Fuzzy Graphs and Fuzzy Hypergraphs

Studies in Fuzziness and Soft Computing

Editor-in-chief
Prof. Janusz Kacprzyk
Systems Research Institute
Polish Academy of Sciences
ul. Newelska 6
01-447 Warsaw, Poland
E-mail: kacprzyk@ibspan.waw.pl
http://www.springer.de/cgi-bin/search_book.pl?series=2941

Vol. 3. A. Geyer-Schulz
Fuzzy Rule-Based Expert Systems and Genetic Machine Learning, 2nd ed. 1996
ISBN 3-7908-0964-0

Vol. 4. T. Onisawa and J. Kacprzyk (Eds.)
Reliability and Safety Analyses under Fuzziness, 1995
ISBN 3-7908-0837-7

Vol. 5. P. Bosc and J. Kacprzyk (Eds.)
Fuzziness in Database Management Systems, 1995
ISBN 3-7908-0858-X

Vol. 6. E.S. Lee and Q. Zhu
Fuzzy and Evidence Reasoning, 1995
ISBN 3-7908-0880-6

Vol. 7. B.A. Juliano and W. Bandler
Tracing Chains-of-Thought, 1996
ISBN 3-7908-0922-5

Vol. 8. F. Herrera and J.L. Verdegay (Eds.)
Genetic Algorithms and Soft Computing, 1996
ISBN 3-7908-0956-X

Vol. 9. M. Sato et al.
Fuzzy Clustering Models and Applications, 1997, ISBN 3-7908-1026-6

Vol. 10. L.C. Jain (Ed.)
Soft Computing Techniques in Knowledge-based Intelligent Engineering Systems, 1997
ISBN 3-7908-1035-5

Vol. 11. W. Mielczarski (Ed.)
Fuzzy Logic Techniques in Power Systems, 1998, ISBN 3-7908-1044-4

Vol. 12. B. Bouchon-Meunier (Ed.)
Aggregation and Fusion of Imperfect Information, 1998
ISBN 3-7908-1048-7

Vol. 13. E. Orłowska (Ed.)
Incomplete Information: Rough Set Analysis, 1998
ISBN 3-7908-1049-5

Vol. 14. E. Hisdal
Logical Structures for Representation of Knowledge and Uncertainty, 1998
ISBN 3-7908-1056-8

Vol. 15. G.J. Klir and M.J. Wierman
Uncertainty-Based Information, 2nd ed., 1999
ISBN 3-7908-1242-0

Vol. 16. D. Driankov and R. Palm (Eds.)
Advances in Fuzzy Control, 1998
ISBN 3-7908-1090-8

Vol. 17. L. Reznik, V. Dimitrov and J. Kacprzyk (Eds.)
Fuzzy Systems Design, 1998
ISBN 3-7908-1118-1

Vol. 18. L. Polkowski and A. Skowron (Eds.)
Rough Sets in Knowledge Discovery 1, 1998
ISBN 3-7908-1119-X

Vol. 19. L. Polkowski and A. Skowron (Eds.)
Rough Sets in Knowledge Discovery 2, 1998
ISBN 3-7908-1120-3

Vol. 20. J.N. Mordeson and P.S. Nair
Fuzzy Mathematics, 1998
ISBN 3-7908-1121-1

Vol. 21. L.C. Jain and T. Fukuda (Eds.)
Soft Computing for Intelligent Robotic Systems, 1998
ISBN 3-7908-1147-5

Vol. 22. J. Cardoso and H. Camargo (Eds.)
Fuzziness in Petri Nets, 1999
ISBN 3-7908-1158-0

Vol. 23. P.S. Szczepaniak (Ed.)
Computational Intelligence and Applications, 1999
ISBN 3-7908-1161-0

Vol. 24. E. Orłowska (Ed.)
Logic at Work, 1999
ISBN 3-7908-1164-5

continued on page 249

John N. Mordeson
Premchand S. Nair

Fuzzy Graphs and Fuzzy Hypergraphs

With 29 Figures
and 10 Tables

Physica-Verlag
A Springer-Verlag Company

John N. Mordeson
Director
Center for Research in Fuzzy Mathematics
and Computer Science
Premchand S. Nair
Associate Professor
Department of Mathematics
and Computer Science
Creighton University
Omaha, Nebraska 68178
USA
E-mail: mordes@creighton.edu
psnair@creighton.edu

ISSN 1434-9922
ISBN 3-7908-1286-2 Physica-Verlag Heidelberg New York

Cataloging-in-Publication Data applied for
Die Deutsche Bibliothek – CIP-Einheitsaufnahme
Mordeson, John N.: Fuzzy graphs and fuzzy hypergraphs: with 10 tables / John N. Mordeson; Premchand S. Nair. – Heidelberg; New York: Physica-Verl., 2000
(Studies in fuzziness and soft computing; Vol. 46)
ISBN 3-7908-1286-2

Physica-Verlag is a company in the specialist publishing group BertelsmannSpringer.

Printed in Germany

Hardcover Design: Erich Kirchner, Heidelberg

SPIN 10761577 88/2202-5 4 3 2 1 0 – Printed on acid-free paper

FOREWORD

In the course of fuzzy technological development, fuzzy graph theory was identified quite early on for its importance in making things work. Two very important and useful concepts are those of granularity and of nonlinear approximations. The concept of granularity has evolved as a cornerstone of Lotfi A.Zadeh's theory of perception, while the concept of nonlinear approximation is the driving force behind the success of the consumer electronics products manufacturing.

It is fair to say fuzzy graph theory paved the way for engineers to build many rule-based expert systems. In the open literature, there are many papers written on the subject of fuzzy graph theory. However, there are relatively books available on the very same topic. Professors' Mordeson and Nair have made a real contribution in putting together a very comprehensive book on fuzzy graphs and fuzzy hypergraphs. In particular, the discussion on hypergraphs certainly is an innovative idea.

For an experienced engineer who has spent a great deal of time in the laboratory, it is usually a good idea to revisit the theory. Professors Mordeson and Nair have created such a volume which enables engineers and designers to benefit from referencing in one place. In addition, this volume is a testament to the numerous contributions Professor John N. Mordeson and his associates have made to the mathematical studies in so many different topics of fuzzy mathematics.

The Center for Research in Fuzzy Mathematics and Computer Science, under the direction of Dr. John N. Mordeson, is one of the earliest of these establishments in the world. The scholarly and academic products that have grown out from the center certainly are very impressive indeed, both in terms of quality as well as quantity.

In a sense, fuzzy mathematics is a generalization of traditional mathematics. In this regard, I have no doubt that Professor John N. Mordeson and his associates will be recognized as important leading researchers and the Center for Research in Fuzzy Mathematics and Computer Science will have its place in the annals of fuzzy theory as an important innovation and institution.

Paul P. Wang
Duke University

PREFACE

In 1965, L. A. Zadeh introduced the concept of a fuzzy subset of a set as a way for representing uncertainty. Zadeh's ideas stirred the interest of researchers worldwide. His ideas have been applied to a wide range of scientific areas. Theoretical mathematics has also been touched by the notion of a fuzzy subset. We consider two areas of mathematics here.

The book deals with fuzzy graph theory and fuzzy hypergraph theory. The book is based on papers that have appeared in journals and conference proceedings. The purpose of this book is to present an up to date account of results from these two areas and to give applications of the results. The book should be of interest to research mathematicians and to engineers and computer scientists interested in applications. For the purpose of a comprehensive presentation of fuzzy graph theory, we include not only much of what appears in volume 20 of this series, but also a greatly expanded version.

In Chapter 1, basic concepts of fuzzy subset theory are given. The notion of a fuzzy relation and its basic properties are presented. The concept of a fuzzy relation is fundamental to many applications given, e. g., cluster analysis and pattern classification. Chapter 1 is based primarily on the work of Rosenfeld and Yeh and Bang.

Chapter 2 presents many concepts and theoretical results of fuzzy graphs. The material from this chapter is the result of the work of many authors including that of the authors of this book. However much of the work is an outgrowth of the ideas of Rosenfeld. We acknowledge the authors at the beginning of each section. This chapter deals with the fuzzification of such concepts as paths, connectedness, bridges, cut vertices, trees, forests, cut sets, chords, cotrees, twigs, 1-chains, cocycles, line graphs, intersection graphs, and interval graphs.

In Chapter 3, applications of fuzzy graph theory are presented. Here again many of the results of this chapter are based on the work of Rosenfeld and Yeh and Bang. Applications of fuzzy graphs to cluster analysis and database theory are presented. Applications of fuzzy graphs to the problem concerning group structure are also given.

In Chapter 4, we present theoretical aspects of fuzzy hypergraph theory with applications to portfolio management, managerial decision making with an example to waste management, and to neural cell-assemblies. The results of this chapter are taken mainly from the work of Goetschel and his coauthors. We have reorganized Goetschel's work and added some examples. This chapter deals with the concepts of fuzzy transversals of fuzzy hypergraphs, colorings of fuzzy hypergraphs, and intersecting fuzzy hypergraphs. In 1982, Z. Pawlak introduced the idea of a rough set in order to provide a systematic approach for the study of indiscernibility of objects. We show how (fuzzy) hypergraphs and rough sets are related in such a way that ideas may be carried back and forth between the two areas.

John N. Mordeson
Premchand S. Nair

ACKNOWLEDGMENTS

The authors are grateful to the editorial and production staffs of Physica-Verlag, especially Janusz Kacpryzk, Martina Bihn, and Gabriele Keidel. We are indebted to Paul Wang and Hu Cheng-ming for their support of fuzzy mathematics. We are also appreciative of the support of Frs. Michael Proterra and Albert Agresti, Deans, Creighton College of Arts and Sciences and Dr. and Mrs. George Haddix. The first author dedicates the book to his children-in-law, James, Brigid, Todd, and Christy. The second author is very grateful to his wife, Suseela, and his parents, Mr. Sukumaran Nair and Ms. Sarada Devi, for supporting his dreams.

CONTENTS

1
FUZZY SUBSETS

In 1965, Lofti Zadeh published his seminal paper "Fuzzy Sets" [11] which described fuzzy set theory and consequently fuzzy logic. The purpose of Zadeh's paper was to develop a theory which could deal with ambiguity and imprecision of certain classes or sets in human thinking, particularly in the domains of pattern recognition, communication of information, and abstraction. This theory proposed making the grade of membership of an element in a subset of a universal set a value in the closed interval $[0, 1]$ of real numbers.

Zadeh's ideas have found applications in computer science, artificial intelligence, decision analysis, information science, system science, control engineering, expert systems, pattern recognition, management science, operations research, and robotics. Theoretical mathematics has also been touched by fuzzy set theory. The ideas of fuzzy set theory have been introduced into topology, abstract algebra, geometry, graph theory, and analysis.

Before introducing the concept of a fuzzy subset, we review briefly some basic properties of sets, relations, and functions. We assume the reader is familiar with the basic ideas from set theory. Let S be a set and let A and B be subsets of S. We use the notation $A \cup B$ and $A \cap B$ to denote the union and intersection of A and B, respectively. We also let $B \setminus A$ denote the relative complement of A in B. The (relative) complement of A in S, $S \setminus A$, is sometimes denoted by A^c when S is understood. Then it is easily verified that $(A \cup B)^c = A^c \cap B^c$ and $(A \cap B)^c = A^c \cup B^c$. These equations are known as DeMorgan's Laws.

Let x be an element of S. We write $x \in A$ if x is an element of A. otherwise we write $x \notin A$. We use the notation $A \subseteq B$ or $B \supseteq A$ to denote that A is a subset of B. If $A \subseteq B$, but there exists $x \in B$ such that $x \notin A$, then we write $A \subset B$ or $B \supset A$ and say that A is a proper subset of B. The cardinality of A is denoted by $|A|$ or card(A). The power set of A, written $\wp(A)$, is defined to be the set of all subsets of A, i. e., $\wp(A) = \{U \mid U \subseteq A\}$. A *partition* of S is a set $\mathcal{P}$ of nonempty subsets of S such that $\forall U, V \in \mathcal{P}$, either (1) $U = V$ or $U \cap V = \emptyset$, the empty set, and (2) $S = \underset{U \in \mathcal{P}}{\cup} U$.

We let $\mathbb{N}$ denote the set of positive integers, $\mathbb{Z}$ the set of integers, and $\mathbb{R}$ the set of real numbers.

Let X and Y be sets. If $x \in X$ and $y \in Y$, then (x, y) denotes the ordered pair of x with y. The *Cartesian cross product* of X with Y is defined to be the set $\{(x, y) \mid x \in X, y \in Y\}$ and is denoted by $X \times Y$. We occasionally write X^2 for $X \times X$. In fact, for $n \in \mathbb{N}, n \geq 2$, we let X^n denote the set of all ordered n-tuples of elements from X. A *relation* R of X into Y is a subset of $X \times Y$. Let R be such a relation. Then the *domain* of R, written Dom(R), is $\{x \in X \mid \exists y \in Y$ such that $(x, y) \in R\}$ and the *image* of R, written Im(R), is $\{y \in Y \mid \exists x \in X$ such that $(x, y) \in R\}$. If $(x, y) \in R$, we sometimes write xRy or $R(x) = y$. If R is a relation from X into X, we say that R is a relation on X. A relation R on X is called

(1) *reflexive* if $\forall x \in X$, $(x, x) \in R$;

(2) *symmetric* if $\forall x, y \in X$, $(x, y) \in R$ implies $(y, x) \in R$;

(3) *transitive* if $\forall x, y, z \in X$, (x, y) and $(y, z) \in R$ implies $(x, z) \in R$.

A relation R on X which is reflexive, symmetric, and transitive is called an *equivalence relation.* If R is an equivalence relation on X, we let $[x]$ denote the equivalence class of x with respect to R. Hence $[x] = \{a \in X \mid aRx\}$. If R is an equivalence relation on X, then $\{[x] \mid x \in X\}$ is a partition of X. Also if $\mathcal{P}$ is a partition of X and R is the relation on X defined by $\forall x, y \in X$, $(x, y) \in R$ if $\exists U \in \mathcal{P}$ such that $x, y \in U$, then R is an equivalence relation on X whose equivalence classes are exactly those members of $\mathcal{P}$.

Let R be a relation on X. Then R is called *antisymmetric* if $\forall x, y \in X$, $(x, y) \in R$ and $(y, x) \in R$ implies $x = y$. If R is a reflexive, antisymmetric, and transitive relation on X, then R is called a *partial order* on X and X is said to be *partially ordered* by R.

Let R be a relation of X into Y and T a relation of Y into a set Z. Then the *composition* of R with T, written $T \circ R$, is defined to be the relation $\{(x, z) \in X \times Z \mid \exists y \in Y$, such that $(x, y) \in R$ and $(y, z) \in T\}$.

Suppose that f is a relation of X into Y such that Dom$(f) = X$ and $\forall x, x' \in X$, $x = x'$ implies $f(x) = f(x')$. Then f is called a *function* of X into Y and we write $f : X \to Y$. Let f be a function of X into Y. Then f is sometimes called a *mapping* and f is said to *map* X into Y. If $\forall y \in Y$, $\exists x \in X$ such that $f(x) = y$, then f is said to be *onto* Y or to map X *onto* Y. If $\forall x, x' \in X$, $f(x) = f(x')$ implies that $x = x'$, then f is said to be *one-to-one* and f is called an *injection.* If f is a one-to-one function of X

onto Y, then f is called a *bijection.* If g is a function of Y into a set Z, then the composition of f with g, $g \circ f$, is a function of X into Z which is one-to-one if f and g are and which is onto Z if f is onto Y and g is onto Z. If $\text{Im}(f)$ is finite, the we say that f is *finite-valued.* We say that an infinite set X is *countable* if there exists a one-to-one function of X onto $\mathbb{N}$; otherwise we call X *uncountable.*

We now introduce the notion of a fuzzy subset of a set S. A *fuzzy subset* of S is a mapping $\mu : S \to [0,1]$, where $[0,1]$ denotes the set $\{t \in \mathbb{R} \mid 0 \leq t \leq 1\}$. We think of μ as assigning to each element $x \in S$ a degree of membership, $0 \leq \mu(x) \leq 1$. Let μ be a fuzzy subset of S. We let $\mu^t = \{x \in S \mid \mu(x) \geq t\}$ for all $t \in [0,1]$. The sets μ^t are called *level sets* or *t-cuts* of μ. We let $\text{supp}(\mu) = \{x \in S \mid \mu(x) > 0\}$. We call $\text{supp}(\mu)$ the *support* of μ. A fuzzy set μ is *nontrivial* is $\text{supp}(\mu) \neq \emptyset$. The set of all fuzzy subsets of S is denoted by and is called the *fuzzy power set* of S. Throughout we use the notation $\vee$ for supremum and $\wedge$ for infimum. Let h be the function of $\mathfrak{F}\wp(S)$ into $[0,1]$ defined by $h(\mu) = \vee\ \{\mu(x) \mid x \in S\} \forall \mu \in \mathfrak{F}\wp(S)$. Then $h(\mu)$ is called the *height* of μ.

Definition 1.1 *Let μ, ν be two fuzzy subsets of S. Then*

(1) $\mu \subseteq \nu$ if $\mu(x) \leq \nu(x)$ for all $x \in S$,

(2) $\mu \subset \nu$ if $\mu(x) \leq \nu(x)$ for all $x \in S$ and there exists at least one $x \in S$ such that $\mu(x) < \nu(x)$,

(3) $\mu = \nu$ if $\mu(x) = \nu(x)$ for all $x \in S$.

Definition 1.2 *Let μ, ν be any two fuzzy subsets of S. Then $\mu \cup \nu$ is the fuzzy subset of S defined by*

$$(\mu \cup \nu)(x) = \mu(x) \vee \nu(x) \text{ for all } x \in S$$

and $\mu \cap \nu$ is the fuzzy subset of S defined by

$$(\mu \cap \nu)(x) = \mu(x) \wedge \nu(x) \text{ for all } x \in S.$$

Definition 1.3 *Let μ be any fuzzy subset of S. Then μ^c is the fuzzy subset of S defined by*

$$\mu^c(x) = 1 - \mu(x) \text{ for all } x \in S.$$

If $\mathcal{S}$ is a collection of fuzzy subsets of S, we define the fuzzy subset $\bigcap_{\xi \in \mathcal{S}} \xi$ (intersection) of $\mathcal{S}$ by $\forall x \in S$, $(\bigcap_{\xi \in \mathcal{S}} \xi)(x) = \wedge\{\xi(x) \mid \xi \in \mathcal{S}\}$ and the fuzzy subset $\bigcup_{\xi \in \mathcal{S}} \xi$ (union) of $\mathcal{S}$ by $\forall x \in S$, $(\bigcup_{\xi \in \mathcal{S}} \xi)(x) = \vee\{\xi(x) \mid \xi \in \mathcal{S}\}$.

Let A be a subset of S. Define $\chi_A : S \to [0,1]$ by $\chi_A(x) = 1$ if $x \in A$ and $\chi_A(x) = 0$ if $x \in S \setminus A$. Then χ_A is called the *characteristic function* of A in S. Now $\chi_S(x) = 1$ for all $x \in S$ and $\chi_\emptyset(x) = 0$ for all $x \in S$. Let μ, ν, and ξ be fuzzy subsets of S. Then we have the following properties.

(1) $\mu \cup \nu = \nu \cup \mu$	(9) $\mu \cup (\nu \cup \xi) = (\mu \cup \nu) \cup \xi$
(2) $\mu \cap \nu = \nu \cap \mu$	(10) $\mu \cap (\nu \cap \xi) = (\mu \cap \nu) \cap \xi$
(3) $\mu \cup \chi_\emptyset = \mu$	(11) $\mu \cap (\nu \cup \xi) = (\mu \cap \nu) \cup (\mu \cap \xi)$
(4) $\mu \cap \chi_\emptyset = \chi_\emptyset$	(12) $\mu \cup (\nu \cap \xi) = (\mu \cup \nu) \cap (\mu \cup \xi)$
(5) $\mu \cup \chi_S = \chi_S$	(13) $(\mu \cup \nu)^c = \mu^c \cap \nu^c$
(6) $\mu \cap \chi_S = \mu$	(14) $(\mu \cap \nu)^c = \mu^c \cup \nu^c$
(7) $\mu \cup \mu = \mu$	(15) $(\mu^c)^c = \mu$
(8) $\mu \cap \mu = \mu$	

It is important to note that the properties $\mu \cap \mu^c = \chi_\emptyset$ and $\mu \cup \mu^c = \chi_S$ do not hold in general. In logic, the former property is known as the law of contradiction while the latter is known as the law of the excluded middle. Additional properties involving fuzzy subsets can be found in [1,2,4,5,6,7,14].

1.1 Fuzzy Relations

Much of the material in the first three sections is based on the work of Rosenfeld, [8]. Let S and T be two sets and let μ and ν be fuzzy subsets of S and T, respectively. Then a *fuzzy relation* ρ from the fuzzy subset μ into the fuzzy subset ν is a fuzzy subset ρ of $S \times T$ such that $\rho(x,y) \leq \mu(x) \wedge \nu(y), \forall x \in S$ and $y \in T$. That is, for ρ to be a fuzzy relation, we require that the degree of membership of a pair of elements never exceed the degree of membership of either of the elements themselves. Also, the restriction $\rho(x,y) \leq \mu(x) \wedge \nu(y), \forall x \in S$ and $y \in T$ allows ρ^t to be a relation from μ^t into ν^t for all $t \in [0,1]$ and for $\text{supp}(\rho)$ to be a relation from $\text{supp}(\mu)$ into $\text{supp}(\nu)$.

There are three special cases of fuzzy relations which are extensively found in the literature:

(1) $S = T$ and $\mu = \nu$. In this case, ρ is said to be a fuzzy relation on μ. Note that ρ is a fuzzy subset of $S \times S$ such that $\rho(x,y) \leq \mu(x) \wedge \mu(y)$.

(2) $\mu(x) = 1.0$ for all $x \in S$ and $\nu(y) = 1.0$ for all $y \in T$. In this case, ρ is said to be a fuzzy relation from S into T.

(3) $S = T, \mu(x) = 1.0$ for all $x \in S$ and $\nu(y) = 1.0$ for all $y \in T$. In this case, ρ is said to be a fuzzy relation on S.

There are many applications in which a fuzzy relation on a fuzzy subset is quite useful. Also, any result we obtain is clearly true for fuzzy relations on a set. We devote this and next two sections of this chapter to fuzzy relations on a fuzzy subset. The last section of this chapter is devoted to the study of cases 2 and 3.

Let ρ be a fuzzy relation on μ. Then ρ is called the *strongest fuzzy relation* on μ if and only if for all fuzzy relations ϖ on $\mu, \forall x, y \in S, \varpi(x,y) \leq \rho(x,y)$. The converse problem may also arise in practice. That is, we know the strength of the pairs and we want to compute the minimum strength required for the elements themselves. For a given fuzzy subset ρ of $S \times S$, the *weakest fuzzy subset* μ of S on which ρ is a fuzzy relation is defined by $\mu(x) = \vee\{\rho(x,y) \vee \rho(y,x) | y \in S\}$ for all $x \in S$. That is, if ν is a fuzzy subset of S and ρ is a fuzzy relation on ν, then $\mu \subseteq \nu$.

We now introduce some important operations on fuzzy relations.

Definition 1.4 *Let $\rho : S \times T \rightarrow [0,1]$ be a fuzzy relation from a fuzzy subset μ of S into a fuzzy subset ν of T and $\varpi : T \times U \rightarrow [0,1]$ be a fuzzy relation from a fuzzy subset ν of T into a fuzzy subset ξ of U. Define $\rho \circ \varpi : S \times U \rightarrow [0,1]$ by*

$$\rho \circ \varpi(x,z) = \vee\{\rho(x,y) \wedge \varpi(y,z) | y \in T\}$$

for all $x \in S, z \in U$. Then $\rho \circ \varpi$ is called the composition *of ρ with ϖ.*

Proposition 1.1 *Let ρ, μ, ϖ and ν as defined in Definition 1.4. Then $\rho \circ \varpi$ is a fuzzy relation from μ into ξ.*

Proof. Let $x \in S$, $y \in T$, and $z \in U$. Then $\rho(x,y) \leq \mu(x) \wedge \nu(y)$ and $\varpi(y,z) \leq \nu(y) \wedge \xi(z)$. Hence $\rho(x,y) \wedge \varpi(y,z) \leq \mu(x) \wedge \nu(y) \wedge \xi(z)$. Thus $(\rho \circ \varpi)(x,z) = \vee\{\rho(x,y) \wedge \varpi(y,z) \mid y \in T\} \leq \mu(x) \wedge \xi(z)$. ■

We see that the composition of ρ with ϖ is a fuzzy relation from a fuzzy subset μ of S into a fuzzy subset ξ of U. A closer look at the definition of the composition operation reveals that $\rho \circ \varpi$ can be computed similar to matrix multiplication, where the addition is replaced by $\vee$ and the multiplication is replaced by $\wedge$. Since composition is associative,we use the notation ρ^2 to denote the composition $\rho \circ \rho$, ρ^k to denote $\rho^{k-1} \circ \rho, k > 1$. Define $\rho^\infty(x,y) = \vee\{\rho^k(x,y) | k = 1, 2, \ldots\}$ for all $x, y \in S$. Finally, it is convenient to define $\rho^0(x,y) = 0$ if $x \neq y$ and $\rho^0(x,y) = \mu(x)$ otherwise, for all $x, y \in S$.We have introduced three binary operations. We now introduce a unary operation on a fuzzy relation. Given a fuzzy relation ρ on a fuzzy subset μ of S, define the fuzzy relation ρ^c on μ by $\rho^c(x,y) = 1 - \rho(x,y)$ for all $x, y \in S$.

Definition 1.5 *Let $\rho : S \times T \rightarrow [0,1]$ be a fuzzy relation from a fuzzy subset μ of S into a fuzzy subset ν of T. Define the fuzzy relation ρ^{-1} : $T \times S \rightarrow [0,1]$ of ν into μ by $\rho^{-1}(y,x) = \rho(x,y)$ for all $(y,x) \in T \times S$.*

Theorem 1.2 *Let τ, π, ρ and ϖ be fuzzy relations on a fuzzy subset μ of a set S. Then the following properties hold.*

(1) $\rho \cup \varpi = \varpi \cup \rho$

(2) $\rho \cap \varpi = \varpi \cap \rho$

(3) $\rho = (\rho^c)^c$

(4) $\pi \cup (\rho \cup \varpi) = (\pi \cup \rho) \cup \varpi$

(5) $\pi \cap (\rho \cap \varpi) = (\pi \cap \rho) \cap \varpi$

(6) $\pi \circ (\rho \circ \varpi) = (\pi \circ \rho) \circ \varpi$

(7) $\pi \cap (\rho \cup \varpi) = (\pi \cap \rho) \cup (\pi \cap \varpi)$

(8) $\pi \cup (\rho \cap \varpi) = (\pi \cup \rho) \cap (\pi \cup \varpi)$

(9) $(\rho \cup \varpi)^c = \varpi^c \cap \rho^c$

(10) $(\rho \cap \varpi)^c = \varpi^c \cup \rho^c$

(11) For all $t \in [0,1], (\rho \cup \varpi)^t = \rho^t \cup \varpi^t$

(12) For all $t \in [0,1], (\rho \cap \varpi)^t = \rho^t \cap \varpi^t$

(13) For all $t \in [0,1], (\rho \circ \varpi)^t \supseteq \rho^t \circ \varpi^t$ and if S is finite, $(\rho \circ \varpi)^t = \rho^t \circ \varpi^t$.

(14) If $\tau \subseteq \rho$ and $\pi \subseteq \varpi$ then $\tau \cup \pi \subseteq \rho \cup \varpi$

(15) If $\tau \subseteq \rho$ and $\pi \subseteq \varpi$ then $\tau \cap \pi \subseteq \rho \cap \varpi$

(16) If $\tau \subseteq \rho$ and $\pi \subseteq \varpi$ then $\tau \circ \pi \subseteq \rho \circ \varpi$

Proof. We provide the proofs for (13) and (16). Let $x, z \in S$.

(13) $(x,z) \in (\rho \circ \varpi)^t \Leftrightarrow (\rho \circ \varpi)(x,z) \geq t \Leftarrow \exists y \in S$ such that $\rho(x,y) \wedge \varpi(y,z) \geq t \Leftrightarrow \exists y \in S$ such that $\rho(x,y) \geq t$ and $\varpi(y,z) \geq t \Leftrightarrow \exists y \in S$ such that $(x,y) \in \rho^t$ and $(y,z) \in \varpi^t \Leftrightarrow (x,z) \in \rho^t \circ \varpi^t$. The implication becomes an equivalence if S is finite.

(16) $(\tau \circ \pi)(x,z) = \vee\{\tau(x,y) \wedge \pi(y,z) \mid y \in S\} \leq \vee\{\rho(x,y) \wedge \varpi(y,z) \mid y \in S\} = (\rho \circ \varpi)(x,z)$ for all $x, z \in S$. ■

1.2 Fuzzy Equivalence Relations

In this section ρ and ϖ are fuzzy relations on a fuzzy subset μ of S. It is quite natural to represent a fuzzy relation in the form of a matrix. We now use the matrix representation of a fuzzy relation to explain the properties of a fuzzy relation. In particular, we shall use the term "diagonal" to represent the principal diagonal of the matrix.

We call ρ *reflexive* (on μ) if $\rho(x,x) = \mu(x)$ for all $x \in S$. If ρ is reflexive, on χ_S, we call ρ *reflexive.* If ρ is reflexive on μ, then $\rho(x,y) \leq \mu(x) \wedge \mu(y) \leq \mu(x) = \rho(x,x)$ and it follows that "any diagonal element of ρ is larger than or equal to any element in its row". Similarly, "any diagonal element is larger than or equal to any element in its column". Conversely, given a fuzzy relation ρ on μ such that "any diagonal element is larger than or equal to any element in its row and column", define a fuzzy subset ν of S as $\nu(x) = \rho(x,x), \forall x \in S$. Then ν is the weakest fuzzy subset of S such that ρ is a fuzzy relation on ν. Further, ρ is reflexive on ν.

Fuzzy reflexive relations have some interesting algebraic properties.

Theorem 1.3 *Let ρ and ϖ be fuzzy relations on a fuzzy subset μ of S. Then the following properties hold.*

(1) If ρ is reflexive, $\varpi \subseteq \varpi \circ \rho$ and $\varpi \subseteq \rho \circ \varpi$.

(2) If ρ is reflexive, $\rho \subseteq \rho^2$.

(3) If ρ is reflexive, $\rho^0 \subseteq \rho \subseteq \rho^2 \subseteq \rho^3 \subseteq \ldots \subseteq \rho^\infty$.

(4) If ρ is reflexive, $\rho^0(x,x) = \rho(x,x) = \rho^2(x,x) = \rho^3(x,x) = \ldots = \rho^\infty(x,x) = \mu(x), \forall x \in S$.

(5) If ρ and ϖ are reflexive, so is $\rho \circ \varpi$ and $\varpi \circ \rho$.

(6) If ρ is reflexive, then ρ^t is a reflexive relation on μ^t for all $t \in [0,1]$.

Proof. Let $x, z \in S$.

(1) $(\rho \circ \varpi)(x,z) = \vee\{\rho(x,y) \wedge \varpi(y,z) \mid y \in S\} \geq \rho(x,x) \wedge \varpi(x,z) = \mu(x) \wedge \varpi(x,z)$. Since $\varpi(x,z) \leq \mu(x) \wedge \mu(z), \mu(x) \wedge \varpi(x,z) = \varpi(x,z)$. Thus $\varpi \subseteq \rho \circ \varpi$. Similarly, $\varpi \subseteq \varpi \circ \rho$.

(2) Choose ϖ as ρ in (1).

(3) Choose ϖ as ρ, ρ^2, ρ^3 and so on in (1).

(4) Note that $\rho(x,x) = \mu(x), \forall x \in S$. Assume that $\rho^n(x,x) = \mu(x), \forall x \in S$. Now for all $x \in S, \rho^{n+1}(x,x) = \vee\{\rho(x,y) \wedge \rho^n(y,x) \mid y \in S\} \leq \vee\{\mu(x) \wedge \mu(x) \mid y \in S\} = \mu(x)$ and $\rho^{n+1}(x,x) = \vee\{\rho(x,y) \wedge \rho^n(y,x) \mid y \in S\} \geq \rho(x,x) \wedge \rho^n(x,x)$. Hence $\rho^{n+1}(x,x) = \mu(x), \forall x \in S$.

(5) $(\rho \circ \varpi)(x,x) = \vee\{\rho(x,y) \wedge \varpi(y,x) \mid y \in S\} \leq \vee\{\mu(x) \wedge \mu(x) \mid y \in S\} = \mu(x)$ and $(\rho \circ \varpi)(x,x) = \vee\{\rho(x,y) \wedge \varpi(y,x) \mid y \in S\} \geq \rho(x,x) \wedge \varpi(x,x) = \mu(x) \wedge \mu(x) = \mu(x)$. The proof that $\varpi \circ \rho$ is reflexive is similar.

(6) If $x \in \mu^t$, then $\rho(x,x) = \mu(x) \geq t$ and thus $(x,x) \in \rho^t$.

We call ρ *symmetric* if $\rho(x,y) = \rho(y,x)$, for all $x, y \in S$. In other words, ρ is symmetric if the matrix representation of ρ is symmetric (with respect to the diagonal).

Theorem 1.4 *Let ρ and ϖ be fuzzy relations on a fuzzy subset μ of S. Then the following properties hold.*

(1) If ρ and ϖ are symmetric, then $\rho \circ \varpi$ is symmetric if and only if $\rho \circ \varpi = \varpi \circ \rho$.

(2) If ρ is symmetric, then so is every power of ρ.

(3) If ρ is symmetric, then ρ^t is a symmetric relation on μ^t for all $t \in [0, 1]$.

Proof. (1) $(\rho \circ \varpi)(x, z) = (\rho \circ \varpi)(z, x) \Leftrightarrow \vee\{\rho(x, y) \wedge \varpi(y, z) \mid y \in S\} = \vee\{\rho(z, y) \wedge \varpi(y, x) \mid y \in S\} \Leftrightarrow$
$\vee\{\rho(x, y) \wedge \varpi(y, z) \mid y \in S\} = \vee\{\varpi(y, x) \wedge \rho(z, y) \mid y \in S\} \Leftrightarrow \rho \circ \varpi = \varpi \circ \rho$.

(2) Assume that ρ^n is symmetric for $n \in \mathbb{N}$. Then $\rho^{n+1}(x, z) = \vee\{\rho(x, y) \wedge \rho^n(y, z) \mid y \in S\} = \vee\{\rho(y, x) \wedge \rho^n(z, y) \mid y \in S\} = \vee\{\rho^n(z, y) \wedge \rho(y, x) \mid y \in S\} = \rho^{n+1}(z, x)$.

(3) Let $0 \leq t \leq 1$. Suppose $(x, z) \in \rho^t$. Then $\rho(x, z) \geq t$. Since ρ is symmetric, $\rho(z, x) \geq t$. Thus $(z, x) \in \rho^t$. ■

We call ρ *transitive* if $\rho^2 \subseteq \rho$. It follows that ρ^∞ is transitive for any fuzzy relation ρ.

Theorem 1.5 *Let π, ρ and ϖ be fuzzy relations on a fuzzy subset μ of S. Then the following properties hold.*

(1) If ρ is transitive and $\pi \subseteq \rho, \varpi \subseteq \rho$, then $\pi \circ \varpi \subseteq \rho$.

(2) If ρ is transitive, then so is every power of ρ.

(3) If ρ is transitive, ϖ is reflexive and $\varpi \subseteq \rho$, then $\rho \circ \varpi = \varpi \circ \rho = \rho$.

(4) If ρ is reflexive and transitive, then $\rho^2 = \rho$.

(5) If ρ is reflexive and transitive, then $\rho^0 \subseteq \rho = \rho^2 = \rho^3 = \ldots = \rho^\infty$.

(6) If ρ and ϖ are transitive and $\rho \circ \varpi = \varpi \circ \rho$, then $\rho \circ \varpi$ is transitive.

(7) If ρ is symmetric and transitive, then $\rho(x, y) \leq \rho(x, x)$ and $\rho(y, x) \leq \rho(x, x)$, for all $x, y \in S$.

(8) If ρ is transitive, then for any $0 \leq t \leq 1, \rho^t$ is a transitive relation on μ^t.

Proof. (1) $(\pi \circ \varpi)(x, z) = \vee\{\pi(x, y) \wedge \varpi(y, z) \mid y \in S\} \leq \vee\{\rho(x, y) \wedge \rho(y, z) \mid y \in S\} = \rho^2(x, z) \leq \rho(x, z)$. Hence $\pi \circ \varpi \subseteq \rho$.

(2) Assume that $\rho^n \circ \rho^n \subseteq \rho^n$. Then $\rho^{n+1} \circ \rho^{n+1} = \rho^{2n+2} = \rho^{2n} \circ \rho^2 \subseteq \rho^n \circ \rho = \rho^{n+1}$.

(3) By (1), taking π to be ρ, $\rho\circ\varpi \subseteq \rho$. $(\rho\circ\varpi)(x,z) = \vee\{\rho(x,y)\wedge\varpi(y,z) \mid y \in S\} \geq \vee\rho(x,z) \wedge \varpi(z,z) = \rho(x,z) \wedge \mu(z) = \rho(x,z)$. Hence $\rho \circ \varpi = \rho$. Similarly, $\varpi \circ \rho = \rho$.

(4) Choose ϖ as ρ in (3).

(5) Note that $\rho = \rho^2$ by (4). Assume that $\rho^n = \rho^{n+1}$, for $n > 1$. Hence $\rho^n \circ \rho = \rho^{n+1} \circ \rho$. That is, $\rho^{n+1} = \rho^{n+2}$.

(6) $(\rho\circ\varpi)\circ(\rho\circ\varpi) = \rho\circ(\varpi\circ\rho)\circ\varpi = \rho\circ(\rho\circ\varpi)\circ\varpi = \rho^2\circ\varpi^2 \subseteq \rho\circ\varpi$. Hence $\rho\circ\varpi$ is transitive.

(7) Since ρ is transitive, $\rho\circ\rho \subseteq \rho$. Hence $(\rho\circ\rho)(x,x) \leq \rho(x,x)$. That is, $\vee\{\rho(x,y)\wedge\rho(y,x) \mid y \in S\} \leq \rho(x,x)$. Since ρ is symmetric, $\vee\{\rho(x,y)\wedge \rho(x,y) \mid y \in S\} \leq \rho(x,x)$. Thus $\rho(x,y) \leq \rho(x,x)$. Since ρ is symmetric, $\rho(y,x) \leq \rho(x,x)$.

(8) Let $0 \leq t \leq 1$. Let $(x,y),(y,z) \in \rho^t$. Hence $\rho(x,y) \geq t$ and $\rho(y,z) \geq t$. Therefore, $\rho(x,z) = \vee\{\rho(x,w)\wedge\rho(w,z) \mid w \in S\} \geq \rho(x,y)\wedge\rho(y,z) \geq t$. Thus $(x,z) \in \rho^t$.

■

A fuzzy relation ρ on S which is reflexive, symmetric, and transitive is called a *fuzzy equivalence relation* on S.

1.3 Pattern Classification

Let S be a set whose elements we think of as patterns. A fuzzy classification ρ on S is a symmetric fuzzy relation on S such that $\rho(x,x) = 1$ for all $x \in S$.

Since ρ is reflexive, $\rho \subseteq \rho^2 \subseteq \rho^3 \subseteq \ldots \subseteq \rho^\infty$. Note that ρ^∞ is a fuzzy equivalence relation. So for any $0 \leq t \leq 1$, $(\rho^\infty)^t$ is an equivalence relation on S. Let P^t the partition of S induced by the equivalence relation $(\rho^\infty)^t$.

Theorem 1.6 *Let ρ be a fuzzy relation on S. Define δ from $S \times S$ into $\mathbb{R}$ by $\forall x,y \in S, \delta(x,y) = 1 - \rho^\infty(x,y)$. Then $\forall x,y,z \in S$:*

(1) $\delta(x,y) = 0$ if and only if $x = y$.

(2) $\delta(x,y) = \delta(y,x)$.

(3) $\delta(x,z) \leq \delta(x,y) + \delta(y,z)$.

That is, δ is a metric on S.

Proof. Let $x,y,z \in S$.

(1) $\rho^\infty(x,y) = 1$ if and only if $x = y \Leftrightarrow 1 - \rho^\infty(x,y) = 0$ if and only if $x = y \Leftrightarrow \delta(x,y) = 0$ if and only if $x = y$.

(2) $\delta(x,y) = 1 - \rho^\infty(x,y) = 1 - \rho^\infty(y,x) = \delta(y,x)$.

(3) ρ^∞ is transitive $\Rightarrow \rho^\infty(x,z) \geq \rho^\infty(x,y) \wedge \rho^\infty(y,z) \geq \rho^\infty(x,y) + \rho^\infty(y,z) - 1 \Rightarrow \rho^\infty(x,z) - 1 \geq \rho^\infty(x,y) - 1 + \rho^\infty(y,z) - 1 \Rightarrow 1 - \rho^\infty(x,z) \leq 1 - \rho^\infty(x,y) + 1 - \rho^\infty(y,z) \Rightarrow \delta(x,z) \leq \delta(x,y) + \delta(y,z)$. ■

Example 1.1 *Let $S = \{x_1, x_2, x_3, x_4, x_5\}$ and define the symmetric fuzzy relation ρ on S as follows:*

	x_1	x_2	x_3	x_4	x_5
x_1	*1.0*				
x_2	*0.8*	*1.0*			
x_3	*0.0*	*0.4*	*1.0*		
x_4	*0.1*	*0.0*	*0.0*	*1.0*	
x_5	*0.2*	*0.9*	*0.0*	*0.5*	*1.0*

Now $\rho^\infty = \rho^3$ is given by

	x_1	x_2	x_3	x_4	x_5
x_1	*1.0*				
x_2	*0.8*	*1.0*			
x_3	*0.4*	*0.4*	*1.0*		
x_4	*0.5*	*0.5*	*0.4*	*1.0*	
x_5	*0.8*	*0.9*	*0.4*	*0.5*	*1.0*

and we have the partitions

$$P^t = \begin{cases} \{\{x_1, x_2, x_3, x_4, x_5\}\} & \text{if } 0 \le t \le 0.4 \\ \{\{x_1, x_2, x_4, x_5\}, \{x_3\}\} & \text{if } 0.4 < t \le 0.5 \\ \{\{x_1, x_2, x_5\}, \{x_4\}, \{x_3\}\} & \text{if } 0.5 < t \le 0.8 \\ \{\{x_1\}, \{x_2, x_5\}, \{x_4\}, \{x_3\}\} & \text{if } 0.8 < t \le 0.9 \\ \{\{x_1\}, \{x_2\}, \{x_5\}, \{x_4\}, \{x_3\}\} & \text{if } 0.9 < t \le 1.0 \end{cases}$$

Thus there are many partitions possible and depending upon the level of detail, one could classify the patterns based on equivalence relations. Note that if $s \ge t$, then P^s is a refinement of P^t.

We now present an experiment done by Tamura, Higuchi and Tanaka [9]. Portraits obtained from 60 families were used in their experiment, each family of which was composed of between four and seven members. They chose portraits because even though parents may not resemble each other, they may be connected through their children, and consequently they could classify the portraits into families. They first divided the 60 families into 20 groups, each of which was composed of 3 families. Each group was, on the average, composed of 15 members. The portraits of each group were presented to a different student to assign the values of the subjective similarity $\rho(x, y)$ between all pairs on a scale of 1 to 5. They used the 5 rank representation instead of a continuous value representation because it has been proved that human beings cannot make distinctions into more than 5 ranks. Twenty students were involved in the experiment. Since the levels of the subjective values are different according to individuals, the threshold was determined in each group as follows. As they lowered the threshold, the number of classes decreased. Hence, under the assumption that the number of classes c to be classified was known to be 3, while lowering the threshold they stopped at the value which divided the patterns into 3 classes (collection of the patterns composed of more than 2 patterns that have a stronger relation than λ with each other) and some nonconnected patterns. However, when some $\rho(x, y)$ are equal, sometimes there is no threshold by which the patterns can be divided into exactly c given classes. In such a case, they divided them into exactly c classes by stopping the threshold at the value where the patterns are divided into less than c classes and separating some connections randomly that have a minimum $\rho(x, y)$ greater than the threshold. The correctly classified rates, the misclassified rates, and the rejected rates of 20 groups were within the range of 50-94 percent, 0-33 percent, and 0-33 percent, respectively, and they obtained the correctly classified rate 75 percent of the time, the misclassified rate 13 percent, and the rejected rate 12 percent as the averages of the 20 groups. Here, since the classes made in this experiment have no label, they calculated these rates by making a one-to-one correspondence between 3 families and 3 classes, so as to have the largest number of correctly classified patterns.

We see that Tamura, Higuchi and Tanaka [9] have studied pattern classification using subjective information and performed experiments involving classification of portraits. The method of classification proposed here is based on the procedure of finding a path connecting 2 patterns. Therefore, this method may be combined with nonsupervised learning and may also be applicable to information retrieval and path detection.

1.4 Similarity Relations

In this section we will show that the concept of a similarity relation introduced by Zadeh [13] is derivable in much the same way as a fuzzy equivalence relation. Throughout this section we shall be dealing with a fuzzy relation on a set. The results in this section are from [9].

Definition 1.6 *Let ρ be a fuzzy relation on a set S. We define the following notions:*

(1) ρ is ϵ-reflexive *if $\forall\ x \in S, \rho(x, x) \geq \epsilon$, where $\epsilon \in [0, 1]$.*

(2) ρ is irreflexive *if $\forall x \in S, \rho(x, x) = 0$.*

(3) ρ is weakly reflexive *if for all x, y in S and for all $\epsilon \in [0, 1]$, $\rho(x, y) = \epsilon \Rightarrow \rho(x, x) \geq \epsilon$.*

Note that the definition of a reflexive relation as a 1-reflexive relation coincides with the definition of a reflexive relation in Section 1.2.

Lemma 1.7 *If ρ is a fuzzy relation from S into T, then the fuzzy relation $\rho \circ \rho^{-1}$ is weakly reflexive and symmetric.*

Proof. $(\rho \circ \rho^{-1})(x, x') = \vee\{\rho(x, y) \wedge \rho^{-1}(y, x') \mid y \in T\} \leq \vee\{\rho(x, y) \wedge \rho(x, y) \mid y \in T\} =$
$\vee\{\rho(x, y) \wedge \rho^{-1}(y, x) \mid y \in T\} = (\rho \circ \rho^{-1})(x, x)$. Hence $\rho \circ \rho^{-1}$ is weakly reflexive.
$(\rho \circ \rho^{-1})(x, x') = \vee\{\rho(x, y) \wedge \rho^{-1}(y, x') \mid y \in T\} = \vee\{\rho^{-1}(y, x) \wedge \rho(x', y) \mid y \in T\}$
$= \vee\{\rho(x', y) \wedge \rho^{-1}(y, x) \mid y \in T\} = (\rho \circ \rho^{-1})(x', x)$. Thus $\rho \circ \rho^{-1}$ is symmetric. ■

Let ρ be a weakly reflexive and symmetric fuzzy relation on S. Define a family of non-fuzzy subsets F^ρ as follows:
$F^\rho = \{K \subseteq S | (\exists 0 < \epsilon \leq 1)(\forall x \in S)[x \in K \Leftrightarrow (\forall x' \in K)[\rho(x, x') \geq \epsilon]]\}$.
Hence if we let
$F^\rho_\epsilon = \{K \subseteq S | (\forall x \in S)[x \in K \Leftrightarrow (\forall x' \in K)[\rho(x, x') \geq \epsilon]]\}$,
then we see that $\epsilon_1 \leq \epsilon_2 \Rightarrow F^\rho_{\epsilon_2} \preccurlyeq F^\rho_{\epsilon_1}$ where "$\preccurlyeq$" denotes a covering relation, i.e., every element in $F^\rho_{\epsilon_2}$ is a subset of an element in $F^\rho_{\epsilon_1}$.

A subset J of S is called *ϵ-complete* with respect to ρ if $\forall\ x, x' \in J, \rho(x, x') \geq \epsilon$. A *maximal ϵ-complete* set is one which is not properly contained in any other ϵ-complete set.

Lemma 1.8 *F^ρ is the family of all maximal ϵ-complete sets with respect to ρ for $0 \leq \epsilon \leq 1$.* ■

Proof. Let $K \in F^\rho$ and $x, x'' \in K$. Then there exists $0 < \epsilon \leq 1$ such that $\forall x' \in K$, $\rho(x, x') \geq \epsilon$. Thus $\rho(x, x'') \geq \epsilon$. Hence K is ϵ-complete. Let J be subset of X such that $K \subseteq J$ and J is ϵ-complete. Let $x \in J$. Since J is ϵ-complete, $\forall x' \in K$, $\rho(x, x') \geq \epsilon$. Since $K \in F^\rho$, $x \in K$. Thus $J \subseteq K$. Hence K is maximal. Now let K be a maximal ϵ-complete set. Let $x \in X$. Then clearly $x \in K \Leftrightarrow \forall\ x' \in K$, $\rho(x, x') \geq \epsilon$. Thus $K \in F^\rho$.

Lemma 1.9 *Whenever $\rho(x, x') > 0$, there is some ϵ-complete set $K \in F^\rho$ such that $\{x, x'\} \subseteq K$.* ■

Proof. If $x = x'$, then $\{x\}$ is certainly ϵ-complete for $\epsilon = \mu_R(x, x)$. Suppose that $x \neq x'$. Then since $\rho(x, x') = \rho(x', x)$ by symmetry, and $\rho(x, x) \geq \rho(x, x')$ and $\rho(x', x') \geq \rho(x, x')$ by weak reflexivity, we see that $\{x, x'\}$ is ϵ-complete, where $\epsilon = \rho(x, x')$. Denote by C_ϵ the family of all ϵ-complete sets C which contain $\{x, x'\}$. Then C_ϵ is not empty since $\{x, x'\} \in C_\epsilon$. It follows easily by Zorn's lemma that C_ϵ has a maximal element K. This element is also maximal in the family of all ϵ-complete sets since any set including K must also include $\{x, x'\}$. Hence $K \in F^\rho$ by Lemma 1.8. ■

We note that sometimes a subclass of F^ρ, satisfying the condition of Lemma 1.9, will cover the set S. For example, let ρ be the fuzzy relation on $S = \{a, b, c, d, e, f\}$ given by the following matrix.

	a	b	c	d	e	f
a	1.0	0.3	0.4	0.0	0.4	0.3
b	0.3	1.0	0.2	0.3	0.0	0.4
c	0.4	0.2	1.0	0.3	0.5	0.0
d	0.0	0.3	0.3	1.0	0.0	0.0
e	0.4	0.0	0.5	0.0	1.0	0.0
f	0.3	0.4	0.0	0.0	0.0	1.0

We see that the family consisting of the three maximal complete sets $\{a, b, f\}$, $\{b, c, d\}$, and $\{a, c, e\}$ satisfy the condition of Lemma 1.9, but it does not contain the maximal complete set $\{a, b, c\}$. Note for example that $\{a, b, f\}$ is maximal ϵ'-complete $\forall\, 0 < \epsilon' \leq \epsilon$ where $\epsilon = 0.3$ since $\rho(a, d) = \rho(b, e) = \rho(f, c) = 0$. We have

$$F^\rho_\epsilon = \begin{cases} \{\{a\}, \{b\}, \{c\}, \{d\}, \{e\}, \{f\}\} & \text{if } 0.5 < \epsilon \leq 1 \\ \{\{c, e\}, \{a\}, \{b\}, \{d\}, \{f\}\}\} & \text{if } 0.4 < \epsilon \leq 0.5 \\ \{\{a, c, e\}, \{b, f\}, \{d\}\} & \text{if } 0.3 < \epsilon \leq 0.4 \\ \{\{a, b, f\}, \{a, c, e\}, \{b, d\}, \{c, d\}\} & \text{if } 0.2 < \epsilon \leq 0.3 \\ \{\{a, b, c\}, \{b, c, d\}, \{a, b, f\}, \{a, c, e\}\} & \text{if } 0 < \epsilon \leq 0.2 \end{cases}$$

We also note that ρ is not transitive: $\rho(b, c) = .2 \not\geq .3 = (.3 \wedge .4) \vee (1 \wedge .2) \vee (.2 \wedge 1) \vee (.3 \wedge .3)\ (0 \wedge .5) \vee (.4 \wedge 0) = \vee\{\rho(b, y) \wedge \rho(y, c) \mid y \in S\} = \rho \circ \rho(b, c)$.

Recall that $\chi_\emptyset$ is the characteristic function of $\emptyset$ in $S \times S$.

Lemma 1.10 *If $\rho \neq \chi_\emptyset$ is a weakly reflexive and symmetric fuzzy relation on S, then there exists a set T and a fuzzy relation π from S into T such that $\rho = \pi \circ \pi^{-1}$.*

Proof. Denote by T the set $\{K^* \mid K \in F^\rho\}$. We define a fuzzy relation π from S to T as follows:

$$\pi(x, K^*) = \begin{cases} t & \text{if } x \in K \text{ and } t \text{ is the largest number such that } K \in F^\rho_t \\ 0 & \text{otherwise.} \end{cases}$$

If $\rho(x, x') = t > 0$, then by Lemma 1.9, there is an t-complete set $K \in F^\rho$ such that $\{x, x'\} \subseteq K$. Since $(\pi \circ \pi^{-1})(x, x') = \vee_{K^*} [\pi(x, K^*) \wedge \pi(x', K^*)] \geq t = \rho(x, x')$, we conclude that $\rho \subseteq \pi \circ \pi^{-1}$.

Suppose now that $(\pi\circ\pi^{-1})(x,x')=s$. Then there exists $K^*\in F_s$ such that $\pi(x,K^*)=\pi(x',K^*)$. This means that $\{x,x'\}\subseteq K$ and hence $\rho(x,x')\geq s$. $(s=(\pi\circ\pi^{-1})(x,x')=\underset{K^*}{\vee}[\pi(x,K^*)\wedge\pi(K^*,x')]=\underset{K^*}{\vee}[\pi(x,K^*)\wedge\pi(x',K^*)]$. There exists K^* such that either $\pi(x,K^*)=s$ and $\pi(x',K^*)\geq s$ or $\pi(x,K^*)$ $\geq s$ and $\pi(x',K^*)=s$. Now s is largest such that $K\in F_s^\rho$. Hence $\pi(x,K^*)$ $=s$ and $\pi(x',K^*)=s$.) Therefore, $\pi\circ\pi^{-1}\subseteq\rho$. ■

Combining Lemmas 1.7 and 1.10, we have the following theorem.

Theorem 1.11 *A fuzzy relation $\rho\neq\chi_\emptyset$ on a set S is weakly reflexive and symmetric if and only if there is a set T and a fuzzy relation π from S into T such that $\rho=\pi\circ\pi^{-1}$.* ■

In the remainder of this section, we shall use the notation ϕ_ρ to denote the fuzzy relation π defined above.

Definition 1.7 *A* cover *$\mathcal{C}$ on a set S is a family of subsets $S_i, i\in I$, of S such that $\bigcup_{i\in I}S_i=S$, where I is a nonempty index set.*

Definition 1.8 *Let ρ be a fuzzy relation from S into T. For $\epsilon\in[0,1]$, we say that:*

(1) ρ is ϵ-determinate *if for each $x\in S$, there exists at most one $y\in T$ such that $\rho(x,y)\geq\epsilon$.*

(2) ρ is ϵ-productive *if for each $x\in S$, there exists at least one $y\in T$ such that $\rho(x,y)\geq\epsilon$.*

(3) ρ is an ϵ-function *if it is both ϵ-determinate and ϵ-productive.*

Lemma 1.12 *If ρ is an ϵ-reflexive fuzzy relation on S, then ϕ_ρ is ϵ-productive and for each $\epsilon'\leq\epsilon$, $F_{\epsilon'}^\rho$ is a cover of S.*

Proof. Let $0<\epsilon'\leq\epsilon$. Since for each $x\in X, \rho(x,x)\geq\epsilon$, and because $\{x\}$ is ϵ -complete, there is some K in $F_{\epsilon'}^\rho$ such that $x\in K$. Hence, $F_{\epsilon'}^\rho$ is a cover of X. Also, by definition of $\phi_\rho, x\in K$ implies that $\phi_\rho(x,K^*)\geq\epsilon$ which implies that ϕ_ρ is ϵ -productive. ■

In the sequel, we use the term *productive (determinate, reflexive, function)* for 1-productive (1-determinate, 1-reflexive, 1-function).

Corollary 1.13 *If ρ is reflexive, then ϕ_ρ is productive and each F_ϵ^ρ $(0<\epsilon\leq 1)$ is a cover of S.* ■

The following result is a consequence of Theorem 1.11 and Corollary 1.13.

Corollary 1.14 *ρ is reflexive and symmetric relation on S if and only if there is a set T and a productive fuzzy relation π from S into T such that $\rho = \pi \circ \pi^{-1}$.* ■

Lemma 1.15 *Let ρ be a weakly reflexive, symmetric and transitive fuzzy relation on S, and let ϕ_ρ^ϵ denote the relation ϕ_ρ whose range is restricted to F_ϵ^ρ. That is, ϕ_ρ^ϵ equals ϕ_ρ on $S \times \{K^* | K \in F_\epsilon^\rho\}$. Then for each $0 < \epsilon \leq 1$, ϕ_ρ^ϵ is ϵ-determinate and the elements of F_ϵ^ρ are pairwise disjoint.*

Proof. Let K and K' be two not necessarily distinct elements of F_ϵ^ρ and assume that $K \cap K' \neq \emptyset$. For any $q_1 \in K \cap K'$, we have $\rho(q, q_1) \geq \epsilon$, for all q in K and $\rho(q_1, q') \geq \epsilon$, for all q' in K'. Since ρ is transitive, we see that $\rho(q, q') \geq \epsilon$, for all $q \in K$, and $q' \in K'$. Since ρ is weakly reflexive and symmetric, we conclude that $K \cup K'$ is ϵ-complete. However, since K and K' are maximal ϵ-complete, we must conclude that $K = K'$. Hence, $K \neq K' \Rightarrow K \cap K' = \emptyset$. Suppose $x \in K$ where $K \in F_\epsilon^\rho$. Then $\phi_\rho(x, K^*) \geq \epsilon$, and since x cannot belong to any other sets in F^R,ϕ_ρ^ϵ is ϵ-determinate. ■

Definition 1.9 *A similarity relation ρ on S is a fuzzy relation on S which is reflexive, symmetric and transitive. ρ is called an* ϵ-similarity relation *if it is ϵ-reflexive for some $0 < \epsilon \leq 1$, symmetric, and transitive.*

Note that a similarity relation on S is merely a fuzzy equivalence relation on S.

Since clearly reflexivity implies weak reflexivity, we have the following consequence of Lemmas 1.12 and 1.15.

Corollary 1.16 *If ρ is a similarity relation on S, then for each $0 < \epsilon \leq 1$, F_ϵ^ρ is a partition of S.* ■

We see that Corollary 1.16 says that every similarity relation ρ can be represented as $\bigcup_t t\rho^t$, where ρ^t is the equivalence relation induced by the partition F_t^ρ. It was noted in [14] that if the $\rho^t, 0 < t \leq 1$, are a nested sequence of distinct equivalence relations on S with $t_1 > t_2$ if and only if $\rho^{t_1} \subseteq \rho^{t_2}$, ρ^{t_1} is nonempty and the domain of ρ^{t_1} is equal to the domain of ρ^{t_2}, then $\rho = \bigcup_t t\rho^t$ is a similarity relation on S, where

$$t\rho^t(x, y) = \begin{cases} t & \text{if } (x, y) \in \rho^t \\ 0 & \text{otherwise.} \end{cases}$$

The following result, which is a straightforward consequence of Theorem 1.11 and Corollary 1.16, yields another characterization of a similarity relation.

Theorem 1.17 *A relation* ρ *is an* ϵ*-similarity* $(0 < \epsilon \leq 1)$ *relation on a set* S *if and only if there is another set* T *and an* ϵ*-function* π *from* S *into* T *such that* $\rho = \pi \circ \pi^{-1}$. ■

Example 1.2 *Let* ρ *be the fuzzy relation on* $S = \{a, b, c, d, e, f\}$ *given by the following matrix,* M_ρ.

	a	b	c	d	e	f
a	1.0	0.5	0.5	0.2	0.2	0.2
b	0.5	1.0	0.5	0.2	0.2	0.2
c	0.5	0.5	1.0	0.2	0.2	0.2
d	0.2	0.2	0.2	1.0	0.4	0.4
e	0.2	0.2	0.2	0.4	1.0	0.4
f	0.2	0.2	0.2	0.4	0.4	1.0

Now ${M_\rho}^2 = M_\rho$. *Thus* ρ *is transitive. Clearly,* ρ *is reflexive and symmetric. We have*

$$F_\epsilon^\rho = \begin{cases} \{\{a\},\{b\},\{c\},\{d\},\{e\},\{f\}\} & \textit{if } 0.5 < \epsilon \leq 1 \\ \{\{a,b,c\},\{d\},\{e\},\{f\}\} & \textit{if } 0.4 < \epsilon \leq 0.5 \\ \{\{a,b,c\},\{d,e,f\}\} & \textit{if } 0.2 < \epsilon \leq 0.4 \\ \{X\}. & \textit{if } 0 < \epsilon \leq 0.2 \end{cases}$$

Let $\epsilon = 0.4$. *Then the* $\epsilon -$ *function* $\pi : X \times \{K^* | K \in F_{0.4}^\rho\} \to [0,1]$, *is defined as follows:* $\pi(a, \{a,b,c\}^*) = \pi(b, \{a,b,c\}^*) = \pi(c, \{a,b,c\}^*) = 0.5, \pi(d, \{d,e,f\}^*) = \pi(e, \{d,e,f\}^*) = \pi(f, \{d,e,f\}^*) = 0.4$, *and* $\pi(x, K^*) = 0$ *otherwise.*

1.5 References

1. Dubois, D. and Prade, H., Fuzzy Sets and Systems: Theory and Applications, *Mathematics in Science and Engineering,* Vol. 144, Academic Press, Inc., Orlando, Florida, 1980.

2. Kandel, A., *Fuzzy Mathematical Techniques with Applications,* Addison-Wesley Pub. Co. 1986.

3. Kaufmann, A., *Introduction to the Theory of Fuzzy Sets,* Vol. 1, Academic Press, Inc., Orlando, Florida, 1973.

4. Klir, G.J., U. St. Clair, U.H., and Yuan, B., *Fuzzy Set Theory, Foundations and Applications,* Prentice Hall, Upper Saddle River, N.J., 1997.

5. Klir, G.J. and Folger, T.A., *Fuzzy Sets, Uncertainty and Information,* Prentice Hall, Englewood Cliffs, N.J., 1988.

6. Klir, G.J. and Yuan, B., *Fuzzy Sets and Fuzzy Logic: Theory and Applications,* Prentice Hall, Upper Saddle River, N.J., 1995.

7. Mordeson, J. N. and Nair, P.S., *Fuzzy Mathematics: An Introduction for Engineers and Scientists,* Studies in Fuzziness and Soft Computing, Physica-Verlag, Heidelberg, Germany, 1998.

8. Rosenfeld, A., Fuzzy graphs. In: L. A. Zadeh, K. S. Fu and M. Shimura, Eds., *Fuzzy Sets and Their Applications,* Academic Press, New York, 77–95, 1975.

9. Tamura, S., Higuchi, S., and Tanaka, K., Pattern Classification Based on Fuzzy Relations, *IEEE Trans.* SMC-1, 61–66, 1971.

10. Yeh, R. T. and Bang, S.Y., Fuzzy graphs, fuzzy relations, and their applications to cluster analysis. In: Zadeh, L. A., K. S. Fu and M. Shimura, Eds., *Fuzzy Sets and Their Applications,* Academic Press, New York, 125–149, 1975.

11. Zadeh, L. A., Fuzzy sets, *Inform. and Control,* **8**:338–353, 1965.

12. Zadeh, L. A., Fuzzy sets as a basis for a theory of possibility, *Fuzzy Sets and Systems,* **1**:3–28, 1978.

13. Zadeh, L.A., Similarity relations and fuzzy orderings, *Inform. Sci.,* **3**:177–200, 1971.

14. Zimmermann, H.J., *Fuzzy Set Theory and Its Applications,* Second Edition, Kluwer Academic Publishers, Boston, Dordrecht/London, 1991.

2

FUZZY GRAPHS

A *graph* is a pair (V, R), where V is a set and R is a relation on V. The elements of V are thought of as vertices of the graph and the elements of R are thought of as the edges. Similarly, any fuzzy relation ρ on a fuzzy subset μ of a set V can be regarded as defining a weighted graph, or fuzzy graph, where the edge $(x, y) \in V \times V$ has weight or strength $\rho(x, y) \in [0, 1]$. In this chapter, we shall use graph terminology and introduce fuzzy analogs of several basic graph-theoretical concepts. For simplicity, we will consider only undirected graphs through out this chapter unless otherwise specified. Therefore, all fuzzy relations are symmetric and all edges are regarded as unordered pairs of vertices. We abuse notation by writing (x, y) for an edge in an undirected graph (V, R), where $x, y \in V$. (We need not consider loops, that is, edges of the form (x, x); we can assume, if we wish, that our fuzzy relation is reflexive.) Formally, a *fuzzy graph* $G = (V, \mu, \rho)$ is a nonempty set V together with a pair of functions $\mu : V \to [0, 1]$ and $\rho : V \times V \to [0, 1]$ such that for all x, y in V, $\rho(x, y) \leq \mu(x) \wedge \mu(y)$. We call μ the *fuzzy vertex set* of G and ρ the *fuzzy edge set* of G, respectively. Note that ρ is a fuzzy relation on μ. We will assume that, unless otherwise specified, the underlying set is V and that it is finite. Therefore, for the sake of notational convenience, we omit V in the sequel and use the notation $G = (\mu, \rho)$. Thus in the most general case, both vertices and edges have membership value. However, in the special case where $\mu(x) = 1$, for all $x \in V$, edges alone have fuzzy membership. So, in this case, we use the abbreviated notation $G = (V, \rho)$. The fuzzy graph $H = (\nu, \tau)$ is called a *partial fuzzy subgraph* of $G = (\mu, \rho)$ if $\nu \subseteq \mu$ and $\tau \subseteq \rho$. Similarly, the fuzzy graph $H = (P, \nu, \tau)$ is called a *fuzzy subgraph* of $G = (V, \mu, \rho)$ induced by P if $P \subseteq V, \nu(x) = \mu(x)$ for all

$x \in P$ and $\tau(x,y) = \rho(x,y)$ for all $x, y \in P$. For the sake of simplicity, we sometimes call H a fuzzy subgraph of G. It is worth noticing that a fuzzy subgraph (P, ν, τ) of a fuzzy graph (V, μ, ρ) is in fact a special case of a partial fuzzy subgraph obtained as follows.

$$\nu(x) = \begin{cases} \mu(x) & \text{if } x \in P \\ 0 & \text{if } x \in V \setminus P \end{cases}$$

$$\tau(x,y) = \begin{cases} \rho(x,y) & \text{if } (x,y) \in P \times P \\ 0 & \text{if } (x,y) \in V \times V \setminus P \times P \end{cases}$$

Hence we see that a fuzzy graph can have only one fuzzy subgraph corresponding to a given subset P of V. Thus we shall use the notation $\langle P \rangle$ to denote the fuzzy subgraph of G induced by P. For any threshold $t, 0 \leq t \leq 1$, $\mu^t = \{x \in V \mid \mu(x) \geq t\}$ and $\rho^t = \{(x,y) \in V \times V \mid \rho(x,y) \geq t\}$. Since $\rho(x,y) \leq \mu(x) \wedge \mu(y)$ for all $x, y \in V$, we have $\rho^t \subseteq \mu^t \times \mu^t$, so that (μ^t, ρ^t) is a graph with the vertex set μ^t and edge set ρ^t for all $t \in [0,1]$.

Proposition 2.1 *Let $G = (\mu, \rho)$ be a fuzzy graph. If $0 \leq u \leq t \leq 1$, then (μ^t, ρ^t) is a subgraph of (μ^u, ρ^u).* ■

Proposition 2.2 *Let $H = (\nu, \tau)$ be a partial fuzzy subgraph of $G = (\mu, \rho)$. For any threshold $t, 0 \leq t \leq 1$, (ν^t, τ^t) is a subgraph of (μ^t, ρ^t).* ■

We say that the partial fuzzy subgraph (ν, τ) *spans* the fuzzy graph (μ, ρ) if $\mu = \nu$. In this case, we call (ν, τ) a spanning fuzzy subgraph of (μ, ρ). For any fuzzy subset ν of V such that $\nu \subseteq \mu$, the partial fuzzy subgraph of (μ, ρ) *induced* by ν is the maximal partial fuzzy subgraph of (μ, ρ) that has fuzzy vertex set ν. This is the partial fuzzy graph (ν, τ), where $\tau(x,y) = \nu(x) \wedge \nu(y) \wedge \rho(x,y)$ for all $x, y \in V$.

2.1 Paths and Connectedness

Let $G = (V, X)$ be a graph. A *path* of G is an alternating sequence of vertices and edges $v_0, e_1, v_1, ..., v_{n-1}, e_n, v_n$, where $v_0, v_i \in V$, $e_i \in X$, $e_i = (v_{i-1}, v_i)$, $i = 1, ..., n$ and all the vertices are distinct except v_0 may be the same as v_n. A path is sometimes denoted by $v_0 v_1 ... v_n$, where the edges are evident by context. Let $v_0 v_1 ... v_n$ be a path. If $n \geq 3$ and $v_0 = v_n$, then the path is called a *cycle*. $G = (V, X)$ is said to be *complete* if $(u, v) \in X$ $\forall u, v \in V$, $u \neq v$. A *clique* of a graph is a maximal complete subgraph.

A *path* P in a fuzzy graph (μ, ρ) is a sequence of distinct vertices $x_0, x_1, ...,$ x_n (except possibly x_0 and x_n) such that $\rho(x_{i-1}, x_i) > 0$, $1 \leq i \leq n$. Here $n \geq 1$ is called the *length* of the path P. The consecutive pairs (x_{i-1}, x_i) are called the *edges* of the path. The diameter of $x, y \in V$, written $diam(x, y)$, is the length of the longest path joining x to y. It is shown in [4] that if $diam(x, y) = 1$, then $\rho^\infty(x, y) = \rho(x, y)$. It is also shown in [4] that if $diam(x, y) = k$, then $\rho^\infty(x, y) = \vee\{\rho^i(x, y) \mid i = 1, 2, \ldots, k\}$. In fact an algorithm to compute ρ^∞ is given in [4]. The *strength* of P is defined as $\wedge_{i=1}^n \rho(x_{i-1}, x_i)$. In other words, the strength of a path is defined to be the weight of the weakest edge of the path. A single vertex x may also be considered as a path. In this case, the path is of length 0. If the path has length 0, it is convenient to define its strength to be $\mu(x_0)$. It may be noted that any path of length $n > 0$ can as well be defined as a sequence of edges $(x_{i-1}, x_i), 1 \leq i \leq n$, satisfying the condition $\rho(x_{i-1}, x_i) > 0$ for $1 \leq i \leq n$. A partial fuzzy subgraph (μ, ρ) is said to be *connected* if $\forall x, y \in \text{supp}(\mu), \rho^\infty(x, y) > 0$.

We call P a *cycle* if $x_0 = x_n$ and $n \geq 3$. Two vertices that are joined by a path are said to be *connected.* It is evident that "connected" is an equivalence relation. In fact, x and y are connected if and only if $\rho^\infty(x, y) > 0$. The equivalence classes of vertices under this relation are called *connected components* of the given fuzzy graph. They are just its maximal connected partial fuzzy subgraphs. A strongest path joining any two vertices x, y has strength $\rho^\infty(x, y)$. We shall sometimes refer to this as the *strength of connectedness* between the vertices.

Proposition 2.3 *If (ν, τ) is a partial fuzzy subgraph of (μ, ρ), then $\tau^\infty \subseteq \rho^\infty$.* ■

Bridges and Cut Vertices

Let $G = (\mu, \rho)$ be a fuzzy graph, let x, y be two distinct vertices, and let G' be the partial fuzzy subgraph of G obtained by deleting the edge (x, y). That is, $G' = (\mu, \rho')$, where $\rho'(x, y) = 0$ and $\rho' = \rho$ for all other pairs. We say that (x, y) is a *bridge* in G if $\rho'^\infty(u, v) < \rho^\infty(u, v)$ for some u, v. In other words, if deleting the edge (x, y) reduces the strength of connectedness between some pair of vertices. Thus, (x, y) is a bridge if and only if there exist vertices u, v such that (x, y) is an edge of every strongest path from u to v.

Theorem 2.4 *The following statements are equivalent:*

(1) (x, y) is a bridge;

(2) $\rho'^\infty(x, y) < \rho(x, y)$;

(3) (x, y) is not the weakest edge of any cycle.

Proof. (2) $\Rightarrow$ (1) If (x, y) is not a bridge, then $\rho'^{\infty}(x, y) = \rho^{\infty}(x, y) \geq \rho(x, y)$.

(1) $\Rightarrow$ (3) If (x, y) is a weakest edge of a cycle, then any path involving edge (x, y) can be converted into a path not involving (x, y) but at least as strong, by using the rest of the cycle as a path from x to y. Thus (x, y) cannot be a bridge.

(3) $\Rightarrow$ (2) If $\rho'^{\infty}(x, y) \geq \rho(x, y)$, there is a path from x to y, not involving (x, y), that has strength $\geq \rho(x, y)$, and this path together with (x, y) forms a cycle of which (x, y) is a weakest edge. ■

Let w be any vertex and let G' be the partial fuzzy subgraph of G obtained by deleting the vertex w. That is, $G' = (\mu', \rho')$ is the partial fuzzy subgraph of G such that $\mu'(w) = 0$, $\mu' = \mu$ for all other vertices, $\rho'(w, z) = 0$ for all z, and $\rho' = \rho$ for all other edges.

We say that w is a *cutvertex* in G if $(\rho')^{\infty}(u, v) < \rho^{\infty}(u, v)$ for some u, v such that $u \neq w \neq v$. In other words, if deleting the vertex w reduces the strength of connectedness between some other pair of vertices. Hence, w is a cutvertex if and only if there exist u, v, distinct from w such that w is on every strongest path from u to v. G' is called *nonseparable* (or sometimes: a *block*) if it has no cut vertices. It should be pointed out that a block may have bridges. However this cannot happen for non-fuzzy graphs. For example, consider the fuzzy graph $G = (V, \rho)$, where $V = \{x, y, z\}$ and $\rho(x, y) = 1$, $\rho(x, z) = \rho(y, z) = 0.5$. Note that the edge (x, y) is a bridge since its deletion reduces the strength of connectedness between x and y from 1 to 0.5. However, it is easily verified that no vertex of this fuzzy graph is a cutvertex.

If between every two vertices x, y of G there exist two strongest paths that are disjoint (except for x, y themselves), G is a block. This is analogous to the "if" of the non-fuzzy graph theorem that G is a block (with at least three vertices) if and only if every two vertices of G lie on a common cycle. The "only if", on the other hand, does not hold in the fuzzy case, as the example shows.

Forests and Trees

A (crisp) graph that has no cycles is called *acyclic* or a *forest*. A connected forest is called a *tree*. We call a fuzzy graph a *forest* if the graph consisting of its nonzero edges is a forest, and a *tree* if this graph is also connected. More generally, we call the fuzzy graph $G = (\mu, \rho)$ a *fuzzy forest* if it has a partial fuzzy spanning subgraph $F = (\mu, \tau)$ which is a forest, where for all edges (x, y) not in F (i.e., such that $\tau(x, y) = 0$), we have $\rho(x, y) < \tau^{\infty}(x, y)$. In other words, if (x, y) is in G but (x, y) is not in F, there is a path in F between x and y whose strength is greater than $\rho(x, y)$. It is clear that a forest is a fuzzy forest.

The fuzzy graphs in Figure 2.1 are fuzzy forests and the fuzzy graphs in Figure 2.2 are not fuzzy forests.

FIGURE 2.1 Fuzzy forests.

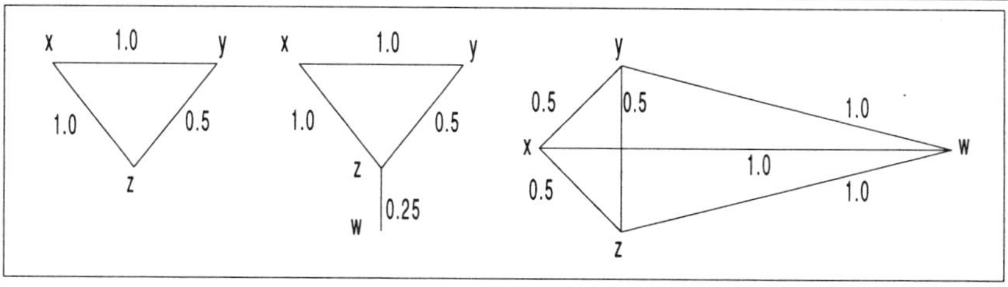

FIGURE 2.2 Fuzzy graphs; but not fuzzy forests.

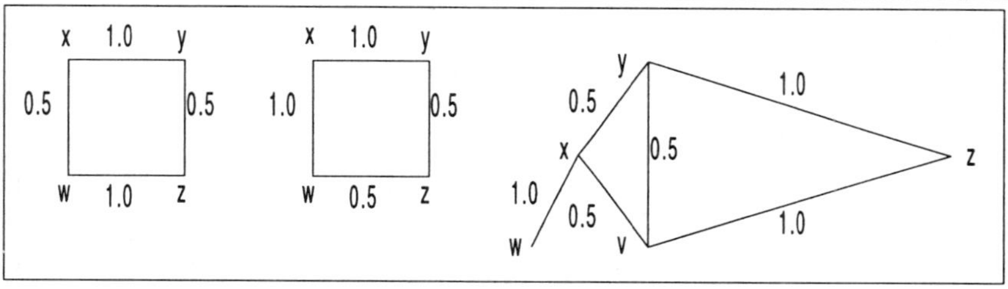

If G is connected, then so is F since any edge of a path in G is either in F, or can be diverted through F. In this case, we call G a *fuzzy tree.* The examples of fuzzy forests given above are all fuzzy trees. Note that if we replaced $<$ by $\leq$ in the definition, then even the fuzzy graph (V, μ, ρ), where $V = \{x, y, z\}$, $\mu(x) = \mu(y) = \mu(z) = 1, \rho(x, y) = \rho(x, z) = \rho(y, z) = 1$, would be a fuzzy forest since it has partial fuzzy spanning subgraphs such as (V, μ, ρ'), where $\rho'(x, y) = \rho'(x, z) = 1$ and $\rho(y, z) = 0$.

Theorem 2.5 *G is a fuzzy forest if and only if in any cycle of G, there is an edge (x, y) such that $\rho(x, y) < \rho'^{\infty}(x, y)$, where $G' = (\mu, \rho')$ is the partial fuzzy subgraph obtained by the deletion of the edge (x, y) from G.*

Proof. Suppose (x, y) is an edge, belonging to a cycle, which has the property of the theorem and for which $\rho(x, y)$ is smallest. (If there are no cycles, G is a forest and we are done.) If we delete (x, y), the resulting partial fuzzy subgraph satisfies the path property of a fuzzy forest. If there are still cycles in this graph, we can repeat the process. Now at each stage, no previously deleted edge is stronger than the edge being currently deleted. Thus the path guaranteed by the property of the theorem involves only edges that have not yet been deleted. When no cycles remain, the resulting partial fuzzy subgraph is a forest F. Let (x, y) not be an edge of F. Then (x, y) is one of the edges that we deleted in the process of constructing F, and there is a path from x to y that is stronger than $\rho(x, y)$ and that does not involve (x, y) nor any of the edges deleted prior to it. If this path involves edges that were deleted later, it can be diverted around them using a path of still stronger edges; if any of these were deleted later, the path can be further diverted; and so on. This process eventually stabilizes with a path consisting entirely of edges of F. Thus G is a fuzzy forest.

Conversely, if G is a fuzzy forest and P is any cycle, then some edge (x, y) of P is not in F. Thus by definition of a fuzzy forest we have $\rho(x, y) < \tau^{\infty}(x, y) \leq \rho'^{\infty}(x, y)$. ∎

We see that if G is connected, then so is F determined by the construction in the first part of the proof.

Proposition 2.6 *If there is at most one strongest path between any two vertices of* G, *then* G *must be a fuzzy forest.*

Proof. Suppose G is not a fuzzy forest. Then by Theorem 2.5, there is a cycle P in G such that $\rho(x, y) \geq \rho'(x, y)$ for all edges (x, y) of P. Thus (x, y) is a strongest path from x to y. If we choose (x, y) to be a weakest edge of P, it follows that the rest of the P is also a strongest path from x to y, a contradiction. ∎

We now show that the converse of Proposition 2.6 does not hold, that is, G can be a fuzzy forest and still have multiple strongest paths between vertices. For example, the fuzzy graph in Figure 2.3 is a fuzzy forest. Here F consists of all edges except (u, y). The strongest paths between x and y have strength $1/4$, due to the edge (x, u); both x, u, v, y and x, u, y are such paths, where the former lies in F but the latter does not.

Proposition 2.7 *If* G *is a fuzzy forest, then the edges of* F *are just the bridges of* G.

Proof. An edge (x, y) not in F cannot be a bridge since $\rho(x, y) < \tau^{\infty}(x, y) \leq \rho'^{\infty}(x, y)$. Suppose that (x, y) is an edge in F. If it were not a bridge, we would have a path P from x to y, not involving (x, y), of strength $\geq \rho(x, y)$. This path must involve edges not in F since F is a forest and has no cycles. However, by definition, any such edge (u_i, v_i) can be replaced by a path P_i in F of strength $> \rho(u, v)$. Now P_i cannot involve (x, y) since all its edges are strictly stronger than $\rho(u, v) \geq \rho(x, y)$. Thus by replacing each (u_i, v_i) by P_i, we can construct a path in F from x to y that does not involve (x, y), giving us a cycle in F, contradiction. ∎

FIGURE 2.3 A fuzzy forest with no multiple strongest paths between vertices.

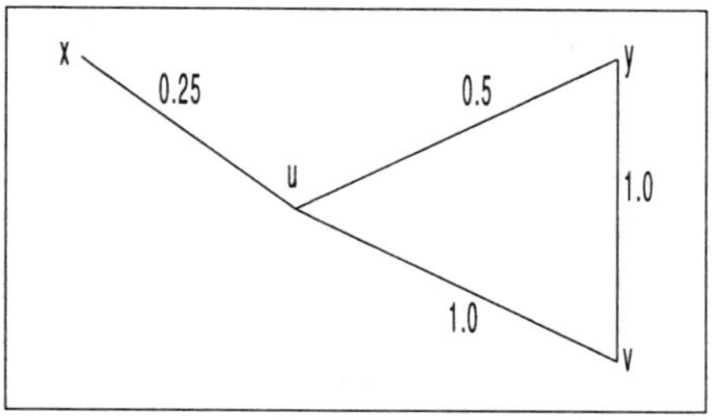

Trees and Cycles

Recall that supp$(\mu) = \{x \in V \mid \mu(x) > 0\}$ and supp$(\rho) = \{(x, y) \in V \times V \mid \rho(x, y) > 0\}$. Since $\rho(x, y) \leq \mu(x) \wedge \mu(y)$, $(x, y) \in$ supp(ρ) implies $x, y \in$ supp(μ). Thus $(supp(\mu),\ supp(\rho))$ is a graph.

We now recall some definitions and give some new ones.

Definition 2.1

(1) $(\mu,\ \rho)$ *is a* tree *if and only if* $(supp(\mu),\ supp(\rho))$ *is a tree.*

(2) $(\mu,\ \rho)$ *is a* fuzzy tree *if and only if* $(\mu,\ \rho)$ *has a fuzzy spanning subgraph* $(\mu,\ \nu)$ *which is a tree such that* $\forall\ (u, v) \in supp(\rho) \backslash\ supp(\nu)$, $\rho(u, v) < \nu^{\infty}(u, v)$, *i. e., there exists a path in* $(\mu,\ \nu)$ *between* u *and* v *whose strength is greater than* $\rho(u, v)$.

Definition 2.2

(1) $(\mu,\ \rho)$ *is a* cycle *if and only if* $(supp(\mu),\ supp(\rho))$ *is a cycle.*

(2) $(\mu,\ \rho)$ *is a* fuzzy cycle *if and only if* $(supp(\mu),\ supp(\rho))$ *is a cycle and* $\nexists$ *unique* $(x, y) \in supp(\rho)$ *such that* $\rho(x, y) = \wedge\{\ \rho(u, v) \mid (u, v) \in supp(\rho)\}$.

Example 2.1 *Let* $V = \{u, v, w, s, t\}$ *and* $X = \{(u, v), (u, w), (v, w), (w, s), (w, t), (s, t)\}$. *Let* $\mu(x) = 1$ *for all* $x \in V$ *and let* ρ *be the fuzzy subset of* X *defined by* $\rho(u, v) = 1/2$, $\rho(u, w) = \rho(v, w) = \rho(w, s) = \rho(w, t) = \rho(s, t) = 1$. *Then* (μ, ρ) *is neither a fuzzy cycle nor a fuzzy tree.*

Example 2.2 *Let* $V = \{w,\ u,\ v\}$ *and* $X = \{(w, u),\ (w, v),\ (u, v)\}$. *Let* $\mu(x) = 1$ *for all* $x \in V$ *and* ρ *and* ρ' *be fuzzy subsets of* X *defined by* $\rho(w, u) = \rho(w, v) = 1$, $\rho(u, v) = 1/2$ *and* $\rho'\ (w, v) = 1$, $\rho'(w, u) = \rho'(u, v) = 1/2$. *Then* $(\mu,\ \rho)$ *is a fuzzy tree, but not a tree and not a fuzzy cycle while* $(\mu,\ \rho')$ *is a fuzzy cycle, but not a fuzzy tree.*

Example 2.2 illustrates the next result.

Theorem 2.8 *Let $(\mu,\ \rho)$ be a cycle. Then $(\mu,\ \rho)$ is a fuzzy cycle if and only if $(\mu,\ \rho)$ is not a fuzzy tree.*

Proof. Suppose that (μ, ρ) is a fuzzy cycle. Then $\exists$ distinct edges (x_1, y_1), $(x_2, y_2) \in \text{supp}(\rho)$ such that $\rho(x_1, y_1) = \rho(x_2, y_2) = \wedge\{\rho(u, v) \mid (u, v) \in \text{supp}(\rho)\}$. If $(\mu,\ \nu)$ is any spanning tree of $(\mu,\ \rho)$, then $\text{supp}(\rho) \setminus \text{supp}(\nu) = \{(u, v)\}$ for some $u,\ v \in V$ since $(\mu,\ \rho)$ is a cycle. Hence $\nexists$ path in (μ, ν) between u and v of greater strength that $\rho(u, v)$. Thus $(\mu,\ \rho)$ is not a fuzzy tree. Conversely, suppose that $(\mu,\ \rho)$ is not a fuzzy tree. Since $(\mu,\ \rho)$ is a cycle, we have $\forall\ (u, v) \in \text{supp}(\rho)$ that $(\mu,\ \nu)$ is a fuzzy spanning subgraph of $(\mu,\ \rho)$ which is a tree and $\nu^\infty(u, v) \le \rho(u, v)$ where $\nu(u, v) = 0$ and $\nu(x, y) = \rho(x, y)\ \forall\ (x, y) \in \text{supp}(\rho) \setminus \{(u, v)\}$. Hence ρ does not attain $\wedge\{\ \rho(x, y) \mid (x, y) \in \text{supp}(\rho)\}$ uniquely. Thus $(\mu,\ \rho)$ is a fuzzy cycle. ■

Theorem 2.9 *If $\exists\ q \in (0, 1]$ such that $(supp(\mu), \rho^q)$ is a tree, ρ^q a $q-cut$, then $(\mu,\ \rho)$ is as fuzzy tree. Conversely, if $(\mu,\ \rho)$ is a cycle and $(\mu,\ \rho)$ is a fuzzy tree, then $\exists\ q \in (0, 1]$ such that $(supp(\mu),\ \rho^q)$ is a tree.*

Proof. Suppose that q exists. Let ν be the fuzzy subset of $V \times V$ such that $\nu = \rho$ on ρ^q and $\nu(x, y) = 0$ if $(x, y) \in V \times V \setminus \rho^q$. Then $(\mu,\ \nu)$ is a spanning fuzzy subgraph of $(\mu,\ \rho)$ such that $(\mu,\ \nu)$ is a fuzzy tree since $(\text{supp}(\mu)$, $\text{supp}(\nu))$ is a tree. Suppose that $(u, v) \in V \times V$ and $(u, v) \notin \rho^q$. Then $\exists$ a path between u and v of strength $\ge q > \rho(u, v)$. Thus $(\mu,\ \rho)$ is a fuzzy tree. For the converse, we note that since $(\mu,\ \rho)$ is a cycle and a fuzzy tree, $\exists$ unique $(x, y) \in \text{supp}(\rho)$ such that $\rho(x, y) = \wedge\{\rho(u, v) \mid (u, v) \in \text{supp}(\rho)\}$. Let q be such that $\rho(x, y) < q \le \wedge\{\rho(u, v) \mid (u, v) \in \text{supp}(\rho) \setminus \{(x, y)\}\}$. Then $(\text{supp}(\mu),\ \rho^q)$ is a tree. ■

We now illustrate Theorem 2.9.

Example 2.3 *Let $V = \{s, t, u, v, w\}$ and $X = \{(s, t), (s, u), (t, u), (u, v), (u, w), (w, v)\}$. Let $\mu(x) = 1$ for all $x \in V$ and let ρ be the fuzzy subset of X defined by $\rho(s, t) = 1/4$, $\rho(s, u) = \rho(t, u) = 3/8$, $\rho(u, v) = 1/2$, and $\rho(u, w) = \rho(w, v) = 1$. Then $\nexists\ q \in (0, 1]$ such that $(supp(\mu),\ \rho^q)$ is a tree. However $(\mu,\ \rho)$ is a fuzzy tree.*

A Characterization of Fuzzy Trees

The results here are taken from [41]. Let $G = (V, \mu, \rho)$ be a fuzzy graph. Recall that $\rho^\infty(u, v)$ denotes the strength of connectedness between u and v in V.

Let $G = (\mu, \rho)$ be a fuzzy graph such that $G^* = (\text{supp}(\mu), \text{supp}(\rho))$ is a cycle and let $t = \wedge\{\rho(u,v) \mid \rho(u,v) > 0\}$. Then all edges (u,v) such that $\rho(u,v) > t$ are bridges of G.

Theorem 2.10 *Let $G = (\mu, \rho)$ be a fuzzy graph such that $G^* = (supp(\mu), supp(\rho))$ is a cycle. Then a vertex is a cutvertex of G if and only if it is a common vertex of two bridges.*

Proof. Let w be a cutvertex of G. Then there exist u and v, other than w, such that w is on every strongest u-v path. Since $G^* = (\text{supp}(\mu), \text{supp}(\rho))$ is a cycle, there exists only one strongest u-v path containing w and all its edges are bridges. Thus w, is a common vertex of two bridges. Conversely, let w be a common vertex of two bridges (u,w) and (w,v). Then both (u,w) and (w,v) are not the weakest edges of G. Also, the path from u to v not containing the edges (u,w) and (w,v) has strength less than $\rho(u,w) \wedge \rho(w,v)$. Hence the strongest u-v path is the path u, w, v and $\rho^\infty(u,v) = \rho(u,w) \wedge \rho(w,v)$. Thus w is a cutvertex. ■

Theorem 2.11 *If w is a common vertex of at least two bridges, then w is a cutvertex.*

Proof. Let (u_1, w) and (w, u_2) be two bridges. Then there exist some u, v such that (u_1, w) is on every strongest u-v path. If w is distinct from u and v it follows that w is a cutvertex. Next, suppose one of v, u is w so that (u_1, w) is on every strongest u-w path or (w, u_2) is on every strongest w-v path. Suppose that w is not a cutvertex. Then between every two vertices there exists at least one strongest path not containing w. In particular, there exists at least one strongest path P, joining u_1 and u_2, not containing w. This path together with (u_1, w) and (w, u_2) forms a cycle.

We now consider two cases. First suppose that u_1, w, u_2 is not a strongest path. Then clearly one of (u_1, w), (w, u_2) or both become the weakest edges of the cycle which contradicts that (u_1, w) and (w, u_2) are bridges.

Second suppose that u_1, w, u_2 is also a strongest path joining u_1 to u_2. Then $\rho^\infty(u_1, u_2) = \rho(u_1, w) \wedge \rho(w, u_2)$, the strength of P. Thus edges of P are at least as strong as $\rho(u_1, w)$ and $\rho(w, u_2)$ which implies that (u_1, w), (w, u_2) or both are the weakest edges of the cycle, which again is a contradiction. ■

That the condition in Theorem 2.11 is not necessary can be observed from the following example. Let $V = \{u, v, w, x\}$ and $X = \{(u,v), (u,w), (u,x), (v,w), (v,x), (w,x)\}$. Let $\mu(s) = 1$ for all $s \in V$ and let ρ be the fuzzy subset of X defined by $\rho(u,w) = \rho(v,x) = 1$, $\rho(v,w) = \rho(w,x) = 0.7$, and $\rho(u,v) = \rho(u,x) = 0.5$. Clearly, w is a cutvertex; (u,w) and (v,x) are the only bridges.

Consider the fuzzy graph $G = (V, \mu, \rho)$ where $V = \{u, v, w, x\}$. Let $X = \{(u, v), (v, w), (w, x), (x, u)\}$. Let $\mu(s) = 1$ for all $s \in V$ and let ρ be the fuzzy subset of X defined by $\rho(u, v) = \rho(w, x) = 0.5$, and $\rho(v, w) = \rho(x, u) = 0.4$. Note that (u, v) and (w, x) are the bridges and no vertex is a cutvertex. This is a significant difference from the crisp graph theory.

Theorem 2.12 *If (u, v) is a bridge, then $\rho^\infty(u, v) = \rho(u, v)$.*

Proof. Suppose that (u, v) is bridge and that $\rho^\infty(u, v) > \rho(u, v)$. Then there exists a strongest u-v path with strength greater than $\rho(u, v)$ and all edges of this strongest path have strength greater than $\rho(u, v)$ and all edges of this strongest path have strength greater than $\rho(u, v)$. Now, this path together with the edge (u, v) forms a cycle in which (u, v) is the weakest edge, contradicting that (u, v) is a bridge. ■

The converse of Theorem 2.12 is not true. The condition for the converse to be true is discussed in Theorem 2.20.

Recall that a connected fuzzy graph $G = (\mu, \rho)$ is a fuzzy tree if it has a fuzzy spanning subgraph $F = (\mu, \tau)$, which is a tree, where for all edges (u, v) not in F, $\rho(u, v) < \tau^\infty(u, v)$.

Equivalently, there is a path in F between u and v whose strength exceeds $\rho(u, v)$.

Lemma 2.13 *If (ν, τ) is a partial fuzzy subgraph of (μ, ρ), then for all $u, v, \tau^\infty(u, v) \leq \rho^\infty(u, v)$.* ■

Theorem 2.14 *If $G = (\mu, \rho)$ is a fuzzy tree and $G^* = (supp(\mu), supp(\rho))$ is not a tree, then there exists at least one edge (u, v) in $supp(\rho)$ for which $\rho(u, v) < \rho^\infty(u, v)$.*

Proof. If G is a fuzzy tree, then by definition there exists a fuzzy spanning subgraph $F = (\mu, \tau)$, which is a tree and $\rho(u, v) < \tau^\infty(u, v)$ for all edges (u, v) not in F. Also $\tau^\infty(u, v) \leq \rho^\infty(u, v)$ by Lemma 2.13. Thus $\rho(u, v) < \rho^\infty(u, v)$ for all (u, v) not in F and by hypothesis there exist at least on edge (u, v) not in F. ■

Definition 2.3 *A complete fuzzy graph is a fuzzy graph $G = (\mu, \rho)$ such that $\rho(u, v) = \mu(u) \wedge \mu(v)$ for all u and v.*

Lemma 2.15 *If G is a complete fuzzy graph, then $\rho^\infty(u, v) = \rho(u, v)$.* ■

Lemma 2.16 *A complete fuzzy graph has no cutvertices.* ■

That the converse of Lemma 2.15 is not true can be observed from the following example.

Example 2.4 *Let* $V = \{u, v, w, x\}$ *and* $X = \{(u, v), (u, w), (u, x), (v, w), (v, x), (w, x)\}$. *Let* $\mu(u) = 0.8$, *and* $\mu(v) = \mu(w) = \mu(x) = 1$. *Let* ρ *be the fuzzy subset of* X *defined by* $\rho(v, w) = 1$, $\rho(v, x) = \rho(w, x) = 0.8$, *and* $\rho(u, v) = \rho(u, w) = \rho(u, x) = 0.6$.

Further, a complete fuzzy graph may have a bridge as illustrated by the next example. Let $V = \{u, v, w, x\}$ and $X = \{(u, v), (u, w), (u, x), (v, w), (v, x), (w, x)\}$. Let $\mu(u) = 0.6$, $\mu(x) = 0.8$, and $\mu(v) = \mu(w) = 1$. Let ρ be the fuzzy subset of X defined by $\rho(v, w) = 1$, $\rho(v, x) = \rho(w, x) = 0.8$, and $\rho(u, v) = \rho(u, w) = \rho(u, x) = 0.6$.

Theorem 2.17 *If* $G = (\mu, \rho)$ *is a fuzzy tree, then* G *is not complete.*

Proof. If possible, let G be a complete fuzzy graph. Then $\rho^\infty(u, v) = \rho(u, v)$ for all u, v by Lemma 2.15. Now G being a fuzzy tree, $\rho(u, v) < \nu^\infty(u, v)$ for all (u, v) not in F. Thus $\rho^\infty(u, v) < \nu^\infty(u, v)$, contradicting Lemma 2.13. ■

Recall that if G is a fuzzy tree, then the edges of F are the bridges of G.

Theorem 2.18 *If* G *is a fuzzy tree, then the internal vertices of* F *are the cutvertices of* G.

Proof. Let w be any vertex in G which is not an end vertex of F. Then w is the common vertex of at least two edges in F which are bridges of G and by Theorem 2.11, w is a cutvertex. Also, if w is an end vertex of F, then w is not a cutvertex; else there would exist u, v distinct from w such that w is on every strongest u-v path and one such path certainly lies in F. But since w is an end vertex of F, this is not possible. ■

Corollary 2.19 *A cutvertex of a fuzzy tree is the common vertex of at least two bridges.* ■

Theorem 2.20 $G = (\mu, \rho)$ *is a fuzzy tree if and only if the following are equivalent for all* u, v :

(1) (u, v) *is a bridge.*

(2) $\rho^\infty(u,v) = \rho(u,v)$.

Proof. Let $G = (\mu, \rho)$ be a fuzzy tree and suppose that (u,v) is a bridge. Then $\rho^\infty(u,v) = \rho(u,v)$ by Theorem 2.12. Now, let (u,v) be an edge in G such that $\rho^\infty(u,v) = \rho(u,v)$. If G^* is a tree, then clearly (u,v) is a bridge; otherwise, it follows from Theorem 2.14 that (u,v) is in F and (u,v) is a bridge.

Conversely, assume that (1) and (2) are equivalent. Construct a maximum spanning tree $T : (\mu, \tau)$ for G [4]. If (u,v) is in T, by an algorithm in [4], $\rho^\infty(u,v) = \rho(u,v)$ and hence (u,v) is a bridge. Now, these are the only bridges of G; for, if possible let (u',v') be a bridge of G which is not in T. Consider a cycle C consisting of (u',v') and the unique u'-v' path in T. Now edges of this u'-v' path are bridges and so they are not weakest edges of C and thus (u',v') must be the weakest edge of C and thus cannot be a bridge.

Moreover, for all edges (u',v') not in T, we have $\rho(u',v') < \tau^\infty(u',v')$; for, if possible let $\rho(u',v') \geq \tau^\infty(u',v')$. But $\rho(u',v') < \rho^\infty(u',v')$, where strict inequality holds since (u',v') is not a bridge. Hence, $\tau^\infty(u',v') < \rho^\infty(u',v')$ which gives a contradiction since $\tau^\infty(u',v')$ is the strength of the unique u'-v' path in T and by an algorithm in [4], $\rho^\infty(u',v') = \tau^\infty(u',v')$. Thus T is the required spanning subgraph F, which is a tree and hence G is a fuzzy tree. ■

For a fuzzy tree G, the spanning fuzzy subgraph F is unique. It follows from the proof of Theorem 2.20 that F is nothing but the maximum fuzzy spanning tree T of G.

Theorem 2.21 *A fuzzy graph is a fuzzy tree if and only if it has a unique maximum fuzzy spanning tree.* ■

For a fuzzy graph which is not a fuzzy tree there is at least one edge in T which is not a bridge and edges not in T are not bridges of G. This observation leads to the following theorem.

Theorem 2.22 *If* $G = (\mu, \rho)$ *is a fuzzy graph with* $supp(\mu) = V$ *and* $|V| = p$ *then* G *has at most* $p-1$ *bridges.* ■

Theorem 2.23 *Let* $G = (\mu, \rho)$ *be a fuzzy graph and let* T *be a maximum fuzzy spanning tree of* G*. Then end vertices of* T *are not cutvertices of* G*.* ■

Corollary 2.24 *Every fuzzy graph has at least two vertices which are not cutvertices.* ■

FIGURE 2.4 Graph of example 2.5.

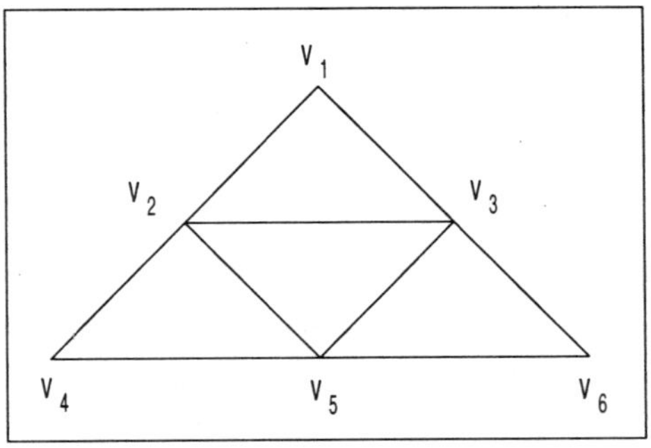

(Fuzzy) Cut Sets

The results in the remainder of the section are from [30]. Before proceeding further, we review some concepts from graph theory. We associate with a graph G, two vector spaces over the field of scalars $\mathbb{Z}_2 = \{0, 1\}$, where addition and multiplication is modulo 2. Then for $1 \in \mathbb{Z}_2$, $1 + 1 = 0$. Let $V = \{v_1, ..., v_n\}$ denote the set of vertices of G and $X = \{e_1, ..., e_m\}$, the set of edges. A *0-chain* of G is a formal linear combination $\sum \epsilon_i v_i$ of vertices and a *1-chain* a formal linear combination of edges $\sum \epsilon_i e_i$, where the $\epsilon_i \in \mathbb{Z}_2$. The *boundary operator* ∂ is a linear function which maps 1-chains to 0-chains such that if $e = (x, y)$, then $\partial(e) = x + y$. The *coboundary operator* δ is a linear function which maps 0-chains to 1-chains such that $\delta(v) = \sum \epsilon_i e_i$ whenever $e_i \in \mathbb{Z}_2$ is incident with v.

Example 2.5 *Let $G = (V, X)$, where $V = \{v_1, ..., v_6\}$ and $X = \{e_1, ..., e_9\}$ and where $e_1 = (v_1, v_2)$, $e_2 = (v_1, v_3)$, $e_3 = (v_2, v_3)$, $e_4 = (v_2, v_4)$, $e_5 = (v_2, v_5)$, $e_6 = (v_3, v_5)$, $e_7 = (v_3, v_6)$, $e_8 = (v_4, v_5)$, and $e_9 = (v_5, v_6)$. The 1-chain $\gamma_1 = e_1 + e_2 + e_4 + e_9$ has boundary $\partial(\gamma_1) = (v_1 + v_2) + (v_1 + v_3) + (v_2 + v_4) + (v_5 + v_6) = v_3 + v_4 + v_5 + v_6$. The 0-chain $\gamma_0 = v_3 + x_4 + v_5 + v_6$ has coboundary $\delta(\gamma_0) = (e_2 + e_3 + e_6 + e_7) + (e_4 + e_8) + (e_5 + e_6 + e_8 - e_9) + (e_7 + e_9)$ $= e_2 + e_3 + e_4 + e_5$.*

A 1-chain with boundary 0 is called a *cycle vector* of G which we can think of as a set of line disjoint cycles. The collection of all cycle vectors, called the *cycle space,* form a vector space over $\mathbb{Z}_2$. A *cut set* of a connected graph is a collection of edges whose removal results in a disconnected graph. A *cocycle* is a minimal cutset. A *coboundary* of G is the coboundary of some 0-chain in G. The coboundary of a subset of V is the set of all edges joining a point in this subset to a point not in the subset. Hence every coboundary is a cutset. Since any minimal cutset is a coboundary, a cocycle is just a minimal nonzero coboundary. The collection of all coboundaries of G is a vector space over $\mathbb{Z}_2$ and is called the *cocycle space* of G. A basis of this spaces which consists entirely of cocycles is called a *cocycle basis* for G.

Let G be a connected graph. Then a *chord* of a spanning tree T is an edge of G which is not in T. The subgraph of G consisting of T and any chord of T has only one cycle. The set $C(T)$ of cycles obtained in this way is independent. Every cycle C depends on the set $C(T)$ since C is the symmetric difference of the cycles determined by the chords of T which lie in C. We define $m(G)$, the *cycle rank*, to be the number of cycles in a basis for the cycle space of G. This discussion yields the following result.

Theorem 2.25 *The cycle rank of a connected graph G is equal to the number of chords of any spanning tree in G.* ■

Similar results can be derived for the cocycle space. Again assume that G is a connected graph. The *cotree* T' of a spanning tree T of G is the spanning subgraph of G containing exactly those edges which are not in T. A cotree of G is the cotree of some spanning tree T. The edges of G which are not in T' are called its *twigs*. The subgraph of G consisting of T' and any one of its twigs contains exactly one cocycle. The collection of cocycles obtained by adding twigs to T', on at a time, is a basis for the cocycle space of G. The cocycle rank $m'(G)$ is the number of cocycles in a basis for the cocycle space of G. A more detailed account can be found in [20].

Theorem 2.26 *The cocycle rank of a connected graph G is the number of twigs in any spanning tree of T.* ■

Let $x \in V$ and let $t \in [0, 1]$. Define the fuzzy subset x_t of V by $\forall y \in V, x_t(y) = 0$ if $y \neq x$ and $x_t(y) = t$ if $y = x$. Then x_t is called a *fuzzy singleton* in V. If $(x, y) \in V \times V$, then $(x, y)_{\rho(x,y)}$ denotes a fuzzy singleton in $V \times V$.

Definition 2.4 *Let E be a subset of supp(ρ).*

(1) $\{(x, y)_{\rho(x,y)} \mid (x, y) \in E\}$ *is a* cut set *of (μ, ρ) if E is a cut set of (supp(μ), supp(ρ)).*

(2) $\{(x, y)_{\rho(x,y)} \mid (x, y) \in E\}$ *is a* fuzzy cut set *of (μ, ρ) if $\exists\, u, v \in$ supp(μ) such that $\rho'^{\infty}(u, v) < \rho^{\infty}(u, v)$ where ρ' is the fuzzy subset of $V \times V$ defined by $\rho' = \rho$ on supp(ρ) $\setminus E$ and $\rho'(x, y) = 0\ \forall (x, y) \in E$, i. e., the removal of E from supp(ρ) reduces the strength of connectedness between some pair of vertices of (μ, ρ).*

When E is a singleton set, a cut set is called a *bridge* and a fuzzy cut set is called a *fuzzy bridge.*

The following is an example of a fuzzy graph (μ, ρ) which has no fuzzy bridges and where ρ is not a constant function.

Example 2.6 *Let* $V = \{t, u, v, w\}$ *and* $X = \{(t, u), (u, v), (v, w), (w, t), (t, v)\}$. *Let* $\mu(x) = 1$ *for all* $x \in V$ *and* $\rho(t, u) = \rho(u, v) = \rho(v, w) = \rho(w, t) = 1$ *and* $\rho(t, v) = 1/2$. *Then* ρ *is not constant, but* (μ, ρ) *does not have a fuzzy bridge since the strength of connectedness between any pair of vertices of* (μ, ρ) *remains* 1 *even after the removal of an edge.*

Theorem 2.27 *Let* $V = \{v_1, ..., v_n\}$ *and* $C = \{(v_1, v_2), (v_2, v_3), ..., (v_{n-1}, v_n), (v_n, v_1)\}$, $n \geq 3$.

(1) Suppose that $supp(\rho) \supseteq C$ *and that* $\forall\ (v_j, v_k) \in supp(\rho) \setminus C$, $\rho(v_j, v_k) < \vee\{\rho(v_i, v_{i+1}) \mid i = 1, ..., n\}$ *where* $v_{n+1} = v_1$. *Then either* ρ *is a constant on* C *or* (μ, ρ) *has a fuzzy bridge.*

(2) Suppose that $\emptyset \neq supp(\rho) \subset C$. *Then* (μ, ρ) *has a bridge.*

Proof. (1) Suppose that ρ is not constant on C. Let $(v_h, v_{h+1}) \in C$ be such that $\rho(v_h, v_{h+1}) = \vee\{\rho(v_i, v_{i+1}) \mid i = 1, ..., n\}$. Since ρ is not constant on C, the strength of the path $C \setminus \{(v_h, v_{h+1})\}$ between v_h and v_{h+1} is strictly less than $\rho(v_h, v_{h+1})$. The strength of any other path P between v_h and v_{h+1} is also strictly less than $\rho(v_h, v_{h+1})$ since P must contain an edge from $\text{supp}(\rho) \setminus C$. Thus $(v_h, v_{h+1})_{\rho(v_h, v_{h+1})}$ is a fuzzy bridge.

(2) Immediate. ■

Theorem 2.28 *Suppose that the dimension of the cycle space of* $(supp(\mu), supp(\rho))$ *is* 1. *Then* (μ, ρ) *does not have a fuzzy bridge if and only if* (μ, ρ) *is a cycle and* ρ *is a constant function.*

Proof. Suppose it is not the case that (μ, ρ) is a cycle and ρ is a constant function. If (μ, ρ) is not a cycle, then $\exists$ edge $(x, y) \in \text{supp}(\rho)$ which is not part of the cycle. Then $(x, y)_{\rho(x,y)}$ is a bridge and hence a fuzzy bridge. Suppose that (μ, ρ) is a cycle, but ρ is not a constant function. Let $(x, y) \in \text{supp}(\rho)$ be such that $\rho(x, y)$ is maximal. Then $(x, y)_{\rho(x,y)}$ is a fuzzy bridge. Conversely, suppose that (μ, ρ) is a cycle and ρ is a constant function. Then the deletion of an edge (v_i, v_{i+1}) yields a unique path between v_i and v_{i+1} of strength equal to $\rho(v_i, v_{i+1})$. Thus $(v_i, v_{i+1})_{\rho(v_i, v_{i+1})}$ is not a fuzzy bridge. ■

(Fuzzy) Chords, (Fuzzy) Cotrees, and (Fuzzy) Twigs

We assume throughout this section that $(supp(\mu), supp(\rho))$ is connected.

Definition 2.5 *Let* $(\mu,\ \nu)$ *be a spanning fuzzy subgraph of* $(\mu,\ \rho)$ *which is a tree. If* $(\mu,\ \nu_f)$ *is a fuzzy tree such that* $\nu \subseteq \nu_f \subseteq \rho$, $\nu_f = \rho$ *on* $supp(\nu_f)$, *and* $\nexists$ *a fuzzy tree* $(\mu,\ \nu')$ *such that* $\nu_f \subset \nu' \subseteq \rho$ *and* $\nu' = \rho$ *on* $supp(\nu')$, *then* $(\mu,\ \nu_f)$ *is called a* fuzzy spanning tree *of* $(\mu,\ \rho)$ *with respect to* ν.

Clearly given $(\mu,\ \nu)$ and $(\mu,\ \rho)$ of Definition 2.5, $(\mu,\ \nu_f)$ exists. We note that $(\mu,\ \nu)$ is a spanning fuzzy subgraph of $(\mu,\ \nu_f)$.

In Example 2.6, let ν, ν_f, and $\nu_f{}'$ be the fuzzy subsets of X defined as follows: $\nu = \rho$ on $\{(t,u),(t,v),(t,w)\}$ and $\nu(w,v) = \nu(u,v) = 0, \nu_f = \rho$ on $\{(t,u),(t,v),(t,w),(u,v)\}$ and $\nu_f(w,v) = 0$, $\nu'_f = \rho$ on $\{(t,u),(t,v),(t,w),$ $(w,v)\}$ and $\nu'_f(u,v) = 0$. Then both (μ,ν_f) and $(\mu,\ \nu'_f)$ are fuzzy spanning trees of (μ,ρ) with respect to ν. That is, given ν, ν_f in Definition 2.5 is not necessarily unique.

Consider the fuzzy graph $(\mu,\ \rho)$ of Example 2.2 . Define the fuzzy subset ν of X by $\nu(w,u) = \nu(u,v) = 1$. Since $(\mu,\ \rho)$ is not a fuzzy cycle, the addition of $(u,v)_{1/2}$ to $(\mu,\ \nu)$ does not create a fuzzy cycle. This fact motivates the definition of a fuzzy chord below.

Definition 2.6 *Let* $(\mu,\ \nu)$ *be a fuzzy spanning subgraph of* $(\mu,\ \rho)$ *which is a tree. Let* $(x,y) \in supp(\rho)$.

(1) $(x,y)_{\rho(x,y)}$ *is a* chord *of* $(\mu,\ \nu)$ *if and only if* $(x,y) \notin supp(\nu)$, *i. e.,* (x,y) *is a chord of* $(supp(\mu),\ supp(\rho))$.

(2) $(x,y)_{\rho(x,y)}$ *is a* fuzzy chord *of* $(\mu,\ \nu_f)$ *if and only if* $(x,y) \notin (x,y)_{\rho(x,y)}$.

Example 2.7 *Let* $V = \{s,t,u,v,w\}$ *and* $X = \{(w,s),(w,t),(w,u),(w,v),$ $(s,t),(u,v)\}$. *Define the fuzzy subsets* μ *of* V *and* ρ, ν *of* X *by* $\mu(x) = 1$ *for all* $x \in V$ *and* $\rho(x,y) = 1$ *for all* (x,y) *in* $X \setminus \{(w,u)\}$ *and* $\rho(w,u) = 1/2$, $\nu(x,y) = 1$ *for all* (x,y) *in* Y, *where* $Y = X \setminus \{(w,s),\ (w,u)\}$. *Then* $\nu_f = \rho$ *on* $X\setminus\{(w,s)\}$ *and* $\nu_f(w,s) = 0$. *Also* $(w,s)_1$ *and* $(w,u)_{1/2}$ *are chords of* $(\mu,\ \nu)$ *and* $(w,s)_1$ *is a fuzzy chord of* $(\mu,\ \nu_f)$. *If we define the fuzzy subset* ν' *of* X *by* $\nu' = \nu_f$ *on* $supp(\nu_f)$ *and* $\nu'(w,s) = t$ *where* $0 < t < 1$, *then* (μ,ν') *is a fuzzy tree such that* $\nu_f \subset \nu'$. *However,* $\nu' \neq \rho$ *on* $supp(\nu')$.

Definition 2.7 *Let* $(\mu,\ \nu)$ *be a spanning fuzzy subgraph of* $(\mu,\ \rho)$ *which is a tree.*

(1) Let ν' *be a fuzzy subset of* $V \times V$. *Then* $(\mu,\ \nu')$ *is the* cotree *of* $(\mu,\ \nu)$ *if and only if* $\forall\ (x,y) \in supp(\rho)$, $\nu'(x,y) = 0$ *if* $\nu(x,y) > 0$ *and* $\nu'(x,y) = \rho(x,y)$ *if* $\nu(x,y) = 0$.

(2) Let $\nu_f{}'$ *be a fuzzy subset of* $V \times V$. *Then* $(\mu,\ \nu_f{}')$ *is the* fuzzy cotree *of* $(\mu,\ \nu_f)$ *if and only if* $\forall\ (x,y) \in supp(\rho)$, $\nu_f{}'(x,y) = 0$ *if* $\nu_f(x,y) > 0$ *and* $\nu_f{}'(x,y) = \rho(x,y)$ *if* $\nu_f(x,y) = 0$.

Let (μ, ν') be a cotree of (μ, ν), where (μ, ν) is a spanning fuzzy subgraph of (μ, ρ) which is a tree. Then $(\text{supp}(\mu), \text{supp}(\nu'))$ is a cotree of $(\text{supp}(\mu),$ $\text{supp}(\nu))$ since $\text{supp}(\nu') \cap \text{supp}(\nu) = \emptyset$, $\text{supp}(\nu') \cup \text{supp}(\nu) = \text{supp}(\rho)$, and $(\text{supp}(\mu), \text{supp}(\nu))$ is a tree.

Definition 2.8 *Let (μ, ν) be a fuzzy spanning subgraph of (μ, ρ) which is a tree and let $(x, y) \in supp(\rho)$.*

(1) Let (μ, ν') be a cotree of (μ, ν). Then $(x, y)_{\rho(x,y)}$ is a twig *of (μ, ν') if and only if $\nu'(x, y) = 0$.*

(2) Let $(\mu, \nu_f{}')$ be a fuzzy cotree of (μ, ν_f). Then $(x, y)_{\rho(x,y)}$ is a fuzzy twig *of $(\mu, \nu_f{}')$ if and only if $\nu_f{}'(x, y) = 0$.*

Example 2.8 *Let (μ, ρ), (μ, ν), and (μ, ν_f) be the fuzzy subgraphs of Example 2.7. Let (μ, ν') be the cotree of (μ, ν) and $(\mu, \nu_f{}')$ be the fuzzy cotree of (μ, ν_f). Then the twigs of (μ, ν') are $(s, t)_1$, $(w, t)_1$, $(w, v)_1$, and $(u, v)_1$. The fuzzy twigs of $(\mu, \nu_f{}')$ are $(s, t)_1$, $(w, t)_1$, $(w, v)_1$, $(u, v)_1$, and $(w, u)_{1/2}$.*

(Fuzzy) 1-Chain with Boundary 0, (Fuzzy) Coboundary, and (Fuzzy) Cocyles

We recall that a pair $(M, *)$ is a semigroup if M is a nonempty set and $*$ is an associative binary operation on M. We let $G = (\mu, \rho)$. If X is a set of fuzzy singletons, we let $foot(X)$ denote $\{x \mid x_t \in X\}$.

Definition 2.9 *Let $(x, y) \in V \times V$. Then (x, y) is called* exceptional *in G if and only if $\exists$ a cycle $C \subseteq V \times V$ such that $(x, y) \in C$ and (x, y) is unique with respect to $\rho(x, y) = \wedge\{\rho(u, v) \mid (u, v) \in C\}$. Let $E = \{(x, y) \in V \times V \mid (x, y)$ is exceptional$\}$. Let ρ_E be the fuzzy subset of $V \times V$ defined by $\rho_E = \rho$ on $V \times V \setminus E$ and $\rho_E(x, y) = 0$ $\forall (x, y) \in E$.*

Let $S_\rho = \{(x, y)_t \mid (x, y) \in \text{supp}(\rho),\ t \in (0, 1]\} \cup \{0_t \mid t \in (0, 1]\}$. Let addition of elements of $\text{supp}(\rho)$ be a formal addition modulo 2, That is, $\forall (x, y), (u, v) \in \text{supp}(\rho)$, we write $(x, y) + (u, v)$ if $(x, y) \neq (u, v)$ and $(x, y) + (u, v) = 0$ if $(x, y) = (u, v)$. Then $\forall (x, y)_t, (u, v)_s \in S_\rho$, $(x, y)_t + (u, v)_s = ((x, y) + (u, v))_r$ where $r = t \wedge s$. Also $\forall (x, y)_t \in S_\rho$, $(x, y)_t + 0_s = (x, y)_r$ and $0_t + 0_s = 0_r$, where $r = t \wedge s$. Clearly $(S_\rho, +)$ is a commutative semigroup with identity, 0_1. If S is a set of fuzzy singletons of a set W, we let $\text{foot}(S) = \{w \in W \mid w_t \in S\}$.

Since $\mathbb{Z}_2 = \{0, 1\}$ is the field of integers modulo 2, $1 + 1 = 0$. We have that $\sum \epsilon_i (x_i, y_i)_{\rho(x_i, y_i)} + \sum \epsilon_i{}'(x_i, y_i)_{\rho(x_i, y_i)} = \sum(\epsilon_i + \epsilon_i{}')(x_i, y_i)_{\rho(x_i, y_i)}$, $\epsilon_i (x_i, y_i)_{\rho(x_i, y_i)} = (x_i, y_i)_{\rho(x_i, y_i)}$ if $\epsilon_i = 1$ and $\epsilon_i (x_i, y_i)_{\rho(x_i, y_i)} = 0_{\rho(x_i, y_i)}$ if ϵ_i

$= 0$, ϵ_i, $\epsilon_i{}' \in \mathbb{Z}_2$. We have that $\sum \epsilon_i(x_i, y_i)_{\rho(x_i,y_i)} = (\sum \epsilon_i(x_i, y_i))_m$ where $m = \bigwedge_i \{\rho(x_i, y_i)\}$.

Definition 2.10

(1) $\sum \epsilon_i(x_i, y_i)_{\rho(x_i,y_i)}$ *is a* 1-chain with boundary 0 *in* $(\mu,\ \rho)$ *where* $(x_i, y_i) \in$ *supp*(ρ) *if and only if* $\sum \epsilon_i(x_i, y_i)$ *is a 1-chain with boundary 0 in* $($*supp*(μ), *supp*$(\rho))$.

(2) $\sum \epsilon_i(x_i, y_i)_{\rho(x_i,y_i)}$ *is a* fuzzy 1-chain with boundary 0 *in* $(\mu,\ \rho)$ *where* $(x_i, y_i) \in$ *supp*(ρ_E) *if and only if* $\sum \epsilon_i(x_i, y_i)$ *is a 1-chain with boundary 0 in* $($*supp*(μ), *supp*$(\rho_E))$.

A (fuzzy) 1-chain with boundary 0 in $(\mu,\ \rho)$ is called a *(fuzzy) cycle vector.*

Definition 2.11

(1) $\sum \epsilon_i(x_i,\ y_i)_{\rho(x_i,y_i)}$ *is a* coboundary *of* $(\mu,\ \rho)$ *where* $(x_i, y_i) \in$ *supp*(ρ) *if and only if* $\sum \epsilon_i(x_i, y_i)$ *is a coboundary of* $($*supp*(μ), *supp*$(\rho))$.

(2) $\sum \epsilon_i(x_i, y_i)_{\rho(x_i,y_i)}$ *is a* fuzzy coboundary *of* $(\mu,\ \rho)$ *where* $(x_i, y_i) \in$ *supp*(ρ_E) *if and only if* $\sum \epsilon_i(x_i, y_i)$ *is a coboundary of* $($*supp*(μ), *supp*$(\rho_E))$.

$\mathcal{S}' \subseteq \mathcal{S}_\rho$ is called a *(fuzzy) cocycle* of $(\mu,\ \rho)$ if and only if foot$(\mathcal{S}')$ is a cocycle of ((supp(μ), supp(ρ_E)))) (supp(μ), supp(ρ)).

(Fuzzy) Cycle Set and (Fuzzy) Cocycle Set

Definition 2.12

(1) The set of all (fuzzy) cycle vectors of $(\mu,\ \rho)$ *is called the* (fuzzy) cycle set *of* $(\mu,\ \rho)$.

(2) The set of all (fuzzy) coboundaries of $(\mu,\ \rho)$ *is called the* (fuzzy) cocycle set *of* $(\mu,\ \rho)$.

The following examples show that the fuzzy cycle, cycle, fuzzy cocycle, and cocycle sets are not and do not necessarily generate vector spaces over $\mathbb{Z}_2$.

Example 2.9 *Let* $V = \{t, u, v, w\}$ *and* $X = \{(t, u), (u, v), (v, w), (w, t), (t, v)\}$. *Define the fuzzy subsets* μ *of* V *and* ρ *of* X *as follows:* $\mu(x) = 1$ *for all* $x \in V$, $\rho(t, u) = \rho(u, v) = 1$, $\rho(v, w) = \rho(w, t) = 1/2$, *and* $\rho(t, v) = 1/4$. *Then the cycle set is* $\{(t, u)_1 + (u, v)_1 + (t, v)_{1/4}, (v, w)_{1/2} + (w, t)_{1/2} +$

$(t,v)_{1/4}, (t,u)_1+(u,v)_1+(v,w)_{1/2}+(w,t)_{1/2}, 0_{1/4}, 0_{1/2}\}$. The fuzzy cycle set is $\{(t,u)_1+(u,v)_1+(v,w)_{1/2}+(w,t)_{1/2}, 0_{1/2}\}$. The cocycle set is $\{(t,u)_1+(t,v)_{1/4}+(w,t)_{1/2}, (u,v)_1+(t,v)_{1/4}+(v,w)_{1/2}, (t,u)_1+(u,v)_1, (v,w)_{1/2}+(w,t)_{1/2}, (w,t)_{1/2}+(t,v)_{1/4}+(u,v)_1, (t,u)_1+(t,v)_{1/4}+(v,w)_{1/2}, (w,t)_{1/2}+(t,u)_1+(u,v)_1+(v,w)_{1/2}, 0_{1/4}, 0_{1/2}\}$. The fuzzy cocycle set is $\{(t,u)_1+(w,t)_{1/2}, (u,v)_1+(v,w)_{1/2}, (t,u)_1+(u,v)_1, (v,w)_{1/2}+(w,t)_{1/2}, (w,t)_{1/2}+(u,v)_1, (t,u)_1+(v,w)_{1/2}, (w,t)_{1/2}+(t,u)_1+(u,v)_1+(v,w)_{1/2}, 0_{1/2}\}$. The cycle set and cocycle set are not and do not generate vector spaces over $\mathbb{Z}_2$ because of the presence of $0_{1/2}$ and $0_{1/4}$. Note also that in the cycle set, $((t,u)_1+(u,v)_1+(t,v)_{1/4})+((v,w)_{1/2}+(w,t)_{1/2}+(t,v)_{1/4}) = (t,u)_1 + (u,v)_1+(v,w)_{1/2}+(w,t)_{1/2}+0_{1/4} \neq (t,u)_1+(u,v)_1+(v,w)_{1/2}+(w,t)_{1/2}$. The fuzzy cycle set is a vector space over $\mathbb{Z}_2$ in this example. The fuzzy cocycle set is not a vector space over $\mathbb{Z}_2$ since $((v,w)_{1/2}+(w,t)_{1/2}) + ((w,t)_{1/2}+(t,u)_1+(u,v)_1+(v,w)_{1/2}) = (t,u)_1+(u,v)_1+0_{1/2} \neq (t,u)_1+(u,v)_1$.

Example 2.10 *Let V, μ, and X be as in Example 2.8. Let $X' = X \cup \{(u,w)\}$. Define the fuzzy subset ρ' of X' by $\rho' = \rho$ on X and $\rho'(u,w) = 1/8$. Then the fuzzy cycle set and the fuzzy cocycle set of (μ, ρ') coincides with the cycle set and the cocycle set of (μ, ρ) of Example 2.8, respectively.*

Examples 2.8 and 2.9 illustrate the results which follow.

Let $\mathcal{CS}(\mu,\rho)$, $\mathcal{FCS}(\mu,\rho)$, $\mathcal{C}o\mathcal{S}(\mu,\rho)$, and $\mathcal{FC}o\mathcal{S}(\mu,\rho)$ denote the cycle set, the fuzzy cycle set, the cocycle set, and the fuzzy cocycle set of (μ,ρ), respectively. When the fuzzy graph (μ,ρ) is understood, we sometimes write $\mathcal{CS}$, $\mathcal{FCS}$, $\mathcal{C}o\mathcal{S}$, and $\mathcal{FC}o\mathcal{S}$ for these sets, respectively.

We now show that even though $\mathcal{CS}$, $\mathcal{FCS}$, $\mathcal{C}o\mathcal{S}$, and $\mathcal{FC}o\mathcal{S}$ are not necessarily vector spaces over $\mathbb{Z}_2$, they are nearly so. In fact, it will be clear by the following results that the concepts of (fuzzy) twigs and (fuzzy) chords introduced here will have consequences similar to what their counterparts have in the crisp case.

For ease of notation, we sometimes use the notation e, f,or g for members of supp(ρ).

Clearly, $\mathcal{CS}$, $\mathcal{FCS}$, $\mathcal{C}o\mathcal{S}$, and $\mathcal{FC}o\mathcal{S}$ are subsets of $S_\rho = \{e_t \mid e \in \text{supp}(\rho), t \in (0,1]\} \cup \{0_t \mid t \in (0,1]\}$. Let $\mathcal{S}$ be a subset of S_ρ. We let $< \mathcal{S} >$ denote the intersection of all subsemigroups of S_ρ which contain $\mathcal{S}$. Then $< \mathcal{S} >$ is the smallest subsemigroup of S_ρ which contains $\mathcal{S}$. Let $\mathcal{S}^+ = \{(e_1)_{t_1} + ... + (e_n)_{t_n} \mid (e_i)_{t_i} \in \mathcal{S}, i = 1,...,n, n \in \mathbb{N}\}$ where $\mathbb{N}$ denotes the set of positive integers. Then $\mathcal{S}^+$ is a subsemigroup of S_ρ.

Theorem 2.29 $< \mathcal{CS} > = (\mathcal{CS})^+ = \mathcal{CS} \cup \{e_a + 0_b \mid e_a \in \mathcal{CS}, 0_b \in (\mathcal{CS})^+\}$. *$< \mathcal{CS} >$ has 0_m as its identity where $m = \vee\{b \mid 0_b \in (\mathcal{CS})^+\}$.*

Proof. Since $(\mathcal{CS})^+$ is a subsemigroup of S_ρ which contains $\mathcal{CS}$, $< \mathcal{CS} > \subseteq (\mathcal{CS})^+$. However, it is clear that $< \mathcal{CS} > \supseteq (\mathcal{CS})^+$ since $< \mathcal{CS} >$ is closed under $+$. Thus $< \mathcal{CS} > = (\mathcal{CS})^+$. Clearly $< \mathcal{CS} > \supseteq \mathcal{CS} \cup \{e_a + 0_b \mid e_a \in \mathcal{CS},\ 0_b \in (\mathcal{CS})^+\}$. Now $foot(\mathcal{CS})$ is a vector space over $\mathbb{Z}_2$. Let $g_t, f_s \in \mathcal{CS}$. Then $g + f \in foot(\mathcal{CS})$. Also $g_t + f_s = (g + f)_r$ where $r = t \wedge s$. Now $g_t = (u_1, v_1)_{t_1} + ... + (u_n, v_n)_{t_n}$ and $f_s = (p_1, q_1)_{s_1} + ... + (p_m, q_m)_{s_m}$, where $(u_i, v_i), (p_j, q_j) \in X$, $i = 1, .., n$ and $j = 1, ..., m$. Let $I = \{(u_i, v_i) \mid i = 1, .., n\} \cap \{(p_j, q_j) \mid j = 1, ..., m\}$. Suppose that $I \neq \emptyset$. By rearranging the summands in the representations of g_t and f_s if necessary, we have $g_t + f_s = (u_1, v_1)_{t_1} + ... + (u_{i-1}, v_{i-1})_{t_{i-1}} + (p_1, q_1)_{s_1} + ... + (p_{j-1}, q_{j-1})_{s_{j-1}} + 0_b$ where $0_b = (u_i, v_i)_{t_i} + ... + (u_n, v_n)_{t_n} + (p_j, q_j)_{s_j} + ... + (p_m, q_m)_{s_m}$, $b = \wedge\{ t_i, ..., t_n,\ s_j, ..., s_m\}$, and $I = \{(u_i, v_i), ..., (u_n, v_n)\} = \{(p_j, q_j), ..., (p_m, q_m)\}$. Now $(u_1, v_1) + ... + (u_{i-1}, v_{i-1}) + (p_1, q_1) + ... + (p_{j-1}, q_{j-1}) \in foot(\mathcal{CS})$ since $foot(\mathcal{CS})$ is a vector space over $\mathbb{Z}_2$. Also $(g + f)_a = (u_1, v_1)_{t_1} + ... + (u_{i-1}, v_{i-1})_{t_{i-1}} + (p_1, q_1)_{s_1} + ... + (p_{j-1}, q_{j-1})_{s_{j-1}}$, where $a = \wedge\{t_1, ..., t_{i-1},\ s_1, ..., s_{j-1}\}$. Now $(g + f)_r + 0_b = (g + f)_a + 0_b$. Hence $g_t + f_s \in \mathcal{CS} \cup \{e_a + 0_b \mid e_a \in \mathcal{CS},\ 0_b \in (\mathcal{CS})^+\}$. That is, the sum of any two elements from $\mathcal{CS}$ is in $\mathcal{CS} \cup \{e_a + 0_b \mid e_a \in \mathcal{CS},\ 0_b \in (\mathcal{CS})^+\}$. From this it follows easily that $(\mathcal{CS})^+ \subseteq \mathcal{CS} \cup \{e_a + 0_b \mid e_a \in \mathcal{CS},\ 0_b \in (\mathcal{CS})^+\}$. The case where $I = \emptyset$ is similar. ■

Corollary 2.30 $< \mathcal{FCS} > = (\mathcal{FCS})^+ = \mathcal{FCS} \cup \{e_a + 0_b \mid e_a \in \mathcal{FCS},\ 0_b \in (\mathcal{FCS})^+\}$. *$< \mathcal{FCS} >$ has 0_m as its identity where $m = \vee\{b \mid 0_b \in (\mathcal{FCS})^+\}$.*

Proof. $\mathcal{FCS}$ is the cycle set of (μ, ρ_E). ■

In a similar manner, we obtain the next two results.

Theorem 2.31 $< \mathcal{C}o\mathcal{S} > = (\mathcal{C}o\ \mathcal{S})^+ = \mathcal{C}o\mathcal{S} \cup \{e_a + 0_b \mid e_a \in \mathcal{C}o\mathcal{S},\ 0_b \in (\mathcal{C}o\mathcal{S})^+\}$. *$< \mathcal{C}o\mathcal{S} >$ has 0_m as its identity where $m = \vee\{b \mid 0_b \in (\mathcal{C}o\mathcal{S})^+\}$.* ■

Corollary 2.32 $< \mathcal{FC}o\mathcal{S} > = (\ \mathcal{FC}o\mathcal{S})^+ = \mathcal{FC}o\mathcal{S} \cup \{e_a + 0_b \mid e_a \in \mathcal{FC}o\mathcal{S},\ 0_b \in (\mathcal{FC}o\mathcal{S})^+\}$. *$< \mathcal{FC}o\mathcal{S} >$ has 0_m as its identity where $m = \vee\{b \mid 0_b \in (\mathcal{FC}o\mathcal{S})^+\}$.* ■

Since $\mathcal{CS}$, $\mathcal{FCS}$, $\mathcal{C}o\ \mathcal{S}$, and $\mathcal{FC}o\mathcal{S}$ are nearly vector spaces over $\mathbb{Z}_2$, we can define the (fuzzy) cycle rank and (fuzzy) cocycle rank of a fuzzy graph in a meaningful way.

Definition 2.13 *The* cycle rank *of (μ, ρ), written $m(\mu, \rho)$, is defined to be $m(\mu, \rho) = \vee\{ \sum_{i=1}^{n} t_i \mid (e_i)_{t_i} \in \mathcal{CS},\ i = 1, ..., n, \{e_1, ..., e_n\}$ is a basis for $foot(\mathcal{CS})\}$. The* fuzzy cycle rank *of (μ, ρ), written $fm(\mu, \rho)$, is defined to be the cycle rank of (μ, ρ_E). If $\{e_1, ..., e_n\}$ is a basis for $foot(\mathcal{CS})$ such that $m(\mu, \rho) = \sum_{i=1}^{n} t_i$, where $(e_i)_{t_i} \in \mathcal{CS}$, $i = 1, ..., n$, then $\{e_1, ..., e_n\}$ is called a* cycle basis *of $< \mathcal{CS} >$. If $\{e_1, ..., e_n\}$ is a basis for $foot(\mathcal{FCS}(\mu, \rho))$*

FIGURE 2.5 Non-isomorphic fuzzy graphs with the same cycle rank.

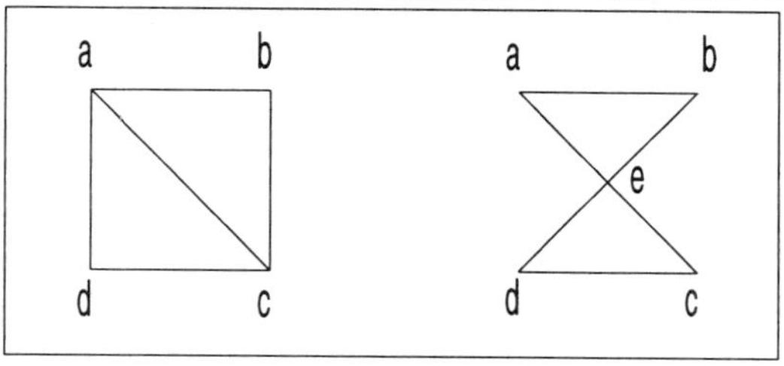

such that $fm(\mu,\rho)=\sum_{i=1}^{n}t_i$, *where* $(e_i)_{t_i}\in\mathcal{FCS}(\mu,\rho)$, $i=1,...,n$, *then* $\{e_1,...,e_n\}$ *is called a* fuzzy cycle basis *of* $<\mathcal{CS}(\mu,\rho)>$.

Theorem 2.33 *Let* $\{e_1,...,e_n\}$ *be a cycle basis of* $<\mathcal{CS}>$. *Then* $\forall\ e_t\in\mathcal{CS}$, *there is a reordering of* $e_1,...,e_n$ *such that* $e_t=(e_1)_{t_1}+\ ...\ +(e_m)_{t_m}$, $m\le n$, *where* $t_i=\rho(e_i)$, $i=1,...,m$.

Proof. Since $\{e_1,...,e_n\}$ is a basis for $foot(\mathcal{CS})$, there is a reordering of $e_1,...,e_n$ such that $e=e_1+...+e_m$, $m\le n$. Suppose that $t>t_1\wedge...\wedge t_m$. Then there is a t_i, $i=1,...,m$, such that $t>t_i$. Now $e\notin<\{e_1+...+e_n\}\backslash\{e_i\}>$. Hence $(\{e_1+...+e_n\}\setminus\{e_i\})\cup\{e\}$ is a basis for $foot(\mathcal{CS})$. However this contradicts the hypothesis that $\{e_1,...,e_n\}$ is a cycle basis of $<\mathcal{CS}>$ since $t>t_i$. Thus $t\le t_1\wedge...\wedge t_m$. Now $e=(u_1,v_1)-...+(u_rv_r)$ where $(u_i,v_i)\in V\times V$, $i=1,...,r$. Hence $e_t=(u_1,v_1)_{a_1}+...+(u_r,v_r)_{a_r}$ and $(e_1)_{t_1}+\ ...\ +(e_m)_{t_m}=(u_1,v_1)_{a_1}+...+(u_r,v_r)_{a_r}+0_a$ for some $a\in(0,1]$ where $\rho(u_i,v_i)=a_i$, $i=1,...,r$, and $t_1\wedge...\wedge t_m=a_1\wedge...\wedge a_r\wedge a$. Now $a_1\wedge...\wedge a_r\wedge a\le a_1\wedge...\wedge a_r=t\le t_1\wedge...\wedge t_m=a_1\wedge...\wedge a_r\wedge a$. Hence $t=t_1\wedge...\wedge t_m$ and so $e_t=(e_1)_{t_1}+\ ...\ +(e_m)_{t_m}$. ■

Corollary 2.34 *Let* $\{e_1,...,e_n\}$ *be a fuzzy cycle basis of* $<\mathcal{CS}(\mu,\rho)>$. *Then* $\forall\ e_t\in\mathcal{FCS}(\mu,\rho)$, *there is a reordering of* $e_1,...,e_n$ *such that* $e_t=(e_1)_{t_1}+\ ...\ +(e_m)_{t_m}$, $m\le n$, *where* $t_i=\rho(e_i)$, $i=1,...,m$.

Proof. $\mathcal{FCS}$ is the cycle set of (μ,ρ_E) and the *fuzzy cycle rank* of (μ,ρ) is the cycle rank of (μ,ρ_E). ■

The following graphs have the same cycle rank, but they are of course not isomorphic since one has four vertices and the other has five vertices.

Definition 2.14 *The* cocycle rank *of* (μ,ρ), *written* $m_c(\mu,\rho)$, *is defined to be*

$m_c(\mu,\rho)=\vee\{\ \sum_{i=1}^{n}t_i\mid(e_i)_{t_i}\in\mathcal{C}o\mathcal{S},\ i=1,...,n,\{e_1,...,e_n\}$ *is a basis for foot*$(\mathcal{C}o\mathcal{S})\}$. *The* fuzzy cocycle rank *of* (μ,ρ), *written* $fm_c(\mu,\rho)$, *is defined to be the cocycle rank of* (μ,ρ_E). *If* $\{e_1,...,e_n\}$ *is a basis for foot*($\mathcal{C}o\mathcal{S}$) *such that* $m_c(\mu, \rho) = \sum_{i=1}^{n} t_i$, *where* $(e_i)_{t_i} \in \mathcal{C}o\mathcal{S}$, $i = 1, ..., n$, *then* $\{e_1, ..., e_n\}$ *is called a* cocycle basis *of* $< \mathcal{C}o\mathcal{S} >$. *If* $\{e_1, ..., e_n\}$ *is a basis for foot*($\mathcal{FC}o\mathcal{S}(\mu, \rho)$) *such that* $fm_c(\mu, \rho) = \sum_{i=1}^{n} t_i$, *where* $(e_i)_{t_i} \in \mathcal{FC}o\mathcal{S}(\mu, \rho)$, $i = 1, ..., n$, *then* $\{e_1, ..., e_n\}$ *is called a* fuzzy cocycle basis *of* $< \mathcal{CS}(\mu, \rho) >$.

In a similar manner, we obtain the next two results.

Theorem 2.35 *Let* $\{e_1, ..., e_n\}$ *be a cocycle basis of* $< \mathcal{C}o\mathcal{S} >$. *Then* $\forall$ $e_t \in \mathcal{C}o\mathcal{S}$, *there is a reordering of* $e_1, ..., e_n$ *such that* $e_t = (e_1)_{t_1} + ... + (e_m)_{t_m}$, $m \leq n$, *where* $t_i = \rho(e_i)$, $i = 1, ..., m$. ■

Corollary 2.36 *Let* $\{e_1, ..., e_n\}$ *be a fuzzy cocycle basis of* $< \mathcal{CS}(\mu, \rho) >$. *Then* $\forall$ $e_t \in \mathcal{FCS}(\mu, \rho)$, *there is a reordering of* $e_1, ..., e_n$ *such that* $e_t = (e_1)_{t_1} + ... + (e_m)_{t_m}$, $m \leq n$, *where* $t_i = \rho(e_i)$, $i = 1, ..., m$. ■

2.2 Fuzzy Line Graphs

The results of this section are taken from [29]. The line graph, $L(G)$, of a graph G is the intersection graph of the set of edges of G. Hence the vertices of $L(G)$ are the edges of G with two vertices of $L(G)$ adjacent whenever the corresponding edges of G are. In this section, the concept of a fuzzy line graph is introduced and its basic properties are developed. We give a necessary and sufficient condition for a fuzzy graph to be isomorphic to its corresponding fuzzy line graph. We also examine when the isomorphism between two fuzzy graphs follows from an isomorphism of their corresponding fuzzy line graphs. A necessary and sufficient condition for a fuzzy line graph to be the fuzzy line graph of some fuzzy graph is also presented in this section. Let $G = (V, X)$ and $G' = (V', X')$ be graphs. If μ is a fuzzy subset of V and ρ is a fuzzy subset of $V \times V$ such that (μ, ρ) is a fuzzy graph and supp$(\rho) \subseteq X$, we call (μ, ρ) a *partial fuzzy subgraph* of G.

Definition 2.15 *Let* (μ, ρ) *and* (μ', ρ') *be partial fuzzy subgraphs of* G *and* G', *respectively. Let* f *be a one-to-one function of* V *onto* V'. *Then*

(1) f *is called a* (weak) vertex-isomorphism *of* (μ, ρ) *onto* (μ', ρ') *if and only if* $\forall v \in V$, $(\mu(v) \leq \mu'(f(v))$ *and* $supp(\mu') = (f(supp(\mu)))$ $\mu(v) = \mu'(f(v))$.

(2) f *is called a* (weak) line-isomorphism *of* (μ, ρ) *onto* (μ', ρ') *if and only if* $\forall (u, v) \in X$, $(\rho(u, v) \leq \rho'(f(u), f(v))$ *and* $supp(\rho') = \{(f(u), f(v)) \mid (u, v) \in supp(\rho)\})$ $\rho(u, v) = \rho'(f(u), f(v))$.

If f is a (weak) vertex-isomorphism and a (weak) line isomorphism of (μ, ρ) onto (μ', ρ'), then f is called a (weak) isomorphism of (μ, ρ) onto (μ', ρ'). If (μ, ρ) is isomorphic to (μ', ρ'), then we write $(\mu, \rho) \simeq (\mu', \rho')$.

Our definition of isomorphism is equivalent to the one in [5, Definition 3.2, p. 160]. For homomorphisms [5, Definition 3.1, p. 159], our definitions of vertex-isomorphism and line-isomorphism are equivalent to the definitions of weak isomorphism and co-weak isomorphism, respectively, which appear in [5, Definition 3.3, 3.5, p.160]. We use the term 'weak' in a different manner than in [5].

Let $G = (V, X)$ be a graph where $V = \{v_{1,}, \ldots, v_n\}$. Let $S_i = \{v_i, x_{i1}, \ldots, x_{iq_i}\}$ where $x_{ij} \in X$ and x_{ij} has v_i as a vertex, $j = 1, \ldots, q_{i;}$ $i = 1, \ldots, n$. Let $S = \{S_1, \ldots, S_n\}$. Let $T = \{(S_i, S_j) \mid S_i, S_j \in S,\ S_i \cap S_j \neq \emptyset, i \neq j\}$. Then $\mathcal{I}(S) = (S, T)$ is an intersection graph and $G \simeq \mathcal{I}(S)$. Any partial fuzzy subgraph (ι, γ) of $\mathcal{I}(S)$ with $\text{supp}(\gamma) = T$ is called a *fuzzy intersection graph.*

Let (μ, ρ) be a partial fuzzy subgraph of G. Let $\mathcal{I}(S)$ be the intersection graph described above. Define the fuzzy subsets ι, γ of S and T, respectively, as follows:

$\forall S_i \in S,\ \iota(S_i) = \mu(v_i)$;

$\forall S_i \in T,\ \gamma(S_i, S_j) = \rho(v_i, v_j)$.

Proposition 2.37 *Let (μ, ρ) be a partial fuzzy subgraph of G . Then*

(1) (ι, γ) is a partial fuzzy subgraph of $\mathcal{I}(S)$;

(2) $(\mu, \rho) \simeq (\iota, \gamma)$.

Proof. (1) $\gamma(S_i, S_j) = \rho(v_i, v_j) \leq \mu(v_i) \wedge \mu(v_j) = \iota(S_i) \wedge \iota(S_j)$

(2) Define $f : V \to S$ by $f(v_i) = S_i, i = 1, \ldots, n$. Clearly, f is a one-to-one function of V onto S. Now $(v_i, v_j) \in X$ if and only if $(S_i, S_j) \in T$ and so $T = \{(f(v_i), f(v_j)) \mid (v_i, v_j) \in X\}$. Also $\iota(f(v_i)) = \iota(S_i) = \mu(v_i)$ and $\gamma(f(v_i), f(v_j)) = \gamma(S_i, S_j) = \rho(v_i, v_j)$. Thus f is an isomorphism of (μ, ρ) onto (ι, γ) . ■

Let $\mathcal{I}(S)$ be the intersection graph of (V, X). Let (ι, γ) be the fuzzy subgraph of $\mathcal{I}(S)$ as defined above. We call (ι, γ) the *fuzzy intersection graph* of (μ, ρ). Proposition 2.37 shows that any fuzzy graph is isomorphic to a fuzzy intersection graph.

Now $L(G)$, the line graph of G, is by definition the intersection graph $\mathcal{I}(X)$. That is, $L(G) = (Z, W)$ where $Z = \{\{x\} \cup \{u_x, v_x\} \mid x \in X, u_x, v_x \in V,\ x = (u_x, v_x)\}$ and $W = \{(S_x, S_y) \mid S_x \cap S_y \neq \emptyset,\ x \in X,\ x \neq y\}$ and where $S_x = \{x\} \cup \{u_x, v_x\}$, $x \in X$. Let (μ, ρ) be a partial fuzzy subgraph of G. Define the fuzzy subsets λ, ω of Z, W, respectively, as follows:

$\forall S_x \in Z, \lambda(S_x) = \rho(x)$;

$\forall (S_x, S_y) \in W,\ \omega(S_x, S_y) = \rho(x) \wedge \rho(y)$.

Proposition 2.38 *(λ, ω) is a fuzzy subgraph of $L(G)$, called the* fuzzy line graph *corresponding to (μ, ρ).*

Proof. $\omega(S_x, S_y) = \rho(x) \wedge \rho(y) = \lambda(S_x) \wedge \lambda(S_y)$. ∎

We recall that a cutpoint of a graph is a vertex whose deletion increases the number of components, and a bridge is such an edge. Every cutpoint of $L(G)$ is a bridge of G which is not an endline, and conversely, [20, p.71].

Recall that $(u, v) \in X$ is defined to be a bridge of (μ, ρ) if and only if deleting (u, v) reduces the strength of connectedness between some pair of vertices. Also, recall that $v \in V$ is defined to be a cutpoint in (μ, ρ) if and only if deleting v reduces the strength of connectedness between some pair of vertices not including v.

The following example shows that the relationship between cutpoints in $L(G)$ and bridges in G does not carry over to the fuzzy case for the above definitions of cutpoints and bridges.

Example 2.11 *Let G_1 be the graph defined in [20, p.72], that is, $G_1 = (V, X)$ where $V = \{v_1, v_2, v_3, v_4\}$ and $X = \{x_1 = (v_1, v_2), x_2 = (v_1, v_3), x_3 = (v_2, v_3), x_4 = (v_3, v_4)\}$. Let $\mu(v_1) = \mu(v_2) = \mu(v_3) = \mu(v_4) = 1$, $\rho(x_1) = \rho(x_3) = 1$, and $\rho(x_2) = \rho(x_4) = 1/2$. Then $\lambda(S_{x_1}) = \lambda(S_{x_3}) = 1, \lambda(S_{x_2}) = \lambda(S_{x_4}) = 1/2$, and $\omega(S_{x_1}, S_{x_2}) = 1, \omega(S_{x_2}, S_{x_3}) = \omega(S_{x_3}, S_{x_4}) = \omega(S_{x_2}, S_{x_4}) = 1/2$. If we delete x_1 from G_1, then the strength of connectedness between v_1 and v_2 is $1/2 = \rho(x_2) \wedge \rho(x_3)$ while the strength of connectedness between v_1 and v_2 before the deletion of x_1 is $1 = \rho(x_1)$. Thus x_1 is a bridge of (μ, ρ) (and not an endline of G_1). However the strength of connectedness between any pair of vertices $S_{x_2}, S_{x_3}, S_{x_4}$ is $1/2$ before and after the deletion of S_{x_1}. Thus S_{x_1} is not a cutvertex of (λ, ω).*

Proposition 2.39 *Let (μ, ρ) and (μ', ρ') be partial fuzzy subgraphs of G and G', respectively. If f is a weak isomorphism of (μ, ρ) onto (μ', ρ'), then f is an isomorphism of (supp(μ), supp(ρ)) onto (supp(μ'), supp(ρ')).*

Proof. $v \in \text{supp}(\mu) \Leftrightarrow f(v) \in \text{supp}(\mu')$ and $(u, v) \in \text{supp}(\rho) \Leftrightarrow (f(u), f(v)) \in \text{supp}(\rho')$. ∎

Proposition 2.40 *If (λ, ω) is the fuzzy line graph of (μ, ρ), then (supp(λ), supp(ω)) is the line graph of (supp(μ), supp(ρ)).*

Proof. (μ, ρ) is a partial fuzzy subgraph of G and (λ, ω) is a partial fuzzy subgraph of $L(G)$. Now $\lambda(S_x) = \rho(x)\ \forall x \in X$ and so $S_x \in \text{supp}(\lambda) \Leftrightarrow x \in \text{supp}(\rho)$. Also $\omega(S_x, S_y) = \rho(x) \wedge \rho(y)\ \forall (S_x, S_y) \in W$ and so $\text{supp}(\omega) = \{(S_x, S_y) \mid S_x \cap S_y \neq \emptyset,\ x, y \in \text{supp}(\rho),\ x \neq y\}$. ∎

We also see in Proposition 2.40 that $(\lambda \mid_{\text{supp}(\lambda)}, \omega \mid_{\text{supp}(\omega)})$ is the fuzzy line graph corresponding to $(\mu \mid_{\text{supp}(\mu)}, \rho \mid_{\text{supp}(\rho)})$.We now give a necessary and sufficient condition for fuzzy graph (μ, ρ) to be isomorphic to its fuzzy line graph (λ, ω) .

Theorem 2.41 *Let (λ, ω) be the fuzzy line graph corresponding to (μ, ρ). Suppose that $(supp(\mu), supp(\rho))$ is connected. Then*

(1) $\exists$ a weak isomorphism of (μ, ρ) onto (λ, ω) if and only if (supp(μ), supp(ρ)) is a cycle and $\forall v \in$ supp(μ), $\mu(v) = \rho(x)$, that is, μ and ρ are constant functions of supp(μ) and supp(ρ), respectively, taking on the same value.

(2) If f is a weak isomorphism of (μ, ρ) onto (λ, ω), then f is an isomorphism.

Proof. Suppose that f is a weak isomorphism of (μ, ρ) onto (λ, ω). By Proposition 2.39, f is an isomorphism of (supp(μ), supp(ρ)) onto (supp(λ), supp(ω)). By Proposition 2.40, (supp(μ), supp(ρ)) is a cycle, [20, Theorem 8.2, p.72]. Let supp(μ) $=$ $\{v_1, \ldots, v_n\}$ and supp(ρ) $= \{(v_1, v_2), (v_2, v_3), \ldots, (v_n, v_1)\}$ where $v_1v_2v_3 \ldots v_nv_1$ is a cycle. Let $\mu(v_i)$ $= s_1$ and $\rho(v_1, v_{1+1}) = r_{i,} i = 1, \ldots, n$ where $v_{n+1} = v_1$. Then for $s_{n+1} = s_1$,

$$r_i \leq s_i \wedge s_{i+1}, i = 1, \ldots, n. \tag{2.1}$$

Now supp(λ) $= \{S_{(v_i, v_{i+1})} \mid i = 1, \ldots, n\}$ and supp(ω) $= \{(S_{(v_i, v_{i+1})}, S_{(v_{i+1}, v_{i+2})}) \mid i = 1, \ldots, n-1\}$. Also for $r_{n+1} = r_1$, $\lambda(S_{(v_i, v_{i+1})}) = \rho(v_i, v_{i+1})$ $= r_i$ and $\omega(S_{(v_i, v_{i+1})}, S_{(v_{i+1}, v_{i+2})}) = \rho(v_i, v_{i+1}) \wedge \rho(v_{i+1}, v_{i+2}) = r_i \wedge r_{i+1}, i =$ $1, \ldots, n$, where $v_{n+2} = v_2$. Since f is an isomorphism of (supp(μ), supp(ρ)) onto (supp(λ), supp(ω)), f maps supp(μ) one to one onto supp(λ) $=$ $\{S_{(v_i, v_2)}, \ldots, S_{(v_n, v_1)}\}$. Also f preserves adjacency. Hence f induces a permutation π of $\{1, \ldots, n\}$ such that $f(v_i) = S_{(v_{\pi(i)}, v_{\pi(i)+1})}$ and $(v_i, v_{i+1}) \rightarrow$ $(f(v_i), f(v_{i+1})) = (S_{(v_{\pi(i)}, v_{\pi(i)+1})}, S_{(v_{\pi(i+1)}, v_{\pi(i+1)+1})}), i = 1, \ldots, n-1$. Now $s_i = \mu(v_i) \leq \lambda(f(v_i)) = \lambda(S_{(v_{\pi(i)}, v_{\pi(i)+1})}) = r_{\pi(i)}$ and $r_i = \rho(v_i, v_{i+1}) \leq$ $\omega(f(v_i), f(v_{i+1})) = \omega(S_{(v_{\pi(i)}, v_{\pi(i)+1})}, S_{(v_{\pi(i+1)}, v_{\pi(i+1)+1})}) = \rho(v_{\pi(i)}, v_{\pi(i)+1}) \wedge$ $\rho(v_{\pi(i+1)}, v_{\pi(i+1)+1}) = r_{\pi(i)} \wedge r_{\pi(i+1)}$, $i = 1, \ldots, n$. That is,

$$s_i \leq r_{\pi(i)} \text{ and } r_i \leq r_{\pi(i)} \wedge r_{\pi(i+1)}, i = 1, \ldots, n. \tag{2.2}$$

By the second part of (2.2), we have that $r_i \leq r_{\pi(i)}$, $i = 1, \ldots, n$, and so $r_{\pi(i)} \leq r_{\pi(\pi(i))}$, $i = 1, \ldots, n$. Continuing, we have that $r_i \leq r_{\pi(i)} \leq \ldots \leq$ $r_{\pi^j(i)} \leq r_i$ and so $r_i = r_{\pi(i)}$, $i = 1, \ldots, n$, where π^{j+1} is the identity map. By (2.2) again, we have $r_i \leq r_{\pi(i+1)} = r_{i+1}, i = 1, \ldots, n$ where $r_{n+1} = r_1$. Hence by (2.1) and (2.2), $r_1 = \ldots = r_n = s_1 = \ldots = s_n$. Thus we have not only proved the conclusion about μ and ρ being constant functions, but we have also shown that (2) holds. Conversely, suppose that (supp(μ), supp(ρ)) is a cycle and $\forall v \in$ supp(μ), $\forall x \in$ supp(ρ), $\mu(v) = \rho(x)$. By Proposition 2.40, (supp(λ), supp(ω)) is the line graph of (supp(μ), supp(ρ)). Since (supp(μ), supp(ρ)) is a cycle, (supp(μ), supp(ρ)) $\simeq$ (supp(λ), supp(ω)) by [20, Theorem 8.2, p.72]. This isomorphism induces an isomorphism of (μ, ρ) onto (λ, ω) since $\mu(v) = \rho(x)$ $\forall v \in V, \forall v \in X$ and so $\mu = \rho = \lambda = \omega$ on their respective domains. ■

Theorem 2.42 *Let (μ, ρ) and (μ', ρ') be the partial fuzzy subgraphs of G and $G^{'}$, respectively, such that (supp(μ), supp(ρ)) and (supp(μ'), supp(ρ')) are connected. Let (λ, ω) and (λ', ω') be the line graphs corresponding to (μ, ρ) and (μ', ρ'), respectively. Suppose that it is not the case that one of (supp(μ), supp(ρ)) and (supp(μ'), supp(ρ')) is K_3 and the other is $K_{1,3}$. If $(\lambda, \omega) \simeq (\lambda', \omega')$, then (μ, ρ) and (μ', ρ') are line-isomorphic.*

Proof. Since $(\lambda, \omega) \simeq (\lambda', \omega')$, (supp($\lambda$), supp($\omega$)) $\simeq$ (supp(λ'), supp(ω')) by Proposition 2.39. Since (supp(λ), supp(ω)) and (supp(λ'), supp(ω')) are line graphs of (supp(μ), supp(ρ)) and (supp(μ'), supp(ρ')) respectively, by Proposition 2.40, we have that (supp(μ), supp(ρ)) $\simeq$ (supp(μ'), supp(ρ')) by [20, Theorem 8.3, p.72]. Let g denote the isomorphism of (λ, ω) onto (λ', ω') and f the isomorphism of (supp(μ), supp(ρ)) onto (supp(μ'), supp(ρ')). Then $\lambda(S_{(u,v)}) = \lambda'(g(S_{(u,v)})) = \lambda'(S_{(f(u),f(v))})$ where the latter equality holds the proof of [20, Theorem 8.3, p.72] and so $\rho(u, v) = \rho'(f(u), f(v))$. Hence (μ, ρ) and (μ', ρ') are line isomorphic. ■

Proposition 2.43 *Let (τ, ν) be a partial fuzzy subgraph of $L(G)$. Then (τ, v) is a fuzzy line graph of some partial fuzzy subgraph of G if and only if $\forall (S_x, S_y) \in W$, $\nu(S_x, S_y) = \tau(S_x) \wedge \tau(S_y)$.*

Proof. Suppose that $\nu(S_x, S_y) = \tau(S_x) \wedge \tau(S_y)$ $\forall (S_x, S_y) \in W$. $\forall x \in X$, define $\rho(x) = \tau(S_x)$. Then $\nu(S_x, S_y) = \tau(S_x) \wedge \tau(S_y) = \rho(x) \wedge \rho(y)$. Any μ that yields the property $\rho(u, v) \leq \mu(u) \wedge \mu(v)$ will suffice, e.g., $\mu(v) = 1$ $\forall v \in V$. The converse is immediate. ■

Not every graph is a line graph of some graph . The following result tells us when a fuzzy graph is a fuzzy line graph of some fuzzy graph.

Theorem 2.44 *(μ, ρ) is a fuzzy line graph if and only if (supp(μ), supp(ρ)) is a line graph and $\forall (u, v) \in$ supp(ρ), $\rho(u, v) = \mu(u) \wedge \mu(v)$.*

Proof. Suppose that (μ, ρ) is a fuzzy line graph. Then the conclusion holds by the Proposition 2.40 and 2.43. Conversely , suppose that (*supp*(μ), *supp*(ρ)) is a line graph and $\forall (u, v) \in$ supp(ρ), $\rho(u, v) = \mu(u) \wedge \mu(v)$. Then the conclusion holds from Proposition 2.43. ■

2.3 Fuzzy Interval Graphs

In this section, we present the results of [8]. Intersection graphs and, in particular, interval graphs are used extensively in mathematical modeling. Applications in archaeology, developmental psychology, ecological modeling, mathematical sociology and organization theory are cited in [33]. These disciplines all have components that are ambiguously defined, require subjective evaluation, or are satisfied to differing degrees. They are active areas of applications of fuzzy methods. It is therefore worthwhile to study the extent that intersection graph results can be extended using fuzzy set theory.

The *intersection graph* of a family (possibly with repeated members) of sets is a graph with a vertex representing each member of the family and an edge connecting two vertices if and only if the two sets have nonempty intersection. Generally, loops are suppressed. If the family is composed of intervals or is the edge set of a hypergraph, then the intersection graph is called an *interval graph* or a *line graph,* respectively.

We prove a fuzzy analog of Marczewski's theorem showing that every fuzzy graph without loops is the intersection graph of some family of fuzzy subsets. We show that the natural generalization of the Fulkerson and Gross characterization of interval graphs fails. We also prove a natural generalization of the Gilmore and Hoffman characterization.

Let $G = (V, \mu, \rho)$ be a fuzzy graph. A *fuzzy digraph* is a triple $D = (V, \mu, \delta)$, where μ is a fuzzy subset of V and δ is a fuzzy subset of $V \times V$ such that $\delta(x, y) \leq \mu(x) \wedge \mu(y)$. We note δ need not be symmetric.

A fuzzy graph (fuzzy digraph) can be represented by an *adjacency matrix,* where the rows and columns are indexed by the vertex set V and the x, y entry is $\rho(x, y)(\delta(x, y))$. Vertex strength can be indicated by adding a column indexed by μ and letting the x, μ entry be $\mu(x)$.

Let $t \in [0, 1]$. Recall that the t level graph of G is the crisp graph $G^t = (\mu^t, \rho^t)$. For a family $\mathcal{F}$ of fuzzy subsets, we define the t level family of $\mathcal{F}$ as $\mathcal{F}^t = \{\alpha^t \mid \alpha \in \mathcal{F}\}$.

Let α be a fuzzy subset of V. Recall that the *height* of α is $h(\alpha) = \vee\{\alpha(x) \mid x \in V\}$. We construct a sequence of crisp level graphs in order to see how a fuzzy subset's structure changes between various levels. Theorems characterizing a fuzzy property in terms of level set properties are significant, in that such theorems demonstrate the extent to which the crisp theory can be generalized. To formalize this sequence of graphs, we define the *fundamental sequence* of a fuzzy graph $G = (\mu, \rho)$ to be the ordered set

$$\mathrm{fs}(G) = \{\mu(x) > 0 \mid x \in V\} \cup \{\rho(x, y) > 0 \mid x, y \in V\},$$

where we use the decreasing order inherited from the real interval $[0, 1]$.

The first element listed in $\mathrm{fs}(G)$ is the maximal vertex strength while the last element listed is the minimal nonzero edge strength.

Fuzzy Intersection Graphs

We now define a fuzzy intersection graph and prove that every fuzzy graph is the fuzzy intersection graph of some collection of fuzzy subsets.

Definition 2.16 *Let $\mathcal{F} = \{\alpha_1, ..., \alpha_n\}$ be a finite family of fuzzy subsets of a set V and consider $\mathcal{F}$ as a crisp vertex set. The fuzzy intersection graph of $\mathcal{F}$ is the fuzzy graph Int($\mathcal{F}$) $= (\mu, \rho)$ where $\mu : \mathcal{F} \to [0,1]$ is defined by $\mu(\alpha_i) = h(\alpha_i)$ and $\rho : \mathcal{F} \times \mathcal{F} \to [0,1]$ is defined by*

$$\rho(\alpha_i, \alpha_j) = \begin{cases} h(\alpha_i \cap \alpha_j) & \textit{if } i \neq j, \\ 0 & \textit{if } i = j. \end{cases}$$

An edge $\{\alpha_i, \alpha_j\}$ in Definition 2.16 has zero strength if and only if $\alpha_i \cap \alpha_j$ is the zero function (empty intersection) or $i = j$ (to preclude loops).

Note that in this section, we have a different notion of a fuzzy intersection graph than in the previous section. For this reason, we use a different notation. Recall that every graph $G = (V, X)$ is an intersection graph: For all $x \in V$, let S_x denote the union of $\{x\}$ with the set of all edges incident with x. It follows that G is isomorphic to the intersection graph of $\{S_x \mid x \in V\}$.

Remark 1. If $\mathcal{F} = \{\alpha_1, ..., \alpha_n\}$ is a family of fuzzy subsets of a set V and $t \in [0,1]$, then $\text{Int}(\mathcal{F}^t) = (\text{Int}(\mathcal{F}))^t$. The graph $\text{Int}(\mathcal{F}^t)$ has a vertex representing $\alpha_i \in \mathcal{F}$ if and only if $h(\alpha_i) \geq t$. The pair $\{(\alpha_i)^t, (\alpha_j)^t\}$ is an edge of $\text{Int}(\mathcal{F}^t)$ if and only if $i \neq j$ and $h(\alpha_i \cap \alpha_j) \geq t$, (so $(\alpha_i)^t \cap (\alpha_j)^t$ is nonempty). These conditions also characterize the graph $(\text{Int}(\mathcal{F}))^t$. In particular, if $\mathcal{F}$ is a family of crisp subsets of V, then the fuzzy intersection graph and crisp intersection graph definitions coincide.

Theorem 2.45 *(Fuzzy analog of Marczewski's theorem [27]). If $G = (\mu, \rho)$ is a fuzzy graph (without loops), then for some family of fuzzy subsets $\mathcal{F}$, $G =$ Int($\mathcal{F}$).*

Proof. Let $G = (\mu, \rho)$ be a fuzzy graph on V. For each $x \in V$ define the anti-reflexive, symmetric fuzzy subset $\alpha_x : V \times V \to [0,1]$ by

$$\alpha_x(y,z) = \begin{cases} \mu(x) & \text{if } y = x \text{ and } z = x, \\ \rho(x,z) & \text{if } y = x \text{ and } z \neq x, \\ \rho(y,x) & \text{if } y \neq x \text{ and } z = x, \\ 0 & \text{if } y \neq x \text{ and } z \neq x. \end{cases}$$

We show G is the fuzzy intersection graph of $\mathcal{F} = \{\alpha_x \mid x \in V\}$. By definition $\alpha_x(x,x) = \mu(x) \geq \rho(x,y)$ and so $h(\alpha_x) = \mu(x)$ as required. For $x \neq y$ a nonzero value of $(\alpha_x \cap \alpha_y)(z,w) = \alpha_x(z,w) \wedge \alpha_y(z,w)$ occurs only if $x = z$ and $y = w$ (or $y = z$ and $x = w$). Thus $h(\alpha_x \cap \alpha_y) = (\alpha_x \cap \alpha_y)(x, y) = \rho(x, y)$ and the desired result holds. ■

Fuzzy Interval Graphs

The families of sets most often considered in connection with intersection graphs are families of intervals of a linearly ordered set. This class of interval graphs is central to many applications.

In both the crisp and fuzzy cases, distinct families of sets can have the same intersection graph. In particular, the intersection properties of a finite family of real intervals (fuzzy numbers) can be characterized by a family of intervals (fuzzy intervals) defined on a finite set. Therefore, as is common in interval graph theory [28], we restrict attention to intervals (fuzzy intervals) with finite support.

We generalize two characterizations of (crisp) interval graphs. Theorem 2.47 gives the Fulkerson and Gross characterization [14] and Theorem 2.49 provides the Gilmore and Hoffman characterization [15]. Both theorems make use of relationships between the finite number of points which define the intervals and the cliques of the corresponding interval graph.

A *clique* of a graph is a maximal (with respect to set inclusion) complete subgraph. It is important to note that we adopt the convention of naming a clique by its vertex set. Clearly, if a vertex z is not a member of a clique K, then there exists an $x \in K$ such that (x, z) is not an edge of the graph. We generalize this concept in Definition 2.18.

Definition 2.17 *Let V be a linearly ordered set. A* fuzzy interval $\mathcal{I}$ *on V is a normal, convex fuzzy subset of V [12]. That is, there exists an $x \in V$ with $\mathcal{I}(x) = 1$ and the ordering $w \leq y \leq z$ implies that $\mathcal{I}(y) \geq \mathcal{I}(w) \wedge \mathcal{I}(z)$. A* fuzzy number *is a fuzzy interval. A* fuzzy interval graph *is the fuzzy intersection graph of a finite family of fuzzy intervals.*

In Definition 2.17, we see that by normality of the fuzzy intervals, the vertex set of a fuzzy interval graph is crisp.

Theorem 2.46 *Let $G = Int(\mathcal{F})$ be a fuzzy interval graph. Then for each $t \in (0, 1]$, the level graph G^t is an interval graph.*

Proof. Let $G = \text{Int}(\mathcal{F})$ for a family of fuzzy intervals $\mathcal{F} = \{\alpha_1, ..., \alpha_n\}$. For each $t \in (0, 1]$, convexity implies that each $(\alpha_i)^t \in \mathcal{F}^t$ is a crisp interval. By Remark 1, $G^t = (\text{Int}(\mathcal{F}))^t = \text{Int}(\mathcal{F}^t)$ and G^t is an interval graph. ■

Example 2.12 *The fuzzy graph G defined by the adjacency matrix of Fig. 2.6 demonstrates that the converse of Theorem 2.46 is false. In Figure 2.6,*

FIGURE 2.6 A fuzzy graph that is not a fuzzy interval graph although each cut level graph has an interval representation.

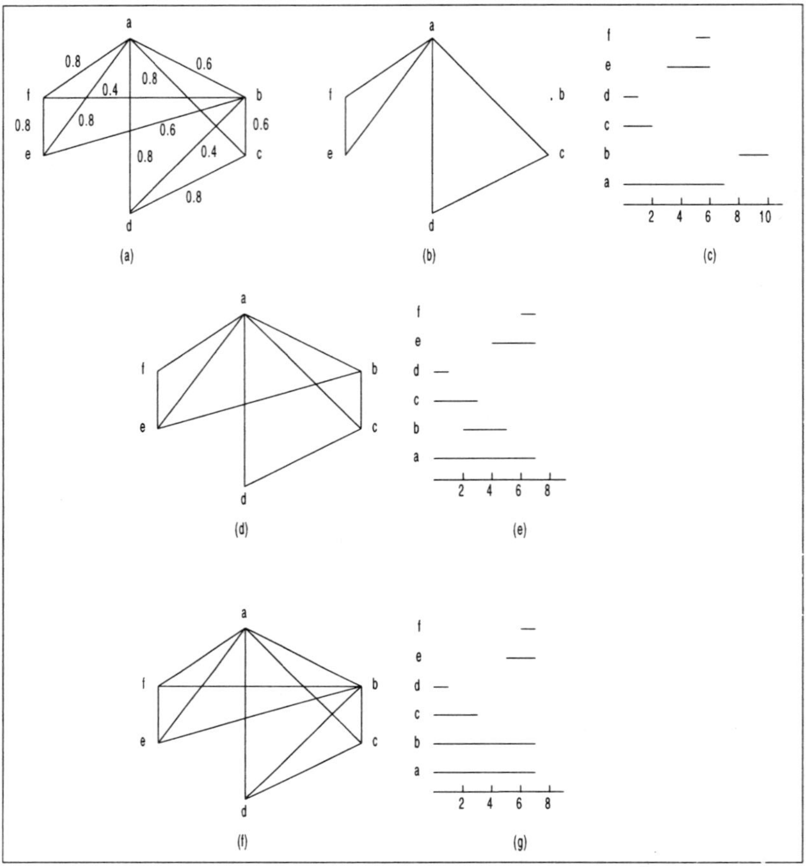

we present a fuzzy graph that is not a fuzzy interval graph although each cut level graph has an interval representation.

Before proceeding by contradiction, we note that $G^{0.8}$ *(see Figure 2.6(b)) has an interval representation. Let* $S_a = \{a\} \cup \{(a,f),(a,e),(a,d),(a,c)\}$, $S_b = \{b\}, S_c = \{c\} \cup \{(a,c),(c,d)\}, S_d = \{d\} \cup \{(c,d),(a,d)\}, S_e = \{e\} \cup \{(e,f),(a,e)\}$ *and* $S_f = \{f\} \cup \{(a,f),(e,f)\}$. *Let* $\{S_a, S_b, S_c, S_d, S_e, S_f\}$ *be our vertex set. Then it is easy to see the intervals of Figure 2.6(c) give an appropriate representation. (For* $G^{0.6}$ *see (d) and (e) and for* $G^{0.4}$ *see (f) and (g).) Suppose that* $G = Int(\mathcal{F})$, *where the fuzzy interval* $v \in \mathcal{F}$ *corresponds to vertex* v *of* G. *Since* $h(c \cap e) = 0$, *we assume without loss of generality that supp(c) lies strictly left of supp(e). An interval graph theorem [13] states that since* $\{a,c,d\}$ *defines a clique of* G^{r_1}, *there exists* x_1 *such that* $x_1 \in a^{0.8} \cap c^{0.8} \cap d^{0.8}$. *Therefore,* $a(x_1) \wedge c(x_1) \wedge d(x_1) \geq 0.8$.

Similarly, there exists an x_5 *such that* $a(x_5) \wedge e(x_5) \wedge f(x_5) \geq 0.8$. *Now* $h(b \cap d) = 0.4$ *and* $h(b \cap f) = 0.4$ *imply* $b(x_1) \leq 0.4$ *and* $b(x_5) \leq 0.4$, *respectively.*

Continuing, $h(b \cap c) = 0.6$ *and* $h(b \cap e) = 0.6$ *imply there exist* x_2 *and* x_4 *with* $b(x_2) \geq 0.6$ *and* $b(x_4) \geq 0.6$. *By the normality of* b *there exists* x_3 *such that* $b(x_3) = 1$. *By the convexity of the fuzzy intervals and the assumption that supp(c) lies strictly to the left of supp(e), the ordering of these points must be* $x_1 < x_2 \leq x_3 \leq x_4 < x_5$, *with* $x_2 < x_4$.

Since $a(x_1) \geq 0.8$, $a(x_5) \geq 0.8$, *and* a *is convex, it follows that* $a(x_3) \geq 0.8$. *Hence* $h(a \cap b) \geq 0.8$. *This contradicts* $h(a \cap b) = 0.6$. *Hence* G *is not a fuzzy interval graph.*

The Fulkerson and Gross Characterization

The Fulkerson and Gross characterization makes use of a correspondence between the set of points on which the family of intervals is defined and the set of cliques of the corresponding interval graph. We provide natural generalizations of the (crisp) definitions and then show that for fuzzy graphs this relationship holds only in one direction.

An interval graph theorem states that any set of intervals defining a clique will have a common point. If one such point is associated with each clique, the linear ordering of these points induces a linear ordering on the cliques of G. Using this ordering the resulting vertex clique incidence matrix has convex rows.

Conversely each convex row naturally defines the characteristic function of a subinterval of the linearly ordered set of cliques. The graph G is the intersection graph of this family of intervals. The following result is a consequence of this argument.

Theorem 2.47 *(Fulkerson and Gross [14]). A (crisp) graph* G *is an interval graph if and only if there exists a linear ordering of the cliques of* G *for which the vertex clique incidence matrix has convex rows.*

Definition 2.18 *Let* $G = (\mu, \rho)$ *be a fuzzy graph. We say that a fuzzy subset* $\mathcal{K}$ *defines a* fuzzy clique *of* G *if for each* $t \in (0,1]$, $\mathcal{K}^t$ *induces a clique of* G^t. *We associate with* G *a vertex clique incidence matrix where the rows are indexed by the domain of* μ, *the columns are indexed by the family of all fuzzy cliques of* G, *and the* x, $\mathcal{K}$ *entry is* $\mathcal{K}(x)$.

Remark 2. Suppose that G is a fuzzy graph with $\mathrm{fs}(G) = \{r_1, ..., r_n\}$ and let $\mathcal{K}$ be a fuzzy clique of G. The level sets of $\mathcal{K}$ define a sequence $\mathcal{K}^{r_1} \subseteq \cdots \subseteq \mathcal{K}^{r_n}$ where each $\mathcal{K}^{r_i}$ is a clique of G^{r_i}. Conversely, any sequence $K_1 \subseteq \cdots \subseteq K_n$ where each K_i is a clique of G^{r_i} defines a fuzzy clique $\mathcal{K}$

where $\mathcal{K}(x) = \vee\{r_i \mid x \in K_i\}$. Therefore, K is a clique of the t-level graph G^t if and only if $K = \mathcal{K}^t$ for some fuzzy clique $\mathcal{K}$.

Theorem 2.48 *(Fuzzy analog of Fulkerson and Gross). Let $G = (V, \rho)$ be a fuzzy graph. Then the rows of any vertex clique incidence matrix of G define a family of fuzzy subsets $\mathcal{F}$ for which $G = Int(\mathcal{F})$. Further, if there exists an ordering of the fuzzy cliques of G such that each row of the vertex clique incidence matrix is convex, then G is a fuzzy interval graph.*

Proof. Let $I = \{\mathcal{K}_1, ..., \mathcal{K}_p\}$ be an ordered family of the fuzzy cliques of G and let M be the vertex clique incidence matrix where the columns are given this ordering. For each $x \in V$ define the fuzzy set $\mathcal{J}_x : I \to [0,1]$ by $\mathcal{J}_x(\mathcal{K}_i) = \mathcal{K}_i(x)$ and let $\mathcal{F} = \{\mathcal{J}_x \mid x \in V\}$. Since each vertex x has strength 1, x is contained in the 1-level cut of some fuzzy clique $\mathcal{K}_i$ in I. Therefore, $\mathcal{J}_x(\mathcal{K}_i) = \mathcal{K}_i(x) = 1$ and $\mathcal{J}_x$ is normal.

We must show for $x \neq y \in V$ that $h(\mathcal{J}_x \cap \mathcal{J}_y) = \rho(x, y)$. Also assuming that each row is convex implies that each $\mathcal{J}_x$ is a fuzzy interval and that G is a fuzzy interval graph. By definition, if $x \neq y$, then $h(\mathcal{J}_x \cap \mathcal{J}_y) = \vee\{(\mathcal{J}_x \cap \mathcal{J}_y)(\mathcal{K}_i) \mid \mathcal{K}_i \in I\} = \vee\{\mathcal{J}_x(\mathcal{K}_i) \wedge \mathcal{J}_y(\mathcal{K}_i) \mid \mathcal{K}_i \in I\} = \vee\{\mathcal{K}_i(x) \wedge \mathcal{K}_i(y) \mid \mathcal{K}_i \in I\} = \vee\{t \in [0,1] \mid \mathcal{K}_i \in I$ and (x, y) is an edge of $(\mathcal{K}_i)^t\}$.

The edge strength $\rho(x, y) = t$ is the maximal value where (x, y) is an edge of G^t so is contained in a clique of G^t. Thus $h(\mathcal{J}_x \cap \mathcal{J}_y) = \rho(x, y)$ as required. ■

Example 2.13 *The fuzzy graph G given by the incidence matrix in Fig. 2.7 shows that the converse of Theorem 2.48 is false. If the set $\mathcal{F}$ of fuzzy intervals is defined by the rows of matrix F, then $G = Int(\mathcal{F})$. Figure 2.7 also shows the cut level graphs of G and a vertex clique incidence matrix M for G. One may verify by exhaustion that no ordering of the fuzzy cliques produces a vertex clique incidence matrix M with convex rows.*

$$G = \begin{array}{c} \\ a \\ b \\ c \\ d \end{array} \begin{array}{c} \begin{array}{cccc} a & b & c & d \end{array} \\ \begin{bmatrix} 0 & 0.5 & 1 & 0 \\ 0.5 & 0 & 0.5 & 0.5 \\ 1 & 0.5 & 0 & 1 \\ 0 & 0.5 & 1 & 0 \end{bmatrix} \end{array} \qquad F = \begin{array}{c} \\ I_a \\ I_b \\ I_c \\ I_d \end{array} \begin{array}{c} \begin{array}{cccc} 1 & 2 & 3 & 4 \end{array} \\ \begin{bmatrix} 1 & 0.5 & 0 & 0 \\ 0.5 & 0.5 & 0.5 & 1 \\ 1 & 1 & 1 & 0.5 \\ 0 & 0 & 1 & 0.5 \end{bmatrix} \end{array}$$

$$M = \begin{array}{c} \\ a \\ b \\ c \\ d \end{array} \begin{array}{c} \begin{array}{cccc} \mathcal{K}_1 & \mathcal{K}_2 & \mathcal{K}_3 & \mathcal{K}_4 \end{array} \\ \begin{bmatrix} 1 & 0.5 & 0 & 0 \\ 0.5 & 1 & 0.5 & 1 \\ 1 & 0.5 & 1 & 0.5 \\ 0 & 0 & 1 & 0.5 \end{bmatrix} \end{array}$$

FIGURE 2.7 A fuzzy interval graph where any vertex clique incidence matrix has a row that is not convex. (a) represents G^1 (b) represents $G^{0.5}$.

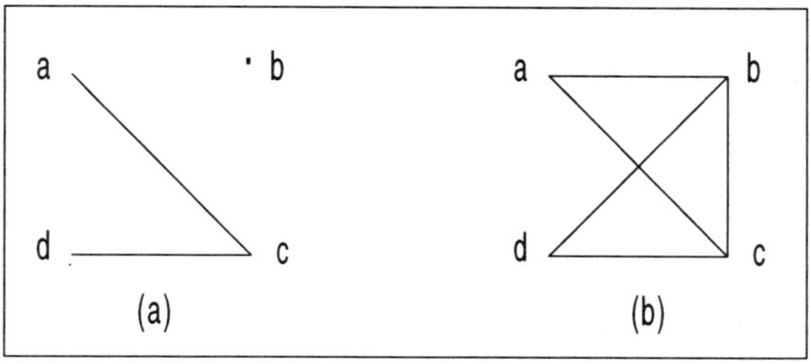

The Gilmore and Hoffman Characterization

Let $G = (V, X)$ be a graph and D be a directed graph. In the remainder of the section, we use the notation (x, y) for an edge of G and $< x, y >$ for a directed edge in D. We begin with several graph theory definitions and state the Gilmore and Hoffman characterization. In the interest of completeness, we provide a reasonably detailed proof of this result. We then give corresponding fuzzy definitions, and conclude with the result that the Gilmore and Hoffman characterization generalizes exactly for fuzzy interval graphs.

Recall that a cycle of length n in a graph $G = (V, X)$ is a sequence $x_0, ..., x_n$ of distinct vertices where $(x_0, x_n) \in X$ and $1 \leq i \leq n$ implies $(x_{i-1}, x_i) \in X$. A graph is *chordal* or *triangulated* if each cycle with $n \geq 4$ has a chord. Formally, if there exist integers $j \neq 0$ or $k \neq n$ with $0 \leq j < k - 1 \leq n$ and $(x_j, x_k) \in X$.

An *orientation* of a graph $G = (V, X)$ is a directed graph $G_A = (V, A)$ that has G as its underlying graph. That is, $(x, y) \in X$ implies that $< x, y > \in A$ or $< y, x > \in A$ but not both. A graph G is *transitively orientable* if there exists an orientation of G for which $< u, v > \in A$ and $< v, w > \in A$ implies $< u, w > \in A$. Further, if $G = (V, X)$ is a graph, the *complement* of G, denoted by G^c, is the graph with vertex set V and edge set consisting of those edges which are not in X. For a fuzzy graph $G = (\mu, \rho)$, we let $G^c = (\mu, 1 - \rho)$.

We merely sketch the proof of the following result.

Theorem 2.49 *(Gilmore and Hoffman [15]). A graph $G = (V, X)$ is an interval graph if and only if it satisfies the following two conditions:*

(1) each subgraph of G induced by four vertices is chordal,

(2) G^c is transitively orientable.

FIGURE 2.8 The clique ordering is well defined.

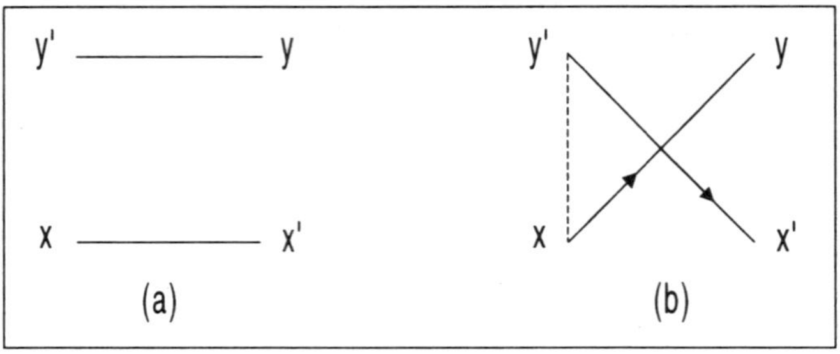

Proof. Further details and examples can be found in [33]. That an interval graph G (hence each four vertex subgraph) is chordal is an immediate consequence of the definitions. To orient G^c, let $< a, b > \in A$ if and only if (a, b) is not an edge of G and the interval corresponding to a lies strictly to the left of the interval corresponding to b.

If each subgraph of G induced by four vertices is chordal and A is a transitive orientation of G^c, we first define a linear ordering $<$ on the set of all cliques of G as follows. For cliques $K \neq L$ of G there exist $x \in K$ with $x \notin L$ and in turn $y \in L$ such that $(x, y) \notin X$; otherwise $\{x\} \cup L$ induces a complete subgraph of G properly containing L. Define $K < L$ if and only if $< x, y > \in A$.

We show $<$ is well defined by contradiction. Suppose that (x, x') is in clique K, (y, y') is in clique L, $< x, y > \in A$ and $< y', x' > \in A$ (Fig. 2.8). Since $xy'yx'x$ would be a cordless 4 cycle, we assume without loss of generality that $(x, y') \notin X$. However, A is transitive so $< x, y' > \in A$ implies $(x, x') \notin X$, a contradiction. Similarly, $< y', x > \in A$ implies $(y', y) \notin X$, a contradiction.

For transitivity let $< x, y > \in A$ define $K < L$ and let $L < M$. If $y \in M$ then $K < M$ as required. If $y \notin M$ there exists $z \in M$ with $(y, z) \notin X$; then $L < M$ implies $< y, z > \in A$. Transitivity of A gives $< x, z > \in A$ and $K < M$ as required.

A well-known graph theory theorem states that any complete transitive relation on a set defines a linear ordering of the set. Therefore, $<$ linearly orders the cliques of G. Since A is anti-symmetric and $< x, y > \in A$, $x \notin M$ whenever $L < M$. Therefore, the vertex clique incidence matrix with the columns ordered by $<$ has convex rows. As in Theorem 2.47, these rows define a family of intervals which have G as its intersection graph. ∎

We now show using natural generalizations of the definitions that fuzzy interval graphs are chordal and have transitively orientable compliments.

The definition of a cycle in the next definition is equivalent to the one previously given. For convenience sake, we use this form in this section.

Definition 2.19 *A cycle of length n in a fuzzy graph is a sequence of distinct vertices $x_0, ..., x_n$ where $\rho(x_0, x_n) > 0$ and if $1 \leq i \leq n$, then $\rho(x_{i-1}, x_i) > 0$. A fuzzy graph $G = (\mu, \rho)$ is* chordal *if for each cycle with $n \geq 4$, there exist integers $j \neq 0$ or $k \neq n$ such that $0 \leq j < k - 1 \leq n$ and $\rho(x_j, x_k) \geq \wedge\{\rho(x_{i-1}, x_i) \mid i = 1, 2, ..., n\} \wedge \rho(x_0, x_n)$.*

A fuzzy graph $G = (\mu, \rho)$ is chordal if and only if for each $t \in (0, 1]$, the t-level graph of G is chordal (triangulated).

Theorem 2.50 *If G is a fuzzy interval graph, then G is chordal.*

Proof. By Theorem 2.46, each cut level graph G^t is an interval graph. As in the proof of Theorem 2.49 any interval graph is chordal. The result then follows from Definition 2.19. ■

To avoid confusion when dealing with cut level graphs, we base an orientation of a fuzzy graph on an orientation of its underlying graph.

Definition 2.20 *Let $G = (\mu, \rho)$ be a fuzzy graph with $fs(G) = \{r_1, ..., r_n\}$ and let A be an orientation of G^{r_n}. Then the* orientation *of G by A is the fuzzy digraph $G_A = (\mu, \rho_A)$ where*

$$\rho_A(< x, y >) = \begin{cases} \rho(x, y) & \text{if} < x, y > \ \in A \\ 0 & \text{if} < x, y > \ \notin A. \end{cases}$$

The fuzzy graph $G = (\mu, \rho)$ is transitively orientable *if there exists an orientation which is transitive. Formally, $\rho_A(< x, y >) \wedge \rho_A(< y, z >) \leq \rho_A(< x, z >)$.*

The t level graph of G_A has arc set $\{< x, y > \mid \rho_A(< x, y >) \geq t\}$. Therefore, an orientation of a fuzzy graph induces consistent orientations on each member of the fundamental sequence of cut level graphs. Conversely, it is possible to have a sequence of transitively oriented subgraphs $G_1 \subseteq G_2 \subseteq G_3$ where the transitive orientation of G_2 does not induce a transitive orientation of G_1, and the transitive orientation of G_2 cannot be extended to a transitive orientation of G_3.

Lemma 2.51 *Suppose that $G = Int(\mathcal{F})$ is a fuzzy interval graph. Then there exists an orientation A that induces a transitive orientation of G^c.*

Proof. Assume (α, β) is a nontrivial edge of G^c. Then $h(\alpha \cap \beta) < r_1 = 1$ and α^{r_1} and β^{r_1} are disjoint. We let $< \alpha, \beta > \in A$ if and only if α^{r_1} lies strictly left of β^{r_1}. Clearly A is a well-defined and transitive orientation of G^c. ■

FIGURE 2.9 A fuzzy graph that is not transitively orientable. (a) represents $(G^{r_3})^c$, (b) represents $(G^{r_2})^c$, and (c) represents $(G^{r_1})^c$.

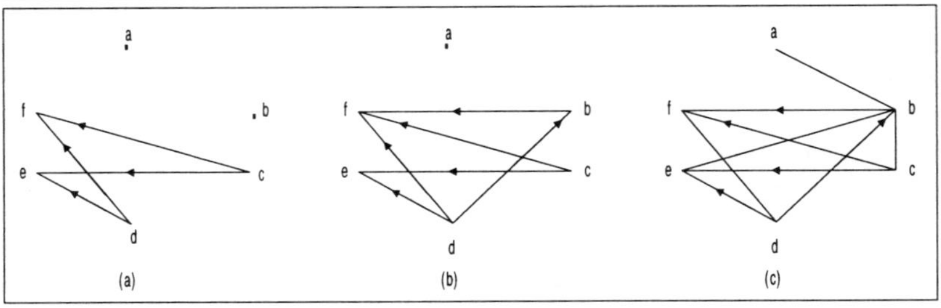

Example 2.14 *The fuzzy graph of Example 2.12. (Fig. 2.6) is not a fuzzy interval graph because any orientation of (d, e) shows there is no transitive orientation of G^c. Fig. 2.9 shows the cut level graphs of G^c with $< d, e > \in A$. Note that $(G^1)^c = (G^c)^{1-r_2}, (G^{r_2})^c = (G^c)^{1-r_3}$ and $(G^{r_3})^c = (G^c)^1$, where $r_1 = 1$ here.*

Theorem 2.52 *(Fuzzy analog of the Gilmore and Hoffman characterization). A fuzzy graph $G = (\mu, \rho)$ is a fuzzy interval graph if and only if*

(1) for all $x \in supp(\mu) = V$, $\mu(x) = 1$;

(2) each fuzzy subgraph of G induced by four vertices is chordal;

(3) G^c is transitively orientable. ■

If G is a fuzzy interval graph the three conditions follow from Definitions 2.16 and 2.17, Theorem 2.50, and Lemma 2.51, respectively.

For the remainder of the section we assume each fuzzy subgraph of $G = (V, \rho)$ induced by four vertices is chordal and that A is a transitive orientation of G^c. Because the proof that G is a fuzzy interval graph is quite involved we first outline the proof. Details are given in Definition 2.21 through Lemma 2.55; the algorithm is applied in Example 2.15. For notational convenience we let $\mathcal{K}_{ij}$ denote the r_j cut level set of the fuzzy set $\mathcal{K}_i$.

By Theorem 2.48 the rows of any vertex clique incidence matrix of G define a family of fuzzy subsets that has G as its fuzzy intersection graph. We define a linear ordering $<$ of the fuzzy cliques of G. If $<$ has a property we call being *cut level consistent,* the rows of the vertex clique incidence matrix will be convex and the result follows immediately from Theorem 2.48.

If $<$ is not level consistent then some row is not convex. We modify this matrix in a "bottom up" construction using the notion of cut level consistent to determine which columns are modified or deleted from the vertex clique incidence matrix. We complete the proof by showing that in the modified matrix each row is normal and convex and that G is the fuzzy intersection graph of the family of fuzzy intervals defined by the rows.

By the discussion following Definitions 2.19 and 2.20 each level graph G^t is chordal and has a transitively orientable complement. Therefore, each G^t is an interval graph and there exists a linear ordering $<_t$ on the family of all cliques of G^t. We now establish definitions that will be used extensively in the discussion below.

Definition 2.21 *Define the relation $<$ on the family of all fuzzy cliques of G as follows. Let $\mathcal{K} < \mathcal{L}$ if and only if $K^t <_t L^t$ where t is the smallest element of $fs(G)$ such that $K^t \neq L^t$. The lexicographic ordering $<$ is clearly well defined, complete, and transitive. Therefore, $<$ defines a linear ordering on the family of all fuzzy cliques of G.*

Definition 2.22 *Let G satisfy the conditions of Theorem 2.52 and let $<$ be the relation of Definition 2.21. Suppose that $t \in fs(G)$ and let $\mathcal{K} \neq \mathcal{L}$ be fuzzy cliques of G. We say $\mathcal{K}$ and $\mathcal{L}$ are consistently ordered by $<$ at level t provided $\mathcal{K}^t <_t \mathcal{L}^t$ if and only if $\mathcal{K} < \mathcal{L}$. We say the linear ordering $<$ is cut level consistent if for each pair of distinct fuzzy cliques of G and for each $t \in fs(G)$ the pair is consistently ordered by $<$ at level t.*

Remark 3. If the linear ordering $<$ is cut level consistent then each row of the vertex clique incidence matrix is convex. That G is a fuzzy interval graph follows immediately from Theorem 2.48. We proceed by contrapositive, assuming there exists a row that is not convex. Suppose that there exist a vertex $x \in V$ and a sequence of fuzzy cliques $\mathcal{K} < \mathcal{L} < \mathcal{M}$ such that $\mathcal{L}(x) < \min\{\mathcal{K}(x), \mathcal{M}(x)\} \equiv t$. Then $x \in \mathcal{K}^t$, $x \notin \mathcal{L}^t$ and $x \in \mathcal{M}^t$. As in Theorem 2.49 there exists $y \in \mathcal{L}^t$ such that $(x, y) \notin \mathcal{E}^t$. If $< x, y > \in A$ then $\mathcal{M}^t <_t \mathcal{L}^t$ with $\mathcal{L} < \mathcal{M}$. If $< y, x > \in A$, then $\mathcal{L}^t <_t \mathcal{K}^t$ with $\mathcal{K} < \mathcal{L}$. In either case the ordering $<$ is not cut level consistent.

By Example 2.13 there exist fuzzy interval graphs where no ordering of the fuzzy cliques is cut level consistent. Therefore, the general construction is considerably more complicated. Our goal is to formalize a process that modifies or deletes "inconsistent" fuzzy cliques (matrix columns). We first give a technical lemma which serves two purposes. First its proof illuminates the "local"structure of noncut level consistent orderings. The lemma is also used to show the construction in Definition 2.23 is well defined.

Lemma 2.53 *Suppose that $\mathcal{K}$ and $\mathcal{L}$ are fuzzy cliques of G and that $s > t$. If $\mathcal{K}^s <_s \mathcal{L}^s$ and $\mathcal{L}^t <_t \mathcal{K}^t$ then there exists a clique M of G^t such that either*

(1) $\mathcal{K}^s \subseteq M$ and $M <_t \mathcal{K}^t$ or

FIGURE 2.10 Basic conditions for inconsistent cut level orderings. (a) represents G^s,(b) represents $(G^s)^c$, (c) represents G^t, and (d) represents $(G^t)^c$.

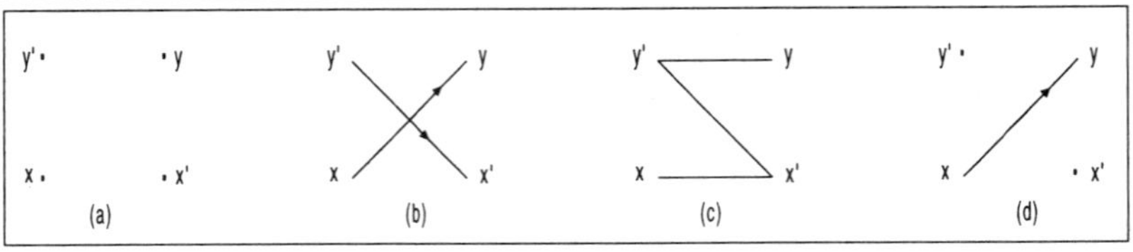

(2) $\mathcal{L}^s \subseteq M$ *and* $\mathcal{L}^t <_t M$.

Proof. We proceed by exhaustion; checking all possible edge configurations. Recall the edge set of the graph G^s is denoted by $\mathcal{E}^s$. Each case shares the general conditions shown in Fig. 2.10. By definition of $<_t$, there exist $x \in \mathcal{K}^t$ and $y \in \mathcal{L}^t$, with $(x, y) \notin \mathcal{E}^t$ and $< x, y > \in A$. Similarly, there exist $x' \in \mathcal{K}^s$ and $y' \in \mathcal{L}^s$, with $(x', y') \notin \mathcal{E}^s$ and $< y', x' > \in A$. Then $s > t$ implies $(x, y) \notin \mathcal{E}^s$, $(x, x') \in \mathcal{E}^s$ (or $x = x'$) and $(y, y') \in \mathcal{E}^t$ (or $y = y'$). As $<_t$ is well defined, $(x', y') \in \mathcal{E}^t$ and either $(x, x') \notin \mathcal{E}^s$ or $(y, y') \notin \mathcal{E}^s$.

If $(y', x) \notin \mathcal{E}^t$ then $(y', x) \notin \mathcal{E}^s$ and transitivity requires $< y', x > \in A$ and $(x, x') \notin \mathcal{E}^s$ (so $x \notin \mathcal{K}^s$). We claim for each $x'' \in \mathcal{K}^s \subseteq \mathcal{K}^t$ that $(y', x'') \in \mathcal{E}^t$. For $(y', x'') \notin \mathcal{E}^t$ with $\mathcal{L}^t <_t \mathcal{K}^t$ implies $< y', x'' > \in A$. However $\mathcal{E}^s \subseteq \mathcal{E}^t$ and $\mathcal{K}^s <_s \mathcal{L}^s$ implies $< x'', y' > \in A$; a contradiction. Therefore, $\{y'\} \cup \mathcal{K}^s$ is a complete subgraph of G^t and is contained in a clique M of G^t. Since $x \notin M$ and $< y', x > \in A$, we have $M <_t \mathcal{K}^t$ and property (1) is satisfied.

Similarly, if $(y, x') \notin \mathcal{E}^t$ we have that $< y, x' > \in A$ and $(y, x') \notin \mathcal{E}^s$. Transitivity requires $\{x'\} \cup \mathcal{L}^s$ to be a complete subgraph of G^t, hence is contained in a clique M of G^t. Then $< y, x' > \in A$ implies $\mathcal{L}^t <_t M$ and property (2) is satisfied.

If $(y', x) \in \mathcal{E}^t$ and $(y, x') \in \mathcal{E}^t$ we show that $\mathcal{K}^s \cup \mathcal{L}^s$ is a complete subgraph of G^t. We need only to show for each $x'' \in \mathcal{K}^s$ and $y'' \in \mathcal{L}^s$ that $(y'', x'') \in \mathcal{E}^t$. Again $(x'', y'') \notin \mathcal{E}^t$ and $\mathcal{L}^t <_t \mathcal{K}^t$ implies $< y'', x'' > \in A$ and $(y'', x'') \notin \mathcal{E}^s$. However, $\mathcal{K}^t <_s \mathcal{L}^s$ implies $< x'', y'' > \in A$; a contradiction.

Therefore, $\mathcal{K}^s \cup \mathcal{L}^s$ induces a complete subgraph of G^t that is contained in some clique M of G^t. If $M <_t \mathcal{L}^t <_t \mathcal{K}^t$, property (1) is satisfied. If $\mathcal{L}^t <_t M$ property (2) is satisfied. These cases exhaust all possibilities. ■

We now construct a directed graph F and in turn a linearly ordered family of fuzzy subsets that define columns of an incidence matrix. These fuzzy subsets will either be fuzzy cliques of G or modifications of fuzzy cliques. The graph theory analogy of a forest with trees allows a good visualization of "vertically growing" cut level sets which define the required fuzzy sets

We use the fuzzy clique ordering $<$ to recursively construct a forest F whose vertex set is the set of all cut level cliques of G and whose arcs connect cut levels of fuzzy sets. We recursively build the forest by "vertically" adding cut level cliques as vertices of F and defining a set of arcs between cut levels. In the recursion let i range from 1 to $n-1$.

Definition 2.23 *Let $G = (V, \rho)$ with $fs(G) = \{r_1, r_2, ..., r_n\}$ be a chordal fuzzy graph and let G^c be* transitively oriented *by A.*

Level r_n : Linearly order the set of all cliques of G^{r_n} by the relation $<_{r_n}$ of Definition 2.22. Each of these cliques of G^{r_n} (vertices of F) represent the root of a tree in the forest.

Level r_{n-i} : Let $s = r_{n-i}$ and $t = r_{n-i+1}$. Linearly order the set of all cliques of G^s by the relation $<_s$. Let X^s be any set of arcs that satisfy:

(1) each clique $\mathcal{K}^s$ of G^s is a vertex of exactly one arc of X^s

(2) if $<\mathcal{K}^t, \mathcal{K}^s> \in X^s$ then $\mathcal{K}_t$ is a clique of G^t, $\mathcal{K}^s$ is a clique of G^s, and $\mathcal{K}^s \subseteq \mathcal{K}^t$. Thus, an arc joins two level sets of (some) fuzzy clique.

(3) For each pair of arcs $<\mathcal{K}^t, \mathcal{K}^s> \in X^s$ and $<\mathcal{L}^t, \mathcal{L}^s> \in X^s$ we have $\mathcal{K}^s <_s \mathcal{L}^s$ if and only if $\mathcal{K}^t <_t \mathcal{L}^t$ or $\mathcal{K}^t = \mathcal{L}^t$.

Thus when viewed as cut levels of a family of fuzzy cliques, the s level ordering is level consistent with the next "lower" level.

We use Lemma 2.53 to demonstrate the existence of at least one such forest, and show in Remark 4 that there may be a number of arc sets that satisfy these conditions. Let $\mathcal{K}^s$ be the minimal (with respect to $<_s$) clique of G^s. Clearly there exists a minimal (with respect to $<_t$) clique $\mathcal{K}^t$ of G^t where $\mathcal{K}^s \subseteq \mathcal{K}^t$. Let $<\mathcal{K}^t, \mathcal{K}^s> \in X^s$.

Next, let $\mathcal{L}^s$ be the successor of $\mathcal{K}^s$ (with respect to $<_s$) and let $\mathcal{L}^t$ be minimal (with respect to $\dot{<}_t$) such that $\mathcal{L}^s \subseteq \mathcal{L}^t$ and $\mathcal{K}^t <_t \mathcal{L}^t$ or $\mathcal{K}^t = \mathcal{L}^t$. If $\mathcal{L}^t$ does not exist, let L be maximal (with respect to $<_t$) with $\mathcal{L}^s \subseteq L$. Now $\mathcal{K}^s <_s \mathcal{L}^s$ and $L <_t \mathcal{K}^t$ are the conditions of Lemma 2.53. However, property (1) contradicts the minimality of $\mathcal{K}^t$ and property (2) contradicts the maximality of L. Therefore, $\mathcal{L}^t$ exists and $<\mathcal{L}^t, \mathcal{L}^s> \in X^s$ is well defined.

Continuing recursively we construct one arc for each clique of G^s. It may be that for some clique M_t of G^t, there is no arc from M_t. We shall call such a clique a *dead branch* of F.

Combining the arc sets $F^{r_{n-i}}$ for $i \in \{1, ..., r-1\}$ defines a forest with arc set $\bigcup_{i=1}^{n-1} F^{r_{n-i}}$. As in Definition 2.22, we lexicographically order the set of paths from a root to a dead branch or a r_1 level clique. For notational convenience we denote the t level vertex of path P_j by P_{jt}. To ensure convex rows in our (still undefined) incidence matrix, we add nonempty vertices "above" dead branches if "adjacent" cliques have nonempty intersection.

More formally let the path P_j end with a dead branch at the t level. For each $s \in \text{fs}(G)$ with $s > t$ we continue the path P_j through the new vertex P_{js}, where $x \in P_{js}$ if and only if there exist $i < j < k$ with $x \in P_{is} \cap P_{ks}$. We call this final forest F. Now each path in F has length n, and it is possible for a vertex P_{js} to be the empty set.

We complete the construction by letting paths in F define a linearly ordered family of fuzzy sets, say I. The fuzzy sets define columns of the *vertex forest matrix* of $(G, <)$; G is the interval graph of the family of rows.

Definition 2.24 *Let G satisfy the conditions of Theorem 2.52, F be a forest for G as defined in Definition 2.23, and P_j be a path in F of length n. Associated with P_j define the fuzzy set $\rho_j \in I$ on the vertex set of G by*

$$\rho_j(x) = \vee\{s \in fs(G) \mid x \text{ is an element of the } s \text{ level vertex of } P_j\}.$$

We construct the *vertex forest matrix of* $(G, <)$ indexing rows by the vertex set of G, columns by the (ordered) fuzzy sets of I and defining the x, ρ_i entry as $\rho_i(x)$. By construction each ρ_j is either a fuzzy clique of G, or has a cut level set that is the intersection of two cut level cliques.

Let $\mathcal{F}$ denote the family of fuzzy sets defined by the rows of the vertex forest matrix. We now complete the proof of Theorem 2.52 by showing that each member of $\mathcal{F}$ is normal and convex (a fuzzy interval) and that $G = \text{Int } \mathcal{F}$.

Lemma 2.54 *We assume the conditions and notation above. For each vertex x of G define $\mathcal{J}_x$ is a fuzzy interval.*

Proof. Let x be a vertex of G. Then x is a vertex in some clique of G^{r_1}, say K. By Definitions 2.23 and 2.24 K is the $r_1 = 1$ level cut of some fuzzy set ρ in I. Therefore, $\mathcal{J}_x(\rho) = \rho(x) = 1$ and $\mathcal{J}_x$ is normal.

Each $\mathcal{J}_x$ is convex if $i < j < k$ implies $\mathcal{J}_x(\rho_i) \wedge \mathcal{J}_x(\rho_k) \leq \mathcal{J}_x(\rho_j)$, or equivalently, if $\rho_i(x) \wedge \rho_k(x) \leq \rho_j(x)$. However, Definition 2.23 clearly provides these conditions. If ρ_i, ρ_j and ρ_k are all fuzzy cliques, the result follows immediately from Remark 3. Otherwise, the result follows by definition of the fuzzy sets ρ_i, ρ_j and ρ_k. ■

We conclude the proof of Theorem 2.52.

Lemma 2.55 *Given the definitions and conditions of Theorem 2.52 through Lemma 2.54, $\mathcal{G} = Int(\mathcal{F})$.*

Proof. There is a clear correspondence between the crisp vertex set V and the family of fuzzy intervals $\mathcal{F}$. Let $x \neq y$ be elements of V. We must show that $\rho(x,y) = h(\mathcal{J}_x \cap \mathcal{J}_y)$. By definition, $h(\mathcal{J}_x \cap \mathcal{J}_y) = \vee\{\mathcal{J}_x(\rho_j) \wedge \mathcal{J}_y(\rho_j) \mid \rho_j \in I\} = \vee\{\rho_j(x) \wedge \rho_j(y) \mid \rho_j \in I\} = \vee\{s \in \text{fs}(G) \mid \{x,y\} \subseteq \rho_j^s\}$.

As $\rho(x,y) = t$ is the maximal value where (x,y) is an edge of G^t, $\rho(x,y)$ is the maximal value where (x,y) is in a clique of G^t. By definition each clique of G^t is the t level set of some fuzzy set $\rho_j \in I$. Therefore, $\rho(x,y) = h(\mathcal{J}_x \cap \mathcal{J}_y)$ as required. ■

We now give a complete example of Theorem 2.52.

Example 2.15 *Consider the fuzzy graph G defined by the incidence matrix G below, where $fs(G) = \{s,t,u\} = \{1, 0.8, 0.5\}$. Fig. 2.11 also shows the cut level graphs of G and a transitive orientation A of G^c.*

$$G = \begin{array}{c} \\ a \\ b \\ c \\ d \\ e \end{array} \begin{array}{c} \begin{array}{ccccc} a & b & c & d & e \end{array} \\ \begin{bmatrix} 0 & 0.8 & 0.8 & 0.5 & 0.8 \\ 0.8 & 0 & 1 & 0 & 0.5 \\ 0.8 & 1 & 0 & 1 & 0.8 \\ 0.5 & 0 & 1 & 1 & 0.8 \\ 0.8 & 0.5 & 0.8 & 0.8 & 0 \end{bmatrix} \end{array}$$

$$M = \begin{array}{c} \\ a \\ b \\ c \\ d \\ e \end{array} \begin{array}{c} \begin{array}{cccccccc} \mathcal{K}_1 & \mathcal{K}_2 & \mathcal{K}_3 & \mathcal{K}_4 & \mathcal{K}_5 & \mathcal{K}_6 & \mathcal{K}_7 & \mathcal{K}_8 \end{array} \\ \begin{bmatrix} 1 & 0.8 & 1 & 0.8 & 1 & 0.8 & 0.5 & 0.5 \\ 0.8 & 1 & 0.5 & 0.5 & 0 & 0 & 0 & 0 \\ 0.8 & 1 & 0.8 & 0.8 & 0.8 & 0.8 & 1 & 0.8 \\ 0 & 0 & 0 & 0 & 0 & 0.5 & 1 & 0.8 \\ 0.5 & 0.5 & 0.8 & 1 & 0.8 & 1 & 0.8 & 1 \end{bmatrix} \end{array}$$

Using Definition 2.22 we linearly order the cut level cliques by:
$s = 1, \quad \{a\} <_s \{b,c\} <_s \{c,d\} <_s \{e\},$
$t = 0.8, \quad \{a,b,c\} <_t \{a,c,e\} <_t \{c,d,e\},$
$u = 0.5, \quad \{a,b,c,e\} <_u \{a,c,d,e\},$

By Remark 2, there are eight fuzzy cliques of G; subscripts indicate the order induced by Definition 2.22. The matrix M is the vertex clique incidence matrix for G. The only convex row is indexed by d; thus the fuzzy clique ordering is not cut level consistent.

Following Definition 2.23 gives the forest F of Fig. 2.12. The paths P_1, P_2, P_4 and P_5 correspond, respectively, to the fuzzy cliques $\mathcal{K}_1$, $\mathcal{K}_2$, $\mathcal{K}_7$ and $\mathcal{K}_8$. The clique $\{a,c,e\}$ is a dead branch; so $P_{3s} = P_{2s} \cap P_{4s} = \{c\}$. The path P_3 is a modification of $\mathcal{K}_4$, the fuzzy cliques $\mathcal{K}_3$, $\mathcal{K}_5$, and $\mathcal{K}_6$ are deleted.

FIGURE 2.11 A chordal fuzzy graph with transitive orientation of G^c. (a) represents G^s,(b) represents $(G^s)^c$, (c) represents G^t, (d) represents $(G^t)^c$ (e) represents G^u, and (f) represents $(G^u)^c$.

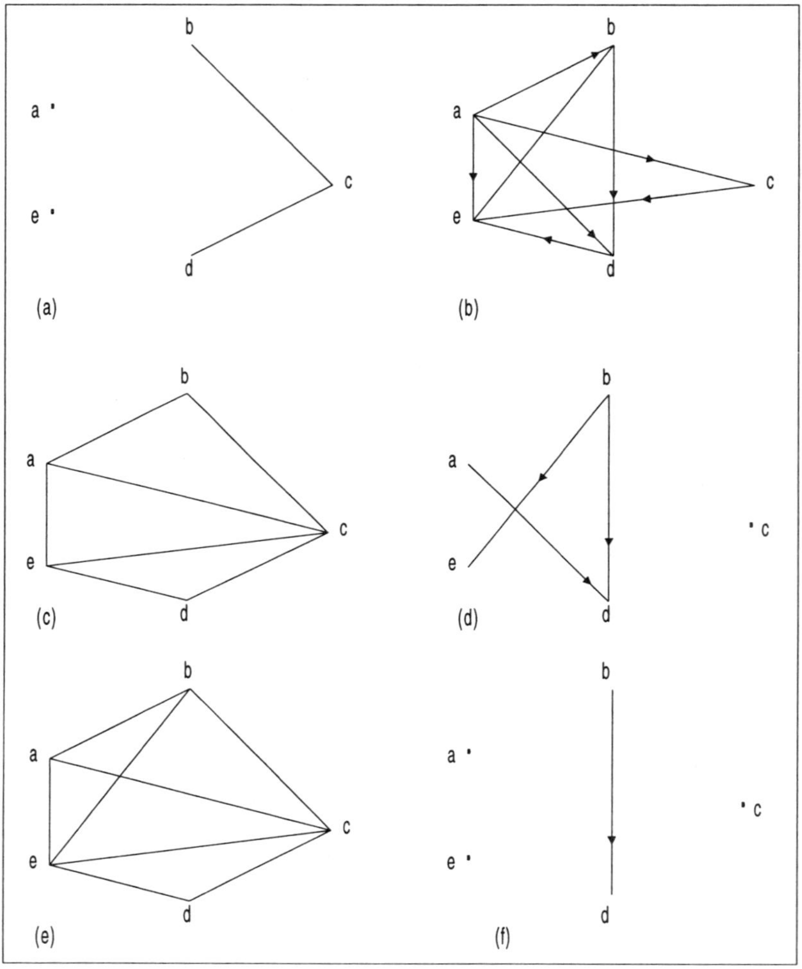

$$V = \begin{array}{c} \\ a \\ b \\ c \\ d \\ e \end{array} \begin{array}{c} \begin{array}{ccccc} \rho_1 & \rho_2 & \rho_3 & \rho_4 & \rho_5 \end{array} \\ \left[\begin{array}{ccccc} 1 & 0.8 & 0.8 & 0.5 & 0.5 \\ 0.8 & 1 & 0.5 & 0 & 0 \\ 0.8 & 1 & 1 & 1 & 0.8 \\ 0 & 0 & 0 & 1 & 0.8 \\ 0.5 & 0.5 & 0.8 & 0.8 & 1 \end{array} \right] \end{array}$$

FIGURE 2.12 A fuzzy interval representation for example 2.15.

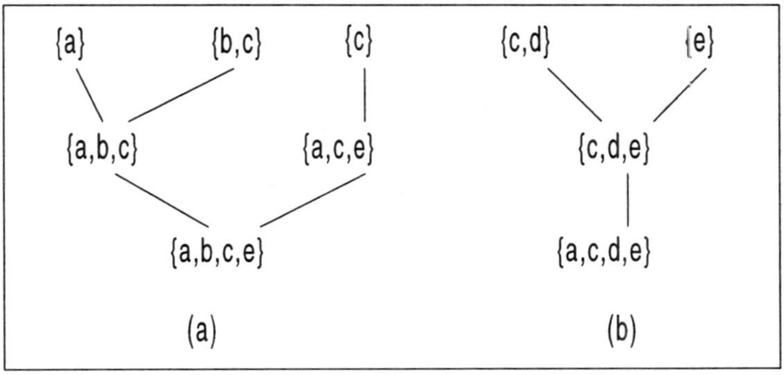

$$V = \begin{array}{c} \\ a \\ b \\ c \\ d \\ e \end{array} \begin{array}{c} \begin{array}{cccccc} \rho_1 & \rho_2 & \rho_3 & \rho_4 & \rho_5 & \rho_6 \end{array} \\ \left[\begin{array}{cccccc} 1 & 0.8 & 0.8 & 0.8 & 0.5 & 0.5 \\ 0.8 & 1 & 0.5 & 0 & 0 & 0 \\ 0.8 & 1 & 1 & 1 & 1 & 0.8 \\ 0 & 0 & 0 & 0.5 & 1 & 0.8 \\ 0.5 & 0.5 & 0.8 & 0.8 & 0.8 & 1 \end{array} \right] \end{array}$$

FIGURE 2.13 Alternate fuzzy interval representation for example 2.15.

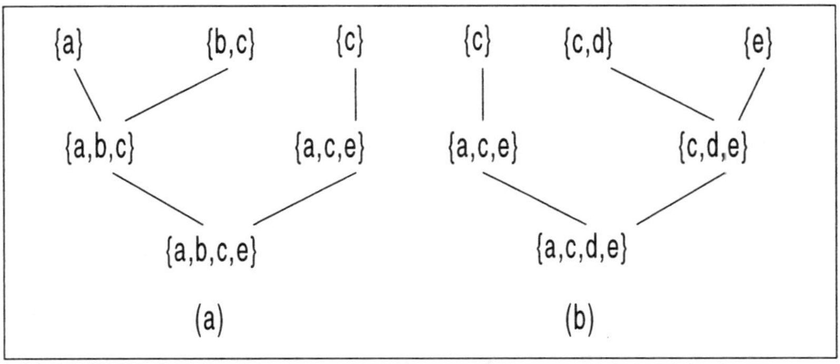

Remark 4. The interval representation of a fuzzy graph G is not in general unique. The construction depends heavily on the orientation of G^c; different orientations can produce different vertex interval matrices.

Slight modifications of Definition 2.23 can produce different vertex intervals matrices. We favored a "left to right" construction of arcs while building the forest F; a "right to left" construction works as well. We also specified each cut level clique be the terminal vertex of only one arc. We could relax this condition as long as cut level consistency is maintained. Figure 2.13 gives an alternative interval representation for the fuzzy graph of Example 2.15.

2.4 Operations on Fuzzy Graphs

The results of this section are taken from [31]. By a partial fuzzy subgraph of a graph (V, X), where X is a set of edges, we mean a partial fuzzy subgraph of the fuzzy graph (χ_V, χ_X). If $G = (V, X)$ is a graph, a partial fuzzy subgraph of G is an ordered pair (μ, ρ) such that μ is a fuzzy subset of V and ρ is a symmetric fuzzy relation on V. However, without any loss of generality, we could have defined ρ as a fuzzy subset of X. Thus it is possible to interpret (μ, ρ) as a partial fuzzy subgraph of G. We follow this interpretation for the remainder of this section for the sake of clarity in presentation. Let (μ_i, ρ_i) be a partial fuzzy subgraph of the graph $G_i = (V_i, X_i)$ for $i = 1, 2$. We define the operations of Cartesian product, composition, union, and join on (μ_1, ρ_1) and (μ_2, ρ_2). Throughout this section we shall denote the edge between two vertices u and v by uv rather than (u, v) because when we take the Cartesian product, a vertex of the graph is, in fact, an ordered pair. If the graph G is formed from G_1 and G_2 by one of these operations, we determine necessary and sufficient conditions in this section for an arbitrary partial fuzzy subgraph of G to also be formed by the same operation from partial fuzzy subgraphs of G_1 and G_2.

Cartesian Product and Composition

Consider the *Cartesian product* $G = G_1 \times G_2 = (V, X)$ of graphs $G_1 = (V_1, X_1)$ and $G_2 = (V_2, X_2)$, [20]. Then

$V = V_1 \times V_2$

and

$X = \{(u, u_2)(u, v_2) \mid u \in V_1, u_2v_2 \in X_2\} \cup \{(u_1, w)(v_1, w) \mid w \in V_2, u_1v_1 \in X_1\}$.

Let μ_i be a fuzzy subset of V_i and ρ_i a fuzzy subset of $X_i, i = 1, 2$. Define the fuzzy subsets $\mu_1 \times \mu_2$ of V and $\rho_1\rho_2$ of X as follows:

$\forall (u_1, u_2) \in V, (\mu_1 \times \mu_2)(u_1, u_2) = \mu_1(u_1) \wedge \mu_2(u_2)$;

$\forall u \in V_1, \forall u_2v_2 \in X_2, \rho_1\rho_2((u, u_2)(u, v_2)) = \mu_1(u) \wedge \rho_2(u_2v_2),$
$\forall w \in V_2, \forall u_1v_1 \in X_1, \rho_1\rho_2((u_1, w)(v_1, w)) = \mu_2(w) \wedge \rho_1(u_1v_1).$

Proposition 2.56 *Let G be the Cartesian product of graphs G_1 and G_2. Let (μ_i, ρ_i) be a partial fuzzy subgraph of $G_i, i = 1, 2$. Then $(\mu_1 \times \mu_2, \rho_1\rho_2)$ is a partial fuzzy subgraph of G.*

Proof. $\rho_1\rho_2((u, u_2)(u, v_2)) = \mu_1(u) \wedge \rho_2(u_2v_2) \leq \mu_1(u) \wedge(\mu_2(u_2) \wedge \mu_2(v_2))$ $= (\mu_1(u) \wedge\mu_2(u_2)) \wedge(\mu_1(u) \wedge \mu_2(u_2)) = (\mu_1 \times \mu_2)(u, u_2) \wedge (\mu_1 \times \mu_2)(u, v_2)$. Similarly, $\rho_1\rho_2((u_1, w)(v_1, w)) \leq (\mu_1 \times \mu_2)(u_1, w) \wedge (\mu_1 \times \mu_2)(v_1, w)$. ∎

The fuzzy graph $(\mu_1 \times \mu_2, \rho_1\rho_2)$ of Proposition 2.56 is called the *Cartesian product* of (μ_1, ρ_1) and (μ_2, ρ_2).

Theorem 2.57 *Suppose that G is the Cartesian product of two graphs G_1 and G_2. Let (μ, ρ) be a partial fuzzy subgraph of G. Then (μ, ρ) is a Cartesian product of a partial fuzzy subgraph of G_1 and a partial fuzzy subgraph of G_2 if and only if the following three equations have solutions for x_i, y_j, z_{jk}, and w_{ih} where $V_1 = \{v_{11}, v_{12}, \ldots, v_{1n}\}$ and $V_2 = \{v_{21}, v_{22}, \ldots, v_{2m}\}$:*

(1) $x_i \wedge y_j = \mu(v_{1i}, v_{2j}), i = 1, \ldots, n; j = 1, \ldots, m;$

(2) $x_i \wedge z_{jk} = \rho((v_{1i}, v_{2j})(v_{1i}, v_{2k})), i = 1, \ldots, n; j, k$ *such that* $v_{2j}v_{2k} \in X_2;$

(3) $y_j \wedge w_{ih} = \rho((v_{1i}, v_{2j})(v_{1h}, v_{2j})), j = 1, \ldots, m; i, h$ *such that* $v_{1i}v_{1h} \in X_1.$

Proof. Suppose that a solution exists. Consider an arbitrary, but fixed, j, k in equations (2) and i, h in equations (3). Let

$$\hat{z}_{jk} = \vee\{\rho((v_{1i}, v_{2j})(v_{1i}, v_{2k})) \mid i = 1, \ldots, n\}$$

and

$$\hat{w}_{ih} = \vee\{\rho((v_{1i}, v_{2j})(v_{1h}, v_{2j})) \mid j = 1, \ldots, m\}.$$

Set
$J = \{(j, k) \mid j, k$ are such that $v_{2j}v_{2k} \in X_2\}$
and
$I = \{(i, h) \mid i, h$ are such that $v_{1i}v_{1h} \in X_1\}$.
Now if $\{x_1, \ldots, x_n\} \cup \{z_{jk} \mid (j, k) \in J\} \cup \{y_1, \ldots, y_m\} \cup \{w_{ih} \mid (i, h) \in I\}$ is any solution to (1), (2), and (3), then $\{x_1, \ldots, x_n\} \cup \{\hat{z}_{jk} \mid (j, k) \in J\} \cup \{y_1, \ldots, y_m\} \cup \{\hat{w}_{ih} \mid (i, h) \in I\}$ is also a solution and, in fact, $\hat{z}_{jk}$ is the smallest possible z_{jk} and $\hat{w}_{ih}$ is the smallest possible w_{ih}. Fix such a solution and define the fuzzy subsets μ_1, μ_2, ρ_1, and ρ_2 of V_1, V_2, X_1, and X_2, respectively, as follows:
$\mu_1(v_{1i}) = x_i$ for $i = 1, \ldots, n;$

$\mu_2(v_{2j}) = y_j$ for $j = 1, \ldots, m$;
$\rho_2(v_{2j}v_{2k}) = \hat{z}_{jk}$ for j, k such that $v_{2j}v_{2k} \in X_2$;
$\rho_1(v_{1i}v_{1h}) = \hat{w}_{ih}$ for i, h such that $v_{1i}v_{1h} \in X_1$.

For any fixed j, k, $\rho((v_{1i}, v_{2j})(v_{1i}, v_{2k})) \leq \mu(v_{1i}, v_{2j}) \wedge \mu(v_{1i}, v_{2k}) = (\mu_1(v_{1i}) \wedge \mu_2(v_{2j})) \wedge (\mu_1(v_{1i}) \wedge \mu_2(v_{2k})) \leq (\mu_2(v_{2j}) \wedge \mu_2(v_{2k})), i = 1, \ldots, n$. Thus $\widehat{z}_{jk} = \vee\{\rho((v_{1i}, v_{2j})(v_{1i}, v_{2k})) \mid i = 1, \ldots, n\} \leq \mu_2(v_{2j}) \wedge \mu_2(v_{2k})$. Hence $\rho_2(v_{2j}v_{2k}) \leq \mu_2(v_{2j}) \wedge \mu_2(v_{2k})$. Thus (μ_2, ρ_2) is a partial fuzzy subgraph of G_2. Similarly, (μ_1, ρ_1) is a partial fuzzy subgraph of G_1. Clearly, $\mu = \mu_1 \times \mu_2$ and $\rho = \rho_1\rho_2$.

Conversely, suppose that (μ, ρ) is the Cartesian product of partial fuzzy subgraphs of G_1 and G_2. Then solutions to equations (1),(2) and (3) exist by definition of Cartesian product. ■

Remark 5. Consider an arbitrary fixed solution to equations (1), (2), and (3) as determined in the proof of Theorem 2.57 (assuming one exists).

(1) Let $(j, k) \in J$ and let $I' = \{i_{jk} \in I \mid \hat{z}_{jk} = \rho((v_{1i_{jk}}, v_{2j})(v_{1i_{jk}}, v_{2k}))\}$ in Theorem 2.57. If $x_{i_{jk}} > \hat{z}_{jk}$ for some $i_{jk} \in I'$, then z_{jk} is unique for these particular $x_1, \ldots, x_n$ and equals $\hat{z}_{jk}$; if $x_{i_{jk}} = \hat{z}_{jk} \forall i_{jk} \in I'$, then $\hat{z}_{jk} \leq z_{jk} \leq 1$ for these particular $x_1, \ldots, x_n$.

(2) Let $(i, h) \in I$ and let $J' = \{j_{ih} \in J \mid \hat{w}_{ih} = \rho((v_{1i}, v_{2j_{ih}})(v_{1h}, v_{2j_{ih}}))\}$ in Theorem 2.57. If $y_{j_{ih}} > \hat{w}_{ih}$ for some $j_{ih} \in J'$, then w_{ih} is unique for these particular $y_1, \ldots, y_m$ and equals $\hat{w}_{ih}$; if $y_{j_{ih}} = \hat{w}_{ih} \forall j_{ih} \in J'$, then $\hat{w}_{ih} \leq w_{ih} \leq 1$ for these particular $y_1, \ldots, y_m$.

Example 2.16 *Let $V_1 = \{v_{11}, v_{12}\}, V_2 = \{v_{21}, v_{22}\}, X_1 = \{v_{11}v_{12}\}$, and $X_2 = \{v_{21}v_{22}\}$. Let $\mu((v_{11}, v_{21})) = 1/4$, $\mu((v_{11}, v_{22})) = 1/2$, $\mu((v_{12}, v_{21})) = 1/8$, and $\mu((v_{12}, v_{22})) = 5/8$. Then (μ, ρ) is not a Cartesian product of partial fuzzy subgraphs of G_1 and G_2 for any choice of ρ since equations (1) of Theorem 2.57 are inconsistent:*

$$x_1 \wedge y_1 = 1/4, x_1 \wedge y_2 = 1/2, x_2 \wedge y_1 = 1/8, x_2 \wedge y_2 = 5/8$$

is impossible.

Examples are easily constructed where equations (1) have a solution, but either equations (2) or (3) are inconsistent.

We now consider the composition of two fuzzy graphs. Let $G_1[G_2]$ denote the composition of graph $G_1 = (V_1, X_1)$ with graph $G_2 = (V_2, X_2)$, [20]. Then $G_1[G_2] = (V_1 \times V_2, X^0)$ where $X^0 = \{(u, u_2)(u, v_2) \mid u \in V_1, u_2v_2 \in X_2\} \cup \{(u_1, w)(v_1, w) \mid w \in V_2, u_1v_1 \in X_1\} \cup \{(u_1, u_2)(v_1, v_2) \mid u_1v_1 \in X_1, u_2 \neq v_2\}$. Let μ_i be a fuzzy subset of V_i and ρ_i a fuzzy subset of $X_i, i = 1, 2$. Define the fuzzy subsets $\mu_1 \circ \mu_2$ and $\rho_1 \circ \rho_2$ of $V_1 \times V_2$ and X^0, respectively, as follows:

$\forall (u_1, u_2) \in V_1 \times V_2$,

$$(\mu_1 \circ \mu_2)(u_1, u_2) = \mu_1(u_1) \wedge \mu_2(u_2);$$

$\forall u \in V_1, \forall u_2v_2 \in X_2$,

$$(\rho_1 \circ \rho_2)((u, u_2)(u, v_2)) = \mu_1(u) \wedge \rho_2(u_2v_2);$$

$\forall w \in V_2, \forall u_1v_1 \in X_1$,

$$(\rho_1 \circ \rho_2)((u_1, w)(v_1, w)) = \mu_2(w) \wedge \rho_1(u_1v_1);$$

$\forall (u_1, u_2)(v_1, v_2) \in X^0 \setminus X$,

$$(\rho_1 \circ \rho_2)((u_1, u_2)(v_1, v_2)) = \mu_2(u_2) \wedge \mu_2(v_2) \wedge \rho_1(u_1v_1),$$

where
$X = \{(u, u_2)(u, v_2) \mid u \in V_1, u_2v_2 \in X_2\} \cup \{(u_1, w)(v_1, w) \mid w \in V_2, u_1v_1 \in X_1\}$.
We see that $\mu_1 \circ \mu_2 = \mu_1 \times \mu_2$ and that $\rho_1 \circ \rho_2 = \rho_1\rho_2$ on X.

Proposition 2.58 *Let G be the composition $G_1 [G_2]$ of graph G_1 with graph G_2. Let (μ_i, ρ_i) be a partial fuzzy subgraph of $G_i, i = 1, 2$. Then $(\mu_1 \circ \mu_2, \rho_1 \circ \rho_2)$ is a partial fuzzy subgraph of $G_1 [G_2]$.*

Proof. We have already seen in the proof of Proposition 2.56 that

$$(\rho_1 \circ \rho_2)((u_1, u_2)(v_1, v_2)) \leq (\mu_1 \circ \mu_2)((u_1, u_2)) \wedge (\mu_1 \circ \mu_2)((v_1, v_2))$$

for all $(u_1, u_2)(v_1, v_2) \in X$. Suppose that $(u_1, u_2)(v_1, v_2) \in X^0 \setminus X$ and so $u_1v_1 \in X_1, u_2 \neq v_2$. Then $(\rho_1 \circ \rho_2)((u_1, u_2)(v_1, v_2)) = \mu_2(u_2) \wedge \mu_2(v_2) \wedge \rho_1(u_1v_1) \leq \mu_2(u_2) \wedge \mu_2(v_2) \wedge \mu_1(u_1) \wedge \mu_1(v_1) = (\mu_1(u_1) \wedge \mu_2(u_2)) \wedge (\mu_1(v_1) \wedge \mu_2(v_2)) = (\mu_1 \circ \mu_2)((u_1, u_2)) \wedge (\mu_1 \circ \mu_2)((v_1, v_2))$. ■

The fuzzy graph $(\mu_1 \circ \mu_2, \rho_1 \circ \rho_2)$ of Proposition 2.58 is called the *composition* of (μ_1, ρ_1) with (μ_2, ρ_2).

Theorem 2.59 *Suppose that G is the composition $G_1 [G_2]$ of two graphs G_1 and G_2. Let (μ, ρ) be a partial fuzzy subgraph of G. Consider the following equations:*

(1) $x_i \wedge y_j = \mu(v_{1i}, v_{2j}), i = 1, \ldots, n; j = 1, \ldots, m;$

(2) $x_i \wedge z_{jk} = \rho((v_{1i}, v_{2j})(v_{1i}, v_{2k})), i = 1, \ldots, n; j, k$ *such that* $v_{2j}v_{2k} \in X_2$;

(3) $y_j \wedge w_{ih} = \rho((v_{1i}, v_{2j})(v_{1h}, v_{2j})), j = 1, \ldots, m; i, h$ *such that* $v_{1i}v_{1h} \in X_1$;

(4) $y_j \wedge y_k \wedge w_{ih} = \rho((v_{1i}, v_{2j})(v_{1h}, v_{2k}))$, *where* $(v_{1i}, v_{2j})(v_{1h}, v_{2k}) \in X^0 \backslash X$;

where X *is defined as above.*

A necessary condition for (μ, ρ) *to be a composition of partial fuzzy subgraphs of* G_1 *and* G_2 *is that a solution to equations (1)–(4) exists.*

Suppose that a solution to equations (1)–(4) exists. If

$$\hat{w}_{ih} \geq \rho((v_{1i}, v_{2j})(v_{1h}, v_{2k}))\forall(i, h) \in I$$

such that $(v_{1i}, v_{2j})(v_{1h}, v_{2j}) \in X^0 \backslash X$, *then* (μ, ρ) *is a composition of partial fuzzy subgraphs of* G_1 *and* G_2.

Proof. The necessary part of the theorem is clear. Suppose that a solution to equations (1)-(4) exists. Then there exists a solution to equations (1)–(4) as determined by in the proof of Theorem 2.57 for equations (1)–(3) because every $w_{ih} \geq \hat{w}_{ih}$ and by the hypothesis concerning the $\hat{w}_{ih}$. Thus if $\mu_i, \rho_i, i = 1, 2$,are defined as in the proof of Theorem 2.57, we have that (μ_i, ρ_i) is a partial fuzzy subgraph of G_i, $i = 1, 2$, and $\mu = \mu_1 \circ \mu_2$ and $\rho = \rho_1 \circ \rho_2$. ■

Example 2.17 *Let* $G_1 = (V_1, X_1)$ *and* $G_2 = (V_2, X_2)$ *be graphs and let* μ_1, μ_2, ρ_1, *and* ρ_2 *be fuzzy subsets of* V_1, V_2, X_1, *and,* X_2, *respectively. Then* $(\mu_1 \times \mu_2, \rho_1\rho_2)$ *is a partial fuzzy subgraph of* $G_1 \times G_2$, *but* (μ_i, ρ_i) *is not a partial fuzzy subgraph of* $G_i, i = 1, 2$: *Let* $V_1 = \{u_1, v_1\}, V_2 = \{u_2, v_2\}, X_1 = \{u_1v_1\}$, *and* $V_2 = \{u_2v_2\}$. *Define the fuzzy subsets* μ_1, μ_2, ρ_1, *and* ρ_2 *as follows:* $\mu_1(u_1) = \mu_1(v_1) = \mu_2(u_2) = \mu_2(v_2) = 1/2$ *and* $\rho_1(u_1v_1) = \rho_2(u_2v_2) = 3/4$. *Then* (μ_i, ρ_i) *is not a partial fuzzy subgraph of* $G_i, i = 1, 2$. *Now for* $x \in V_1$ *and* $y \in V_2, \rho_1\rho_2((x, u_2)(x, v_2)) = \mu_1(x) \wedge \rho_2(u_2v_2) = 1/2 = \mu_1(x) \wedge \mu_2(u_2) \wedge \mu_2(v_2) = (\mu_1 \times \mu_2)((x, u_2)) \wedge (\mu_1 \times \mu_2)((x, v_2))$ *and similarly,* $\rho_1\rho_2((u_1, y)(v_1, y)) = (\mu_1 \times \mu_2)((u_1, y) \wedge (\mu_1 \times \mu_2)((v_1, y))$. *Thus* $(\mu_1 \times \mu_2, \rho_1\rho_2)$ *is a partial fuzzy subgraph of* $G_1 \times G_2$. *Note that for* $x_1y_1 \in X_1$ *and* $x_2, y_2 \in V_2$, $(\rho_1 \circ \rho_2)((x_1, x_2)(y_1, y_2)) = \mu_2(x_2) \wedge \mu_2(y_2) \wedge \rho_1(x_1y_1)$ $= 1/2 = (\mu_1 \times \mu_2)((x_1, x_2)) \wedge (\mu_1 \times \mu_2)((y_1, y_2))$. *Thus* $(\mu_1 \circ \mu_2, \rho_1 \circ \rho_2)$ *is a partial fuzzy subgraph of* $G_1[G_2]$.

We note that in Example 2.17, $(\mu_1 \times \mu_2, \rho_1\rho_2)$ satisfies the conditions in Theorem 2.57. Hence $(\mu_1 \times \mu_2, \rho_1\rho_2)$ is the Cartesian product of partial fuzzy subgraphs (ν_i, τ_i) of $G_i, i = 1, 2$. In fact, these ν_i and τ_i $(i = 1, 2)$ are constant membership functions with membership value 1/2.

Union and Join

Consider the *union* $G = G_1 \cup G_2$ of two graphs $G_1 = (V_1, X_1)$ and $G_2 = (V_2, X_2)$, [20]. Then $G = (V_1 \cup V_2, X_1 \cup X_2)$. Let μ_i be a fuzzy subset of V_i

and ρ_i a fuzzy subset of $X_i, i = 1, 2$. Define the fuzzy subsets $\mu_1 \cup \mu_2$ of $V_1 \cup V_2$ and $\rho_1 \cup \rho_2$ of $X_1 \cup X_2$ as follows:

$(\mu_1 \cup \mu_2)(u) = \mu_1(u)$ if $u \in V_1 \setminus V_2$, $(\mu_1 \cup \mu_2)(u) = \mu_2(u)$ if $u \in V_2 \setminus V_1$, and $(\mu_1 \cup \mu_2)(u) = \mu_1(u) \vee \mu_2(u)$ if $u \in V_1 \cap V_2$;

$(\rho_1 \cup \rho_2)(uv) = \rho_1(uv)$ if $uv \in X_1 \setminus X_2$, $(\rho_1 \cup \rho_2)(uv) = \rho_2(uv)$ if $uv \in X_2 \setminus X_1$, and $(\rho_1 \cup \rho_2)(uv) = \rho_1(uv) \vee \rho_2(uv)$ if $uv \in X_1 \cap X_2$.

Proposition 2.60 *Let G be the union of the graphs G_1 and G_2. Let (μ_i, ρ_i) be a partial fuzzy subgraph of $G_i, i = 1, 2$. Then $(\mu_1 \cup \mu_2, \rho_1 \cup \rho_2)$ is a partial fuzzy subgraph of G.*

Proof. Suppose that $uv \in X_1 \setminus X_2$. We have three different cases to consider: (1) $u, v \in V_1 \setminus V_2$, (2) $u \in V_1 \setminus V_2, v \in V_1 \cap V_2$ and (3) $u, v \in V_1 \cap V_2$.

(1) Let $u, v \in V_1 \setminus V_2$. Then $(\rho_1 \cup \rho_2)(uv) = \rho_1(uv) \leq \mu_1(u) \wedge \mu_1(v) = (\mu_1 \cup \mu_2)(u) \wedge (\mu_1 \cup \mu_2)(v)$.

(2) Let $u \in V_1 \setminus V_2$ and $v \in V_1 \cap V_2$. Then $(\rho_1 \cup \rho_2)(uv) \leq (\mu_1 \cup \mu_2)(u) \wedge (\mu_1(v) \vee \mu_2(v)) = (\mu_1 \cup \mu_2)(u) \wedge (\mu_1 \cup \mu_2)(v)$.

(3) Let $u, v \in V_1 \cap V_2$. Then

$(\rho_1 \cup \rho_2)(uv) \leq (\mu_1(u) \vee \mu_2(u)) \wedge (\mu_1(v) \vee \mu_2(v)) = (\mu_1 \cup \mu_2)(u) \wedge (\mu_1 \cup \mu_2)(v)$.

Similarly, if $uv \in X_2 \setminus X_1$, then $(\rho_1 \cup \rho_2)(uv) \leq (\mu_1 \cup \mu_2)(u) \wedge (\mu_1 \cup \mu_2)(v)$. Suppose that $uv \in X_1 \cap X_2$. Then $(\rho_1 \cup \rho_2)(uv) = \rho_1(uv) \vee \rho_2(uv) \leq (\mu_1(u) \wedge \mu_1(v)) \vee (\mu_2(u) \wedge \mu_2(v)) \leq (\mu_1(u) \vee \mu_2(u)) \wedge (\mu_1(v) \vee \mu_2(v)) = (\mu_1 \cup \mu_2)(u) \wedge (\mu_1 \cup \mu_2)(v)$. ■

The fuzzy subgraph $(\mu_1 \cup \mu_2, \rho_1 \cup \rho_2)$ of Proposition 2.60 is called the *union* of (μ_1, ρ_1) and (μ_2, ρ_2).

Theorem 2.61 *If G is a union of two fuzzy subgraphs G_1 and G_2, then every partial fuzzy subgraph (μ, ρ) is a union of a partial fuzzy subgraph of G_1 and a partial fuzzy subgraph of G_2.*

Proof. Define the fuzzy subsets μ_1, μ_2, ρ_1, and ρ_2 of V_1, V_2, X_1, and X_2, respectively, as follows: $\mu_i(u) = \mu(u)$ if $u \in V_i$ and $\rho_i(uv) = \rho(uv)$ if $uv \in X_i, i = 1, 2$. Then $\rho_i(u_i v_i) = \rho(u_i v_i) \leq \mu(u_i) \wedge \mu(v_i) = \mu_i(u_i) \wedge \mu_i(v_i)$ if $u_i v_i \in X_i, i = 1, 2$. Thus (μ_i, ρ_i) is a partial fuzzy subgraph of $G_i, i = 1, 2$. Clearly, $\mu = \mu_1 \cup \mu_2$ and $\rho = \rho_1 \cup \rho_2$. ■

Consider the join $G = G_1 + G_2 = (V_1 \cup V_2, X_1 \cup X_2 \cup X')$ of graphs $G_1 = (V_1, X_1)$ and $G_2 = (V_2, X_2)$ where X' is the set of all edges joining the vertices of V_1 and V_2 and where we assume $V_1 \cap V_2 = \emptyset$, [20]. Let μ_i be a fuzzy subset of V_i and ρ_i a fuzzy subset of $X_i, i = 1, 2$. Define the fuzzy subsets $\mu_1 + \mu_2$ of $V_1 \cup V_2$ and $\rho_1 + \rho_2$ of $X_1 \cup X_2 \cup X'$ as follows:

$(\mu_1 + \mu_2)(u) = (\mu_1 \cup \mu_2)(u) \forall u \in V_1 \cup V_2$;
$(\rho_1 + \rho_2)(uv) = (\rho_1 \cup \rho_2)(uv)$ if $uv \in X_1 \cup X_2$ and $(\rho_1 + \rho_2)(uv) = \mu_1(u) \wedge \mu_2(v)$ if $uv \in X'$.

Proposition 2.62 *Let G be the join of two graphs G_1 and G_2. Let (μ_i, ρ_i) be a partial fuzzy subgraph of $G_i, i = 1, 2$. Then $(\mu_1+\mu_2, \rho_1+\rho_2)$ is a partial fuzzy subgraph of G.*

Proof. Suppose that $uv \in X_1 \cup X_2$. Then the desired result follows from Proposition 2.60. Suppose that $uv \in X'$. Then $(\rho_1 + \rho_2)(uv) = \mu_1(u) \wedge \mu_2(v) = ((\mu_1 \cup \mu_2)(u) \wedge (\mu_1 \cup \mu_2)(v)) = ((\mu_1 + \mu_2)(u) \wedge (\mu_1 + \mu_2)(v))$. ■

The fuzzy subgraph $(\mu_1 + \mu_2, \rho_1 + \rho_2)$ of Proposition 2.62 is called the *join* of (μ_1, ρ_1) and (μ_2, ρ_2).

Definition 2.25 *Let (μ, ρ) be a partial fuzzy subgraph of a graph $G = (V, X)$. Then (μ, ρ) is called a* strong partial fuzzy subgraph *of G if $\rho(uv) = \mu(u) \wedge \mu(v)$ for all $uv \in X$.*

Theorem 2.63 *If G is the join of two subgraphs G_1 and G_2, then every strong partial fuzzy subgraph (μ, ρ) of G is a join of a strong partial fuzzy subgraph of G_1 and a strong partial fuzzy subgraph of G_2.*

Proof. Define the fuzzy subsets $\mu_1, \mu_{\prime 2}, \rho_1$, and ρ_2 of V_1, V_2, X_1, and X_2 as follows: $\mu_i(u) = \mu(u)$ if $u \in V_i$ and $\rho_i(uv) = \rho(uv)$ if $uv \in X_i, i = 1, 2$. Then (μ_i, ρ_i) is a fuzzy partial subgraph of $G_i, i = 1, 2$, and $\mu = \mu_1 + \mu_2$ as in the proof of Theorem 2.61. If $uv \in X_1 \cup X_2$, then $\rho(uv) = (\rho_1 + \rho_2)(uv)$ as in the proof of Theorem 2.61. Suppose that $uv \in X'$ where $u \in V_1$ and $v \in V_2$. Then $(\rho_1 + \rho_2)(uv) = \mu_1(u) \wedge \mu_2(v) = \mu(u) \wedge \mu(v) = \rho(uv)$ where the latter equality hold since (μ, ρ) is strong. ■

Example 2.18 *Let $G_1 = (V_1, X_1)$ and $G_2 = (V_2, X_2)$ be graphs and let μ_1, μ_2, ρ_1, and ρ_2 be fuzzy subsets of V_1, V_2, X_1, and X_2, respectively. Then $(\mu_1 \cup \mu_2, \rho_1 \cup \rho_2)$ is a partial fuzzy subgraph of $G_1 \cup G_2$, but (μ_i, ρ_i) is not a partial fuzzy subgraph of $G_i, i = 1, 2$: Let $V_1 = V_2 = \{u, v\}$ and $X_1 = X_2 = \{uv\}$. Define the fuzzy subsets $\mu_1, \mu_2, \rho_1, \rho_2$ of V_1, V_2, X_1, X_2, respectively, as follows: $\mu_1(u) = 1 = \mu_2(v), \mu_1(v) = 1/4 = \mu_2(u), \rho_1(uv) = 1/2 = \rho_2(uv)$. Then (μ_i, ρ_i) is not a partial fuzzy subgraph of $G_i, i = 1, 2$. Now $(\rho_1 \cup \rho_2)(uv) = \rho_1(uv) \vee \rho_2(uv) = 1/2 < 1 = (\mu_1(u) \vee \mu_2(u)) \wedge (\mu_1(v) \vee \mu_2(v)) = ((\mu_1 \cup \mu_2)(u) \wedge (\mu_1 \cup \mu_2)(v))$. Thus $(\mu_1 \cup \mu_2, \rho_1 \cup \rho_2)$ is a partial fuzzy subgraph of $G_1 \cup G_2$.*

The above example can be extended to the case where $V_1 \not\subseteq V_2, V_2 \not\subseteq V_1$, $X_1 \not\subseteq X_2$, and $X_2 \not\subseteq X_1$ as follows: Let $V_1 = \{u, v, w\}, V_2 = \{u, v, z\}$, $X_1 = \{uv, uw\}$, $X_2 = \{uv, vz\}$, and $\mu_1(w) = \mu_2(z) = 1 = \rho_1(uw) = \rho_2(uz)$.

Theorem 2.64 *Let $G_1 = (V_1, X_1)$ and $G_2 = (V_2, X_2)$ be graphs. Suppose that $V_1 \cap V_2 = \emptyset$. Let $\mu_1, \mu_2, \rho_1, \rho_2$ be fuzzy subsets of V_1, V_2, X_1, X_2, respectively. Then $(\mu_1 \cup \mu_2, \rho_1 \cup \rho_2)$ is a partial fuzzy subgraph of $G_1 \cup G_2$ if and only if (μ_1, ρ_1) and (μ_2, ρ_2) are partial fuzzy subgraphs of G_1 and G_2, respectively.*

Proof. Suppose that $(\mu_1 \cup \mu_2, \rho_1 \cup \rho_2)$ is a partial fuzzy subgraph of $G_1 \cup G_2$. Let $uv \in X_1$. Then $uv \notin X_2$ and $u, v \in V_1 \setminus V_2$. Hence $\rho_1(uv) = (\rho_1 \cup \rho_2)(uv) \leq ((\mu_1 \cup \mu_2)(u) \wedge (\mu_1 \cup \mu_2)(v)) = (\mu_1(u) \wedge \mu_1(v))$. Thus (μ_1, ρ_1) is a partial fuzzy subgraph of G_1. Similarly, (μ_2, ρ_2) is a partial fuzzy subgraph of G_2. The converse is Proposition 2.60. ■

The following result follows from the proof of Theorem 2.64 and Proposition 2.62.

Theorem 2.65 *Let $G_1 = (V_1, X_1)$ and $G_2 = (V_2, X_2)$ be graphs. Suppose that $V_1 \cap V_2 = \emptyset$. Let $\mu_1, \mu_2, \rho_1, \rho_2$ be fuzzy subsets of V_1, V_2, X_1, X_2, respectively. Then $(\mu_1 + \mu_2, \rho_1 + \rho_2)$ is a partial fuzzy subgraph of $G_1 + G_2$ if and only if (μ_1, ρ_1) and (μ_2, ρ_2) are partial fuzzy subgraphs of G_1 and G_2, respectively.* ■

Definition 2.26 *Let (μ, ρ) be a partial fuzzy subgraph of (V, T). Define the fuzzy subsets μ' of V and ρ' of T as follows: $\mu' = \mu$ and $\forall uv \in T, \rho'(uv) = 0$ if $\rho(u, v) > 0$ and $\rho'(uv) = \mu(u) \wedge \mu(v)$ if $\rho(u, v) = 0$.*

Clearly, $G' = (\mu', \rho')$ is a fuzzy graph.

Definition 2.27 *(μ, ρ) is said to be* complete *if $X = T$ and $\forall uv \in X, \rho(uv) = \mu(u) \wedge \mu(v)$.*

We use the notation $C_m(\mu, \rho)$ for a complete fuzzy graph where $| V |= m$.

Definition 2.28 *(μ, ρ) is called a* fuzzy bigraph *if and only if there exists partial fuzzy subgraphs $(\mu_i, \rho_i), i = 1, 2$, of (μ, ρ) such that (μ, ρ) is the join $(\mu_1, \rho_1) + (\mu_2, \rho_2)$ where $V_1 \cap V_2 = \emptyset$ and $X_1 \cap X_2 = \emptyset$. A fuzzy bigraph is said to be* complete *if $\rho(uv) > 0$ for all $uv \in X'$.*

We use the notation $C_{m,n}(\mu, \rho)$ for a complete fuzzy bigraph such that $| V_1 |= m$ and $| V_2 |= n$.

Proposition 2.66 $C_{m,n}(\mu, \rho) = C_m(\mu_1, \rho_1)' + C_n(\mu_2, \rho_2)'$. ■

2.5 On Fuzzy Tree Definition

In this section, we consider other definitions of a fuzzy tree that can be found in the literature. The results are taken from [10]. The first definition of a fuzzy graph was introduced by Kaufmann [22], based on fuzzy relations (see Section 1.1). A more elaborate definition is due to Rosenfeld [34]. In this section, some concepts and properties, such as path, connectedness and fuzzy tree, are presented.

We may interpret a fuzzy graph (μ, ρ) as a network of roads such that

(1) its vertices are towns which can be classified in several ways, e.g., population size and a corresponding value assigned via μ;

(2) its edges are the roads joining the towns and the roads can be of different categories and a weight is associated to each category via ρ.

The above network road can be interpreted as a fuzzy graph. For a such network road the most important problems arise in relation with the connectedness among vertices.

As developed previously, the existence of a chain joining a pair of vertices guarantees the connectedness between both vertices, regardless of the category to which these vertices and edges belongs. From a practical point of view, in a network road like the above one it is very important to analyze the connectedness by levels in order to know whether vertices of a given category are connected by chains formed without edges of a lower category than the aforementioned vertices. If is clear the existence of important towns joined by roads of low category only, reveals defects in the design of the network.

The same question arises when the concept of a fuzzy tree is considered. It is important to remark that the situation above explained is the subject of a work which has been analyzed by Delgado, Verdegay and Vila on the network road of Andalusia (Spain). The objectives of this section are the definition and study of some structural properties of finite fuzzy graphs in order to find a tool that allows us to solve some Operational Research problems, like the above mentioned. Therefore, on the basis of the initial definitions, we will specify the concept of connectedness, by way of t-cuts. The acyclicity problem in a fuzzy graph is also treated, giving the definitions of a cyclomatic function and an acyclicity level. Finally, we will give some definition for fuzzy trees and we will analyze their relations.

Recall that the t-cut, $t \in [0,1]$, of a fuzzy graph $G = (V, \mu, \rho)$ is the classical graph $G^t = (V^t, U^t)$ with

$\mu^t = \{x \in V \mid \mu(x) \geq t\}$,

$\rho^t = \{(x, y) \in V \times V \mid \rho(x, y) \geq t\}$.

We now examine connectedness for fuzzy graphs. From its underlying structure, some definitions of connected fuzzy graphs can be established. First, we recall our previous definition.

FIGURE 2.14 Connected fuzzy graphs.

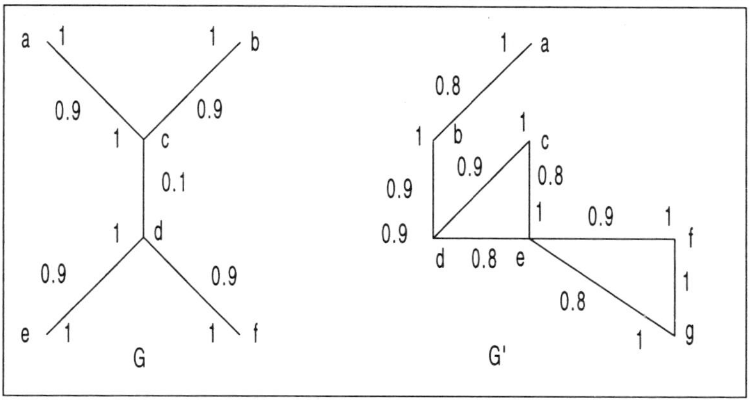

Let $G = (V, \mu, \rho)$ be a fuzzy graph. Then vertices x and y are connected if and only if there is a path with positive strength joining both, that is, $\rho^\infty(x, y) > 0$. With the assumption of reflexivity, that is, every vertex is connected with itself, "connectedness" is an equivalence relation.

We also recall that a fuzzy graph G is called connected if and only if it has only one connected component, which is itself. This definition is equivalent to saying that a fuzzy graph is connected if and only if there exists a path with non-zero strength joining every pair of different vertices.

We now present some examples which will lead us to a different notion of a connected fuzzy graph.

Example 2.19 *We consider the two fuzzy graphs G and G' in Figure 2.14. The two fuzzy graphs are connected, but they do not satisfy this property in the same way. The smallest strength of a chain in G' is* 0.8, *however there are chains which connect the vertices in G with strength* 0.1 *at most. Thus G' seems to be "more connected" than G.*

Example 2.20 *Consider the fuzzy graphs F and F' (Figure 2.15). Neither F nor F' are connected, but the t-cuts of F are connected (in the classical sense) for $t \in (0.2, 0.8]$, whereas no t-cut of F' is connected. F presents "some kind of connectedness" which does not appear obviously in F'.*

These examples explain the following point of view:

(a) The property of connectedness on a fuzzy structure is a matter of degree.

(b) There is a kind of connectedness in fuzzy graphs, characterized by means of t-cuts, which can be used to treat some fuzzy graphs.

This leads us to the following definition.

FIGURE 2.15 Nonconnected fuzzy graphs.

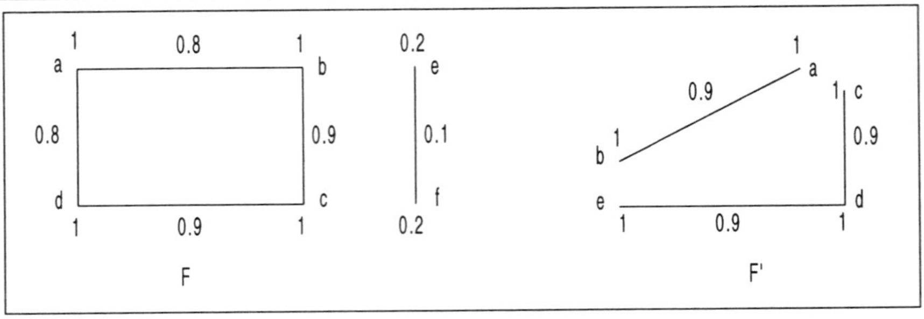

Definition 2.29 *Let* $G = (V, \mu, \rho)$ *be a fuzzy graph. Then the* connectedness level *of* G *is the value* $C(G) = \wedge\{\rho^{\infty}(x, y) \mid x, y \in V, x \neq y\}$.

Obviously, G is connected if and only if $C(G) > 0$. Moreover, if $C(G) > 0$ then $\forall t \in (0, 1]$, $t \leq C(G) \Rightarrow G^t$ is connected.

Definition 2.30 *Let* $G = (V, \mu, \rho)$ *be a fuzzy graph. We say* G *is* weakly connected *if there is some* t*-cut of* G *which is connected.*

We see that a fuzzy graph is weakly connected if and only if $\exists t \in [0, 1]$ such that $\wedge\{\rho^{\infty}(x, y) \mid x, y \in \mu^t\} \geq t$.

Clearly, connectedness implies weak connectedness, but not conversely.

We see that weak connectedness is only meaningful if $\mu(x) < 1$ for some $x \in V$, as in this situation the set of vertices changes with the variations of membership degree.

A fuzzy graph $G = (V, \mu, \rho)$ is called an *acyclic* fuzzy graph if there is a fuzzy subgraph $F = (V, \mu, \tau)$ of G such that F is a forest and $\forall x, y \in V$, $\rho(x, y) > 0$ and $\tau(x, y) > 0$ implies $\tau^{\infty}(x, y) > \rho(x, y)$.

The concept of an acyclic graph plays a very important role in general graph theory, as it is related with connectedness problems. Further, it is closely related to the concept of a tree. For example, one may define a fuzzy tree as an acyclic and connected fuzzy graph. The following example will lead us to a new notion of an acyclic fuzzy graph.

Example 2.21 *Consider the graphs* G *and* F *in Figure 2.16.* G *is a fuzzy tree (*F *is the fuzzy subgraph which appears immediately above); however, there is no* t*-cut of* G*, which is a tree in the classical sense.*

The difficulty arises from the definition of an acyclic fuzzy graph and so we present some alternative definitions of it. For that we must introduce some concepts such as the cyclomatic function.

FIGURE 2.16 Fuzzy tree with no t-cuts as trees.

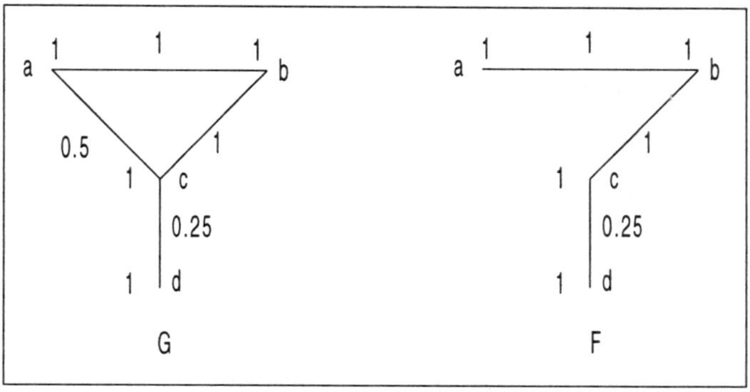

Given a graph G, the *cyclomatic number* of G is defined as $m - n + p$, where n, m, and p denote the number of vertices, edges and connected components of G, respectively.

Definition 2.31 *Let $G = (V, \mu, \rho)$ be a fuzzy graph with card$(V) = n$. We call $h(G, \cdot) : [0,1] \to \mathbb{N} \cup \{0\}$ defined by $h(G,t) =$ cyclomatic number of G^t, the* cyclomatic function *of G.*

If G is a fuzzy graph and $t \in [0,1]$, we let n^t, m^t, and p^t denote the number of vertices, edges, and connected components of G^t, respectively. Then $h(G,t) = m^t - n^t + p^t$.

The following properties of $h(G, \cdot)$ are important for further definitions and developments.

Proposition 2.67 $\forall t \in [0,1]$, $h(G,t) \geq 0$.

Proposition 2.68 *$h(G, \cdot)$ is a piecewise constant function with finite jumps.*

These two properties follow directly from the definition of $h(G, \cdot)$

Let $G = (V, \mu, \rho)$ be a fuzzy graph. If we remove an edge from G to obtain a fuzzy graph G', then $m' = m - 1, n' = n$, and $p' \leq p + 1$, where n', m', and p' are the number of nodes, edges, and connected components of G', respectively. Hence $h(G', \cdot) = m' - n' + p' \leq (m-1) - n + (p+1) = h(G, \cdot)$. Now suppose a node v and edges connected to v are removed from G. Suppose the number of such edges is k. If we remove the edges one at a time, but not v, then the resulting graph has less than or equal to $p + k$ connected components. Finally, removing v, the resulting graph G' has less than or equal to $p + k - 1$ connected components. Hence $h(G', \cdot) =$

$m' - n' + p' \leq (m-k) - (n-1) + (p+k-1) = h(G,\cdot)$. This reasoning leads to the following result.

Proposition 2.69 *$h(G,\cdot)$ is non-increasing in t, that is,*

$$\forall t, t^{'} \in [0,1]\ t \geq t^{'} \Rightarrow h(G,t) \leq h(G,t^{'}).$$

Proof. By properties of t-cuts
$m^t \leq n^t \Rightarrow m^t = m^{t'} - k_1,$
$k_1 \in \mathbb{Z}, k_1 \geq 0,$
$n^t \leq n^{t'} \Rightarrow n^t = n^t - k_2,$
$k_2 \in \mathbb{Z}, k_2 \geq 0.$

We cannot conclude anything about the variation of p^t with t, because the number of connected components of G^t may increase, decrease or remain the same as t ranges in $[0,1]$. Thus, for $t \geq t'$, $p^t = p^{t'} - k_3$ for some $k_3 \in \mathbb{Z}$.

In this situation

$$\begin{aligned} h(G,t) &= (m^{t'} - k_1) - (n^{t'} - k_2) + (p^{t'} - k_3) \\ &= h(G,t') + (k_2 - k_1 - k_3) \\ &= h(G,t') + k. \end{aligned}$$

We will prove $k > 0$ is impossible. Two possibilities must be considered:

(a) $k_2 = 0$. This assumption implies the vertex set does not change from $G^{t'}$ to G^t. Thus $k_3 \leq 0$, because connected components cannot decrease by a possible edge suppression. Moreover, since a new connected component appears only by edge suppression
$-k_3 \leq k_1 \Rightarrow k \leq 0.$

(b) $k_2 > 0$. If we denote h the number of eliminated connected components of $G^{t'}$, obviously $k_3 \leq h$ and $k_2 \geq h$. Let us write $h = k_2 - s$; $s \in \{0, \ldots, k_2\}$. From the definition of s, we can derive $k_1 \geq s$ and thus
$k = k_2 - k_3 - k_1 \leq k_2 - k_2 + s - k_1 \Rightarrow k \leq s - k_1 \leq 0.$ ■

Let $H = \{t \in [0,1] \mid h(G,t) = 0\}$. By Propositions 2.68 and 2.69 we can assure only two possibilities for H :

(1) $H = \emptyset$.

(2) $H = (0,1]$.

Definition 2.32 *The acyclic level of an fuzzy graph G is*
$S(G) = \wedge\{t \mid t \in H\};$
$S(G) = \infty$ *if* $H = \emptyset$.

The following result can be easily proved from this definition and the properties of the cyclomatic function.

Proposition 2.70 *There are no cycles in G^t if and only if $S(G) < \infty$ and $t > S(G)$.*

Two definitions of acyclic fuzzy graphs can be formulated by means of $S(G)$.

Definition 2.33 *The fuzzy graph $G = (V, \mu, \rho)$ is said to be* fully acyclic *if $S(G) = 0$.*

Actually, a fully acyclic fuzzy graph is a forest and conversely, since $S(G) = 0$, it is equivalent to say the graph formed by the edges with nonzero membership degree must be acyclic. However, we introduce this nomenclature to emphasize the acyclic situation underlying in a such fuzzy graph.

Definition 2.34 *We shall say an fuzzy graph $G = (V, \mu, \rho)$ is acyclic by t-cuts if there exists a $t \in [0, 1]$ such that G^t has no cycles.*

Obviously, G will be acyclic by t-cuts if and only if $S(G) \neq \infty$.

Proposition 2.71 *Every acyclic fuzzy graph is acyclic by t-cuts.*

Proof. Let us assume that G is an acyclic fuzzy graph such that $S(G) = \infty$. This implies there is a cycle L in G such that $\rho(x, y) = 1$ for every edge (x, y) belonging to L. Let $(\bar{x}, \bar{y})$ be an edge of L. Let $\zeta(\cdot, \cdot)$ denote the membership function of the fuzzy set of edges in the fuzzy subgraph of G which appears when $(\bar{x}, \bar{y})$ is suppressed. Then
$\zeta^\infty(\bar{x}, \bar{y}) = 1 \Rightarrow \zeta^\infty(\bar{x}, \bar{y}) = \rho(\bar{x}, \bar{y})$.
Hence, by Theorem 2.5, G cannot be an acyclic fuzzy graph and thus we conclude that $S(G) \neq \infty$. ■

To see the converse of Proposition 2.71 does not hold, we can consider the example shown in Figure 2.17. Obviously, $G^{3/4}$ is an acyclic graph, but the definition of an acyclic fuzzy graph never holds for the edges (a, d) and (b, c).

We now consider several definitions of a fuzzy tree. By using the concepts of connectedness and acyclicity, some fuzzy tree definitions can be introduced.

Definition 2.35 *The fuzzy graph $G = (V, \mu, \rho)$ is called a* full fuzzy tree *if it satisfies the conditions $C(G) > 0$ and $S(G) = 0$.*

Obviously, this is the tree definition for fuzzy graphs. We introduce the adjective "full" to emphasize that Definition 2.33 is used.

Definition 2.36 *The fuzzy graph $G = (V, \mu, \rho)$ is a* complete fuzzy tree *if there exists $t \in [0, 1]$ such that G^t is a tree and $\mu^t = V$.*

FIGURE 2.17 Acyclic by t-cuts fuzzy graph.

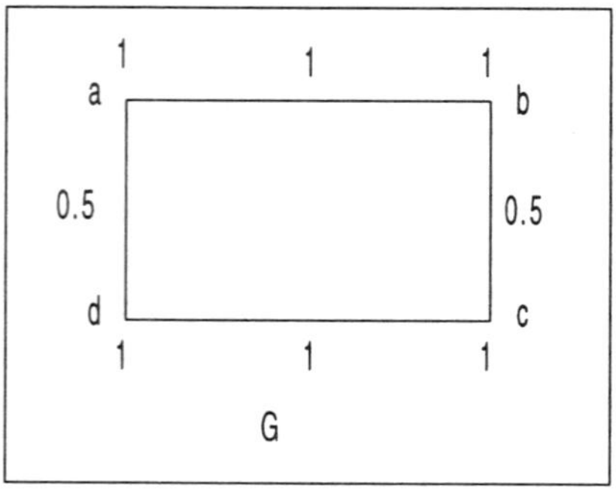

Another characterization of a complete fuzzy tree can be established by means of the following lemma.

Lemma 2.72 *$G = (V, \mu, \rho)$ is a complete fuzzy tree if and only if it satisfies the conditions $C(G) > 0$ and $S(G) < C(G)$.*

Proof. Let $G^{\bar{t}}$ be the tree which appears in Definition 2.36. It is a connected acyclic graph so that $C(G) > \bar{t}$ and $S(G) < \bar{t}$. Therefore $S(G) < \bar{t} < C(G)$.

Let us assume the above conditions hold for G. For every $\bar{t} \in (S(G), C(G)]$, $G^{\bar{t}}$ is a tree. Therefore to prove it is a complete fuzzy tree, it suffices to prove $\mu^{\bar{t}} = V$, that is, $\mu(x) \geq \bar{t}$, $\forall x \in V$. Let $x \in V$. By the definition of $C(G)$, we have $\forall y \in V$, $x \neq y \Rightarrow \rho^{\infty}(x, y) \geq C(G)$. Since $\rho^{\infty}(x, y)$ is the strength of the strongest chain joining x and y, we can assure that $\exists z \in V$ such that $\rho(x, z) \geq \rho^{\infty}(x, y)$. Moreover $\rho(x, z) \leq \mu(x) \wedge \mu(z)$ and thus $\mu(x) \geq \rho(x, z) \geq \rho^{\infty}(x, y) \geq C(G) \geq \bar{t}$. ■

The following property arises from this characterization.

Proposition 2.73 *If G is a complete fuzzy tree, then $\forall t, t' \in (S(G), C(G)]$, $G^t = G^{t'}$.*

Proof. By Lemma 2.72, $\mu^t = \mu^{t'} = V$, therefore $n = n^{t'} = \text{card}(V)$. Moreover, G^t and $G^{t'}$ are trees and so $\text{card}(\rho^t) = \text{card}(\rho^{t'}) = \text{card}(V) - 1$, i.e., both trees have the same number of edges. Now if $t \leq t'$, then $\mu^t \supseteq \mu^{t'}$ and $\rho^t \supseteq \rho^{t'}$. Hence $\mu^t = \mu^{t'}$ and $\rho^t = \rho^{t'}$. Thus $G^t = G^{t'}$. ■

Definition 2.37 *$G = (V, \mu, \rho)$ is called* weak fuzzy tree *if there is $t' \in (0, 1]$ such that $G^{t'}$ is a tree.*

Another characterization of weak fuzzy tree is: G is a weak fuzzy tree if and only if the following conditions hold:

FIGURE 2.18 A complete, but not full fuzzy tree.

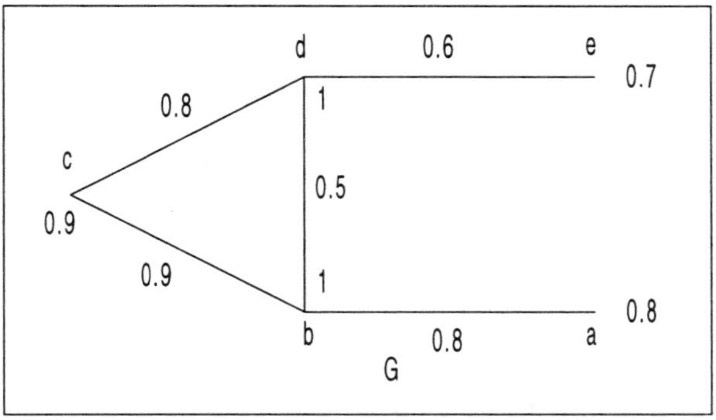

(1) G is weakly connected,

(2) $S(G) < \bar{t}$, $\bar{t}$ being some level such that $G^{\bar{t}}$ is connected.

The proof of this characterization is like one of Lemma 2.72, by using $\bar{t}$ instead of $C(G)$.

Obviously, all these definitions are related, as is shown in the following result.

Proposition 2.74 *The following implications hold.*

(1) If G is a full fuzzy tree, then G is a complete fuzzy tree.

(2) If G is a complete fuzzy tree, then G is a fuzzy tree and in fact, G is a weak fuzzy tree.

Proof. (1) This statement follows from Definition 2.35 and Lemma 2.72.

(2) Let G be a complete fuzzy tree. From Lemma 2.72, $C(G) > 0$ and $S(G) < C(G)$. For every $t \in (S(G), C(G)]$ we can define the fuzzy subgraph of G, $F = (V, \mu, \nu)$ where

$$\nu(x,y) = \begin{cases} t & \text{if } (x,y) \in \rho^t, \\ 0 & \text{otherwise.} \end{cases}$$

Since G^t is a tree, obviously F is a full fuzzy tree. Moreover, $\nu^\infty(x,y) = t$ if $x \neq y$. Thus F is acyclic and G is connected. Therefore G is a fuzzy tree. That G is a complete fuzzy tree implies that G is a weak fuzzy tree follows from Lemma 2.72 and Definition 2.37 with $\bar{t} = C(G)$. ■

It may be noted that weak connectedness does not imply connectedness. The converse of (1) does not hold as it is shown in counter example, illustrated in Figure 2.18: $S(G) = 0.5$, $C(G) = 0.6$ and $V^{0.6} = V$, that is, G is a complete fuzzy tree, but it is not a full fuzzy tree. Example 2.21 shows that a fuzzy tree need not be complete.

2.6 References

1. Arya, S.P. and Hazarika, D., Functions with closed fuzzy graph, *J. Fuzzy Math.* **2**:593–600, 1994.

2. Berge, C., *Theorie des Graphes et ses Applications,* Dunod, Paris, 1958.

3. Bhattacharya, P., Some remarks on fuzzy graphs, *Pattern Recognition Letters* **6**:297–302, 1987.

4. Bhattacharya, P. and Suraweera, An algorithm to compute the supremum of max-min powers and a property of fuzzy graphs, *Pattern Recognition Letters* **12**:413–420, 1991.

5. Bhutani, K.R., On automorphisms of fuzzy graphs, *Pattern Recognition Letters* **9**:159–162, 1989.

6. Cerruti, U., Graphs and fuzzy graphs, *Fuzzy Information and Decision Processes* 123–131, North-Holland, Amsterdam-New York, 1982.

7. Chen, Q. J., Matrix representations of fuzzy graphs (Chinese), *Math. Practice Theory* **1**:41–46, 1990.

8. Craine, W.L., Characterization of fuzzy interval graphs, *Fuzzy Sets and Systems* **68**:181–193, 1994.

9. De Luca, A., and Termini, S., A definition of a non-probabilistic entropy in the setting of fuzzy sets theory, *Inform. and Control* **20**:301–312, 1972.

10. Delgado, M. and Verdegay, J.L., and Vila, M.A., On fuzzy tree definition, *European J. Operational Res.* **22**:243–249, 1985.

11. Dodson, C.T.J., A new generalization of graph theory, *Fuzzy Sets and Systems* **6**:293–308, 1981.

12. Dubois, D. and H. Prade, H., *Fuzzy Sets and Systems, Theory and Applications,* Academic Press, New York,1980.

13. Fishburn, P.C., *Interval Orders and Interval Graphs,* Wiley, New York, 1985.

14. Fulkerson, D.R. and O.A. Gross, O.A., Incidence matrices and interval graphs, *Pacific J. Math.* **5**:835–855, 1965.

15. Gilmore, P.C. and Hoffman, A.J., A characterization of comparability graphs and of interval Graphs, *Canad. J. Math.* **16**:539–548,1964.

16. El-Ghoul, M., Folding of fuzzy graphs and fuzzy spheres, *Fuzzy Sets and Systems* **58:**355–363, 1993.

17. El-Ghoul, M., Folding of fuzzy torus and fuzzy graphs, *Fuzzy Sets and Systems* **80:**389–396, 1996.

18. Halpern, J., Set adjacency measures in fuzzy graphs, *J. Cybernet.* **5:**77-87, 1976.

19. Harary, F., On the group of the composition of two graphs, *Duke Math. J.* **26:**29–34, 1959.

20. Harary, F., *Graph Theory*, Addison Wesley, Third printing, October 1972.

21. Jacobson, M.S., McMorris, F.R. and Scheinerman, E.R., General results on tolerance intersection graphs, *J. Graph Theory* **15**:573–577, 1991.

22. Kaufmann, A., Introduction a la Theorie des sons-ensembles flous, Vol. 1, Masson Paris, 41–189, 1973.

23. Leenders, J.H., Some remarks on an article by Raymond T. Yeh and S.Y. Bang dealing with fuzzy relations: Fuzzy relations, fuzzy graphs, and their applications to clustering analysis, Fuzzy sets and their applications to cognitive and decision processes (Proc. U.S.-Japan Sem., Univ. Calif., Berkeley, Calif., 1974), 125–149, *Simon Stevin* **51:**93–100, 1977/78.

24. Luo, C.S., Decomposition theorems and representation theorems in fuzzy graph theory (Chinese), *J. Xinjiang Univ. Nat. Sci.* **3:**27–33, 1986.

25. Luo, C.S., The theorems of decomposition and representation for fuzzy graphs, *Fuzzy Sets and Systems* **42:**237–243, 1991.

26. McAllister, L.M.N., Fuzzy intersection graphs, *Intl. J. Computers in Mathematics with Applications* **5**:871–886, 1988.

27. Marczewski, E., Sur deux Propriétés des Classes d'ensembles, *Fund. Math.* **33**:303–307,1945.

28. McKee,T.A., Foundations of intersection graph theory, *Utilitas Math.* **40**:77–86,1991.

29. Mordeson, J. N., Fuzzy line graphs, *Pattern Recognition Letters* **14:** 381–384, 1993.

30. Mordeson, J.N. and Nair, P.S, Cycles and cocycles of fuzzy graphs, *Inform. Sci.* **90:**39–49, 1996.

31. Mordeson, J.N. and Peng, C-S, Operations on fuzzy graphs, *Inform. Sci.* **79:**159–170, 1994.

32. Mori, M. and Kawahara, Y., Fuzzy graph rewritings, *Theory of rewriting systems and its applications (Japanese)* **918:**65–71, 1995.

33. Roberts, F., *Discrete Mathematical Models,* Prentice Hall, Englewood Cliffs, NJ,1976.

34. Rosenfeld, A., Fuzzy graphs, In: L. A. Zadeh, K. S. Fu, M. Shimura, Eds., *Fuzzy Sets and Their Applications*, 77–95, Academic Press, 1975.

35. Roubens, M. and Vincke, P., Linear fuzzy graphs, *Fuzzy Sets and Systems* **10:**79–86, 1983.

36. Sabidussi, G., The composition of graphs, *Duke Math. J.* **26:**693–696, 1959.

37. Sabidussi, G., Graph multiplication, *Math. Z.* **72:**446–457, 1960.

38. Sabidussi, G., The lexicographic product of graphs, *Duke Math. J.* **28:**573–578, 1961.

39. Sarma, R. D. and Ajmal, N., Category $\mathcal{N}$ and functions with closed fuzzy graph, *Fuzzy Sets and Systems* **63:**219–226, 1994.

40. Shannon, A., Atanassov, K., Intuitionistic fuzzy graphs from t-, β- and $t\beta$-levels, *Notes IFS* **1:**32–35, 1995.

41. Sunitha, M.S. and Vijayakumar, A., A characterization of fuzzy trees, *Inform. Sci.* **113:**293–300, 1999.

42. Sunouchi, H. and Morioka, M., Some properties on the connectivity of a fuzzy graph (Japanese), *Bull. Sci. Engrg. Res. lab. Waseda Univ.* no. 132, 70–78, 1991.

43. Takeda, E., Connectvity in fuzzy graphs, *Tech. Rep. Osaka Univ.* **23:**343–352, 1973.

44. Tong, Z. and Zheng, D., An algorithm for finding the connectedness matrix of a fuzzy graph, *Congr. Numer.* **120:**189–192, 1996.

45. Wu, L. G. and Chen, T.P., Some problems concerning fuzzy graphs (Chinese), *J. Huazhong Inst. Tech.* no 2, Special issue on fuzzy math, iv, 58–60, 1980.

46. Yeh, R.T. and Bang, S.Y., Fuzzy relations, fuzzy graphs, and their applications to clustering analysis, In: L. A. Zadeh, K. S. Fu, M. Shimura, Eds., *Fuzzy Sets and Their Applications*, 125–149, Academic Press, 1975.

47. Zadeh, L.A., Similarity relations and fuzzy ordering. *Inform. Sci.* **3**:177-200, 1971.

48. Zimmermann, H.J., *Fuzzy Set Theory and its Applications,* Kluwer-Nijhoff, Boston,1985.

3

APPLICATIONS OF FUZZY GRAPHS

Let (V, μ, ρ) be a fuzzy graph. We now provide two popular ways of defining the distance between a pair of vertices. One way is to define the "distance" $dis(x, y)$ between x and y as the length of the shortest strongest path between them. This "distance" is symmetric and is such that $dis(x, x) = 0$ since by our definition of a fuzzy graph, no path from x to x can have strength greater than $\mu(x)$, which is the strength of the path of length 0. However, it does not satisfy the triangle property, as we see from the following example. Let $V = \{u, v, x, y, z\}$, $\rho(x, u) = \rho(u, v) = \rho(v, z) = 1$ and $\rho(x, y) = \rho(y, z) = 0.5$. Here any path from x to y or from y to z has strength $\leq 1/2$ since it must involve either edge (x, y) or edge (y, z). Thus the shortest strongest paths between them have length 1. On the other hand, there is a path from x to z, through u and v, that has length 3 and strength 1. Thus $dis(x, z) = 3 > 1 + 1 = dis(x, y) + dis(y, z)$ in this case.

A better notion of distance in a fuzzy graph can be defined as follows: For any path $P = x_0, ..., x_n$, define the *ρ-length* of P as the sum of the reciprocals of P's edge weights, that is,

$$l(P) = \sum_{i=1}^{n} \frac{1}{\rho(x_{i-1}, x_i)}.$$

If $n = 0$, we define $l(P) = 0$. Clearly, for $n \geq 1$, we have $l(P) \geq 1$. For any two vertices x, y, we can now define their *ρ-distance* $\delta(x, y)$ as the smallest ρ-length of any path from x to y. Thus $\delta(x, y) = \wedge\{l(P) | P$ is a path between x and $y\}$ if x and y are connected. We define $\delta(x, y) = \infty$ if x and y are not connected.

Proposition 3.1 *δ is a metric on V. That is, $\forall x, y, z \in V$,*

(1) $\delta(x,y) = 0 \Leftrightarrow x = y$,

(2) $\delta(x,y) = \delta(y,x)$,

(3) $\delta(x,z) \leq \delta(x,y) + \delta(y,z)$.

Proof. (1) Since $l(P) = 0$ if $x = y$ and $\delta(x,y) \geq 1$ if $x \neq y, \delta(x,y) = 0 \Leftrightarrow x = y$.

(2) Since reversal of a path from x to y is a path from y to x and vice-versa, $\delta(x,y) = \delta(y,x)$.

(3) Since the concatenation of two paths, a path from x to y and a path from y to z yields a path from x to z, $\delta(x,z) \leq \delta(x,y) + \delta(y,z)$. ■

In the crisp case, $l(P)$ is just the length n of P since all the ρ's are 1. Hence, $\delta(x,y)$ becomes the usual definition of distance, that is, the length of the shortest path between x and y.

Proposition 3.2 *Consider a fuzzy graph $G = (\mu, \rho)$ with the vertex set V and let $x, y \in V$. Let j $(1 \leq j \leq diam(x,y))$ be the smallest possible length of a path joining x to y and let $\rho^\infty(x,y)$ be its strength. Suppose further that there exists at least one path of length j whose strength is less than*

$$\left\{ j \left(\frac{1}{\rho^\infty(x,y)} - 1 \right) + 1 \right\}^{-1}.$$

Then $l(P) = \delta(x,y)$.

Proof. Let P be a path joining x to y with strength $\rho^\infty(x,y)$ and having the smallest path length j where $j \leq \text{diam}(x,y)$. Let the vertices of V be $x = x_0, \ldots, x_i, \ldots, x_{j,} = y$. Let (u,v) be the edge which gives the strength of P. In the sum corresponding to the definition of $l(P)$, the contribution by the edge (u,v) is equal to $1/\rho^\infty(u,v)$. It can now be shown that

$$\left\{ (j-1) + \frac{1}{\rho^\infty(x,y)} \right\} \leq l(P) \leq \frac{j}{\rho^\infty(x,y)}.$$

Note that the upper bound is achieved when each edge weight is equal to $\rho^\infty(x,y) = s(P)$, where $s(P) = \wedge\{\rho(x_{i-1}, x_i) \mid i = 1, 2, \ldots, n\}$. The lower bound is achieved when the edge (u,v) has weight $\rho^\infty(x,y)$ and the rest of the edges forming the path P have weight equal to 1 each.

If $j = 1$, then

$$l(P) = \frac{1}{s(P)} = \frac{1}{\rho^\infty(x,y)},$$

As $1/\rho^\infty(x,y)$ is the smallest possible ρ-length of any path from x to y, the result follows in this case. Now, consider the case $j > 1$. Let P' be another path joining x to y with vertices $x = x'_0, \ldots, x'_j = y$. Since P is the smallest path from x to y having strength $\rho^\infty(x,y)$, it follows that

$$\rho(x_{i-1}, x_i) \geq \rho^\infty(x,y), 1 \leq i \leq j.$$

so, $s(P') \leq s(P)$ implies that

$$1/s(P') \geq 1/s(P) = 1/\rho^\infty(x,y).$$

Now $l(P) = \delta(x,y)$ if and only if the maximum value of $l(P)$ is less than or equal to $l(P')$. That is,

$$\frac{j}{\rho^\infty(x,y)} \leq \frac{1}{s(P')} + (j' - 1).$$

Suppose $j' = fj$. Substituting in the above, we get

$$\frac{j}{\rho^\infty(x,y)} - fj \leq \frac{1}{s(P')} - 1.$$

Now, let $f = 1$. This implies that the two paths P and P' have the same length. Simplifying the above inequality, we get that $s(P')$ is less than or equal to the expression $\left\{j\left(\frac{1}{\rho^\infty(x,y)} - 1\right) + 1\right\}^{-1}$ and hence the desired result holds. ■

Example 3.1 *Consider the fuzzy graph defined by Figure 3.1. Here,*

FIGURE 3.1 ρ-length equals ρ-distance.

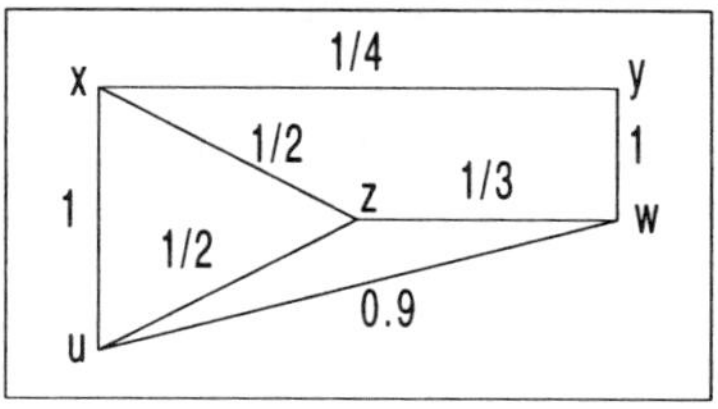

$\rho^\infty(x,y) = 0.9$, *diam*$(x,y) = 4$ *and* $j = 3$. *Let* P *be the path* x, u, w, y. *Here, it is easy to see that* $\left\{j\left(\frac{1}{\rho^\infty(x,y)} - 1\right) + 1\right\}^{-1} = 0.75$. *Therefore,* $s(P') \leq 0.75$. *In this example we see that* $s(P') = 1/3$ *which is indeed less that* 0.75, *and thus we have illustrated Proposition 3.2.*

Corollary 3.3 *If there exists a path of length $j(>1)$ whose strength is less than*

$$\left\{ j\left(\frac{1}{\rho^{\infty}(x,y)} - 1 \right) + 1 \right\}^{-1}.$$

Then

$$\delta(x,y)\rho^{\infty}(x,y) \leq diam(x,y).$$

3.1 Clusters

In graph theory, there are several ways of defining "clusters" of vertices. One approach is to call a subset C of V a *cluster* of order k if the following two conditions hold:

(a) for all vertices x, y in C, $d(x,y) \leq k$;

(b) for all vertices $z \notin C$, $d(z,w) > k$ for some $w \in C$;

where $d(u,v)$ is the length of a shortest path between two vertices u and v.

Thus in a k-cluster C, all pairs of vertices are within distance k of each other, and C is maximal with respect to this property. That is, no vertex outside of C is within distance k of every vertex in C.

A 1-cluster is called a *clique*; it is a maximal complete subgraph. That is, a maximal subgraph in which each pair of vertices is joined by an edge. At the other extreme, if we let $k \to \infty$, a k-cluster becomes a connected component, that is, a maximal subgraph in which each pair of vertices is joined by a path (of any length).

These ideas can be generalized to fuzzy graphs as follows: In $G = (\mu, \rho)$, we can call $C \subseteq V$ a *fuzzy cluster* of order k if

$$\wedge\{\rho^k(x,y) \mid x, y \in C\} > \vee\{\wedge\{\rho^k(w,z) \mid w \in C\} \mid z \notin C\}.$$

Note that C is an ordinary subset of V, not a fuzzy subset. If G is an ordinary graph, we have $\rho^k(a,b) = 0$ or 1 for all a and b. Hence this definition reduces to

(1) $\rho^k(x,y) = 1$ for all x, y in C,

(2) $\rho^k(w,z) = 0$ for all $z \notin C$ and some $w \in C$.

Property (1) implies that for all x, y in C, there exists a path of length $\leq k$ between x and y and property (2) implies that for all $z \notin C$ and some $w \in C$, there does not exist a path of length $\leq k$. This is the same as the definition of a cluster of order k.

In fact, the k-clusters obtained using this definition are just ordinary cliques in graphs obtained by thresholding the kth power of the given fuzzy graph. Indeed, let C be a fuzzy k-cluster, and let $\wedge\{\rho^k(x,y) \mid x,y \in C\} = t$. If we threshold ρ^k (and μ) at t, we obtain an ordinary graph in which C is now an ordinary clique.

Example 3.2 *Let*

$$V = \{x, y, z, u, v\}$$

and

$$X = \{(x,y), (x,z), (y,z), (z,u,), (u,v)\}.$$

Let $\mu(x) = \mu(y) = \mu(z) = \mu(u) = \mu(v) = 1$ *and* $\rho(x,y) = \rho(x,z) = \rho(y,z) = 1/2$, $\rho(z,u,) = \rho(u,v) = 1/4$. *Let* $C = \{x,y,z\}$. *Then* $\bigwedge_{c,d\in C} \rho^k(c,d) = 1/2$ *for* $k = 1, 2, \ldots$, $\bigvee_{e\notin C} (\bigwedge_{c\in C} \rho^1(c,e)) = (1/4 \wedge 0 \wedge 0) \vee (0 \wedge 0 \wedge 0) = 0$, $\bigvee_{e\notin C} (\bigwedge_{c\in C} \rho^2(c,e)) = (1/4 \wedge 1/4 \wedge 1/4) \vee (1/4 \wedge 0 \wedge 0) = 1/4$, *and* $\bigvee_{e\notin C} (\bigwedge_{c\in C} \rho^k(c,e)) = (1/4 \wedge 1/4 \wedge 1/4) \vee (1/4 \wedge 1/4 \wedge 1/4) = 1/4$ *for* $k \geq 3$. *Hence* $C = \{x,y,z\}$ *is a fuzzy cluster of order* k *for all* $k \geq 1$.

Now let $\rho(x,y) = \rho(x,z) = \rho(y,z) = 1/8, \rho(z,u) = \rho(u,v) = 1/4$. *Then* $\bigwedge_{c,d\in C} \rho^k(c,d) = 1/8$ *for* $k = 1, 2, \ldots$. $\bigvee_{e\notin C} (\bigwedge_{c\in C} \rho^1(c,e)) = (1/4 \wedge 0 \wedge 0) \vee (0 \wedge 0 \wedge 0) = 0$, $\bigvee_{e\notin C} (\bigwedge_{c\in C} \rho^2(c,e)) = (1/4 \wedge 1/8 \wedge 1/8) \vee (1/4 \wedge 0 \wedge 0) = 1/8$, *and* $\bigvee_{e\notin C} (\bigwedge_{c\in C} \rho^k(c,e)) = (1/4 \wedge 1/8 \wedge 1/8) \vee (1/4 \wedge 1/8 \wedge 1/8) = 1/8$ *for* $k \geq 3$. *Hence* C *is a fuzzy cluster of order* 1, *but not of order* k *for* $k \geq 2$.

3.2 Cluster Analysis

In this section, we analyze fuzzy graphs from the viewpoint of connectedness. We apply results to cluster analysis. We do not assume our (fuzzy) graphs are necessarily undirected in this section.

Let $G = (\mu, \rho)$ be a fuzzy graph. We denote by M_ρ the corresponding fuzzy matrix of a fuzzy graph G. In other words, $(M_\rho)_{ij} = \rho(v_i, v_j)$.

Theorem 3.4 *Let* $G = (V, \mu, \rho)$ *be a fuzzy graph such that cardinality of* V *is* n. *Then*

(1) if ρ *is reflexive, there exists* $k \leq n$ *such that* $M_\rho < M_\rho^2 < \ldots < M_{\rho^3}^k = M_{\rho^3}^{k+1}$;

(2) if ρ *is irreflexive, the sequence* M_ρ, M_ρ^2, $\ldots$ *is eventually periodic.* ■

Definition 3.1 *Let G be a fuzzy graph. Let $0 \leq \epsilon \leq 1$. A vertex v is said to be* ϵ-reachable *from another vertex u if there exists a positive integer k such that $\rho^k(u,v) \geq \epsilon$. The reachability matrix of G, denoted by M_{ρ^∞}, is the matrix of the fuzzy graph (μ, ρ^∞). The ϵ-reachability matrix of G, denoted by $M^\epsilon_{\rho^\infty}$, is defined as follows: $M^\epsilon_{\rho^\infty}(u,v) = 1$ if $\rho(u,v) \geq \epsilon$ and $M^\epsilon_{\rho^\infty}(u,v) = 0$, otherwise.*

The following algorithm can determine the reachability between any pair of vertices in a fuzzy graph G.

Algorithm 2.1. Determination of M_{ρ^∞}

1. Let $R_i = (a_{i1}, ..., a_{in})$ denote the ith row.
2. Obtain the new R_i by the following procedure:

$$a_{ij}(new) = \vee_j \{\vee_k \{a_{kj} \wedge a_{ik}(old)\}\}, a_{ij}(old)\}.$$

3. Repeat Step 2 until no further changes occur.
4. $M_{\rho^\infty}(i,j) = a_{ij}(new)$.

Note that a similar algorithm can be constructed for the determination of $M^\epsilon_{\rho^\infty}$, $0 \leq \epsilon \leq 1$.

Definition 3.2 *Let G be a fuzzy graph. The* connectivity *of a pair of vertices u and v, denoted by $C(u,v)$ is defined to be $\rho^\infty(u,v) \wedge \rho^\infty(v,u)$. The* connectivity matrix *of G, denoted by C_G, is defined such that $C_G(u,v) = C(u,v)$. For $0 \leq \epsilon \leq 1$, the ϵ-connectivity matrix of G, denoted by C^ϵ_G, is defined as follows: $C^\epsilon_G(u,v) = 1$ if $C(u,v) \geq \epsilon$ and $C^\epsilon_G(u,v) = 0$ otherwise.*

Algorithm 2.2. Determination of C_G.

1. Construct M_{ρ^∞}.
2. $C_G(i,j) = C_G(j,i) = M_{\rho^\infty}(i,j) \wedge M_{\rho^\infty}(j,i)$.

An algorithm for determining C^ϵ_G is similar to Algorithm 2.2.

Definition 3.3 *Let G be a fuzzy graph. G is called* strongly ϵ-connected *if every pair of vertices are mutually ϵ-reachable. G is said to be* initial ϵ-connected *if there exists $v \in V$ such that every vertex u in G is ϵ-reachable from v. A* maximal strongly ϵ-connected fuzzy subgraph $(MS\epsilon CS)$ *of G is a strongly ϵ-connected fuzzy subgraph not properly contained in any other $MS\epsilon CS$.*

Clearly strongly ϵ-connectedness implies initial ϵ-connectedness. Also, the following result is straightforward.

Theorem 3.5 *A fuzzy graph G is strongly ϵ-connected if and only if there exists a vertex u such that for any other vertex v in $G, \rho^\infty(u,v) \geq \epsilon$ and $\rho^\infty(v,u) \geq \epsilon$.* ■

TABLE 3.1 Fuzzy matrix and connectivity matrix of a fuzzy graph.

$$M_\rho = \begin{bmatrix} 1.0 & 0.6 & 0.4 & 0.0 & 0.0 \\ 0.0 & 1.0 & 0.2 & 0.6 & 0.3 \\ 0.0 & 0.8 & 1.0 & 0.0 & 0.9 \\ 0.2 & 0.7 & 0.3 & 1.0 & 0.2 \\ 0.4 & 0.0 & 0.5 & 0.3 & 1.0 \end{bmatrix} \qquad C_G^{0.5} = \begin{bmatrix} 1 & 0 & 0 & 0 & 0 \\ 0 & 1 & 0 & 1 & 0 \\ 0 & 0 & 1 & 0 & 1 \\ 0 & 1 & 0 & 1 & 0 \\ 0 & 0 & 1 & 0 & 1 \end{bmatrix}$$

Algorithm 2.3. Determination of all $MS\epsilon CS$ in G.

1. Construct C_G^ϵ.

2. The number of $MS\epsilon CS$ in G is given by the number of distinct row vectors in C_G^ϵ. For each row vector α in C_G^ϵ, the vertices contained in the corresponding $MS\epsilon CS$ are the nonzero elements of the corresponding columns of α.

Example 3.3 *Let G be a fuzzy graph whose corresponding fuzzy matrices M_ρ and $C_G^{0.5}$ are given in Table 3.1. We see that the $MS0.5CS$'s of G contain the following vertex sets,* $\{1\}, \{2,4\}, \{3,5\}$, *respectively.*

We now apply the results obtained to clustering analysis. We assume that a data fuzzy graph $G = (V, \rho)$ is given, where V is a set of data and $\rho(u, v)$ is a quantitative measure of the similarity of the two data items u and v. For $0 < \epsilon \leq 1$, an ϵ-cluster in V is a maximal subset W of V such that each pair of elements in W is mutually ϵ-reachable. Therefore, the construction of ϵ-clusters of V is equivalent to finding all maximal strongly ϵ-connected fuzzy subgraphs of G.

Algorithm 2.4. Construction of ϵ-clusters

1. Compute $\rho, \rho^2, ..., \rho^k$, where k is the smallest integer such that $\rho^k = \rho^{k+1}$;

2. Let $\varsigma = \bigcup_{i=1}^{k} \rho^i$.

3. Construct F_ϵ^ς.

Then, each element in F_ϵ^ς is an ϵ-cluster.

We may also define an ϵ-cluster in V as a maximal subset W of V such that every element of W is ϵ-reachable from a special element v in W. In this case, the construction of ϵ-clusters is equivalent to finding all maximal initial ϵ-connected fuzzy subgraphs of G. However, the relation induced by initial ϵ-connected fuzzy subgraphs is not, in general, a similarity relation.

Another application is the use of fuzzy graphs to model information networks. Such a model was proposed in [27] utilizing the concepts of a directed graph. In [27] a measure of flexibility of a network was introduced.

Let $G = (V, \rho)$ be a fuzzy graph. Define the *degree* of a vertex v to be $d(v) = \sum_{u \neq v} \rho(v, u)$. The *minimum degree* of G is $\delta(G) = \wedge\{d(v) \mid v \in V\}$, and the *maximum degree* of G is $\Delta(G) = \vee\{d(v) \mid v \in V\}$.

Definition 3.4 *Let $G_i = (V_i, \rho_i), i = 1, 2$ be two fuzzy subgraphs of $G = (V, \rho)$. The union of G_1 and G_2, denoted by $G_1 \cup G_2$, is the fuzzy graph (V', ρ'), where $V' = V_1 \cup V_2$ and*

$$\rho'(u, v) = \begin{cases} \rho(u, v) & \text{if } \{u, v\} \subseteq V_1 \cup V_2 \\ 0 & \text{if } \{u, v\} \not\subseteq V_1 \cup V_2 \end{cases}$$

Lemma 3.6 *Let $G = (V, \rho)$ be a fuzzy graph and $G_i = (V_i, \rho_i), i = 1, ..., n$, be fuzzy subgraphs of G such that $V_i \cap V_j = \emptyset$ for $i \neq j, 1 \leq i, j \leq n$, and $\bigcup_{i=1}^{n} G_i$ is connected. Then*

(1) $\delta(\bigcup_{i=1}^{n} G_i) \geq \wedge\{\delta(G_i) \mid i = 1, \ldots, n\}$,

(2) $\Delta(\bigcup_{i=1}^{n} G_i) \geq \vee\{\delta(G_i) \mid i = 1, \ldots, n\}$. ■

Recall that G is said to be connected if for each pair of vertices u and v in V, there exists a $k > 0$ such that $\rho^k(u, v) > 0$.

Definition 3.5 *Let $G = (V, \rho)$ be a fuzzy graph. G is called* τ-degree connected, *for some $\tau \geq 0$, if $\delta(G) \geq \tau$ and G is connected. A* τ-degree component *of G is a maximal τ-degree connected fuzzy subgraph of G.*

Theorem 3.7 *For any $\tau > 0$, the τ-degree components of a fuzzy graph are disjoint.*

Proof. Let G_1 and G_2 be two τ-degree components of G such that their vertex sets have at least one common element. Since $\delta(G_1 \cup G_2) \geq \delta(G_1) \wedge \delta(G_2)$ by Lemma 3.6, $G_1 \cup G_2$ is τ-degree connected. Since G_1 and G_2 are maximal with respect to τ-degree connectedness, we have that $G_1 = G_2$. ■

Algorithm 2.5. Determination of τ-degree components of a finite fuzzy graph G.

1. Calculate the row sums of M_ρ.

2. If there are rows whose sums are less than τ, then obtain a new reduced matrix by eliminating those vertices, and go to 1.

3. If there is no such row, then stop.

4. Each disjoint fuzzy subgraph of the graph induced by the vertices in the last matrix as well as each eliminated vertex is a maximal τ-degree connected fuzzy subgraph.

Definition 3.6 *Let G be a fuzzy graph, and $\{V_1, V_2\}$ be a partition of its vertex set V. The set of edges joining vertices of V_1 and vertices of V_2 is called a* cut-set *of G, denoted by (V_1, V_2), relative to the partition $\{V_1, V_2\}$. The* weight *of the cut-set (V_1, V_2) is defined to be*

$$\sum_{u \in V_1, v \in V_2} \rho(u, v).$$

Definition 3.7 *Let G be a fuzzy graph. The* edge connectivity *of G, denoted by $\lambda(G)$, is defined to be the minimum weight of cut-sets of G. G is called* τ-edge connected *if G is connected and $\lambda(G) \geq \tau$. A* τ-edge component *of G is a maximal τ-edge connected subgraph of G.*

Example 3.4 *Consider the fuzzy graph G given below.*

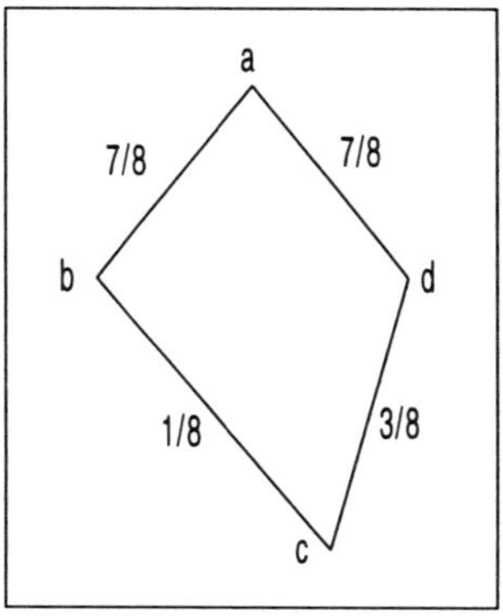

We summarize different cut-sets along with their weights in Table 3.2. We see that $\lambda(G) = 1/2$.

The following results can be proved similar to that of Lemma 3.6 and Theorem 3.7.

Lemma 3.8 *Let G be a fuzzy graph and $G_i, i = 1, ..., n$, be fuzzy subgraphs of G such that $V_i \cap V_j = \emptyset$ for all $i, j, i \neq j, 1 \leq i, j \leq n$ and $\bigcup_{i=1}^{n} G_i$ is connected. Then $\lambda(\bigcup_{i=1}^{n} G_i) \geq \bigwedge_{i=1}^{n} (\lambda(G_i))$.* ■

TABLE 3.2 Cut sets and their weights.

V_1	V_2	*weight*
$\{a\}$	$\{b,c,d\}$	$\frac{7}{8}+\frac{7}{8}=\frac{7}{4}$
$\{b\}$	$\{a,c,d\}$	$\frac{7}{8}+\frac{1}{8}=1$
$\{c\}$	$\{a,b,d\}$	$\frac{1}{8}+\frac{3}{8}=\frac{1}{2}$
$\{d\}$	$\{a,b,c\}$	$\frac{7}{8}+\frac{3}{8}=1\frac{1}{4}$
$\{a,b\}$	$\{c,d\}$	$\frac{7}{8}+\frac{1}{8}=1$
$\{a,c\}$	$\{b,d\}$	$\frac{7}{8}+\frac{7}{8}+\frac{1}{8}+\frac{3}{8}=2\frac{1}{4}$
$\{a,d\}$	$\{b,c\}$	$\frac{7}{8}+\frac{3}{8}=1\frac{1}{4}$

Theorem 3.9 *For $\tau > 0$, the τ-edge components of a fuzzy graph are disjoint.*

The algorithm for determining τ-edge components is based on a result of Matula [22]. In order to understand the algorithm we need to introduce the concept of a cohesive matrix and that of narrow slicing. ■

Cohesiveness

Let $G = (V, \rho)$ be a fuzzy graph. An element of G is defined to be either a vertex or edge. That is, e either a member of V or e is a pair of members of V such that $\rho(e) > 0$.

Definition 3.8 *Let e be an element of a fuzzy graph G. The* cohesiveness *of e, denoted by $h(e)$, is the maximum value of edge-connectivity of the subgraphs of G containing e.*

Lemma 3.10 *For any fuzzy graph G and element e and $0 < \tau \leq h(e)$, there exists a unique τ-edge component of G containing e.* ■

The unique τ-edge component of G, for $\tau = h(e) > 0$, containing the element e has the highest order of the maximum edge-connectivity subgraphs of G containing e, and will be termed the *$h(e)$-edge component* of e, denoted by H_e.

Example 3.5 *Consider the fuzzy graph G given Example 2.3. We summarize the τ-edge components of G in the form a table. Recall that if V_1 is a subset of the set of vertices of G, $\langle V_1\rangle$ denotes the fuzzy subgraph induced by V_1.*

τ	τ-edge components
$(7/8, 1]$	$\langle\{a\}\rangle, \langle\{b\}\rangle, \langle\{c\}\rangle, \langle\{d\}\rangle$
$(1/2, 7/8]$	$\langle\{c\}\rangle, \langle\{a, b, d\}\rangle$
$[0, 1/2]$	$\langle\{a, b, c, d\}\rangle$

The cohesiveness of an element may be determined from the knowledge of any subgraph of maximum edge-connectivity containing that given element, and clearly knowledge of the τ-edge components of G for all $\tau > 0$ is sufficient to determine $h(e)$ for all elements e of G. The following theorem shows an important converse relation, that by utilizing the cohesiveness function it is possible to readily determine H_e for any element e with $h(e) > 0$.

Theorem 3.11 *Let e be an element of the fuzzy graph G with $h(e) > 0$. Let M_e be a maximal connected fuzzy subgraph of G containing e such that all elements of M_e have cohesiveness at least $h(e)$. Then $M_e = H_e$.* ■

Corollary 3.12 *For any fuzzy graph G and any $\tau > 0$, the elements of G of cohesiveness at least τ form a fuzzy graph whose components are τ-edge components of G.* ■

Corollary 3.13 *If G' is an τ-edge component of the fuzzy graph G for some $\tau > 0$, then $G' = H_e$ for some element e of G.* ■

Slicing in Fuzzy Graphs

An ordered partition of the edges of the fuzzy graph $G, (C_1, C_2, \ldots, C_m)$, is a *slicing* of G if each member

$$C_i \text{ is a cut-set } (A_i, \bar{A}_i) \text{ of } \begin{cases} G & \text{for } i = 1 \\ G \setminus \bigcup_{j=1}^{i-1} C_j & \text{for } 2 \leq i \leq m \end{cases}$$

A member of the slicing will also be termed a *cut of the slicing.* A slicing of G, which is minimal in the sense that there is no subpartition which is a slicing of G, is called a *minimal slicing* of G. Clearly each cut C_i of a minimal slicing must be a minimal cut of some component of $G \setminus \bigcup_{j=1}^{i-1} C_j$. Further, a slicing of G is a *narrow slicing* of G, if each cut C_i is a minimum cut of some component of $G \setminus \bigcup_{j=1}^{i-1} C_j$. Note that the notion of slicing pertains only to graphs with at least one edge.

A slicing may be given a dynamic interpretation as a sequence of nonvoid cuts which separates G into isolated vertices and a minimal (narrow) slicing effects this separation using only minimal (minimum) cuts at each step. This provides a way to compute the minimal (narrow) slicing. However, we want to make the observation that a narrow slicing is a minimal slicing but not vice versa.

Algorithm 2.6. Narrow slicing of connected fuzzy graph G.
1. $Z = \emptyset, G_1 = G, i = 1$.
2. While $G_i \neq \emptyset$ do
   $V =$ the vertex set of G_i.
   $v =$ a vertex in G_i with minimum degree.
   $C_i = (\{v\}, V \setminus \{v\})$
   $Z = Z \cup \{C_i\}$
   $i = i + 1$
   $G_i =$ the fuzzy subgraph induced by $V \setminus \{v\}$.
3. Z is a narrow slicing of G.

The following result is an important link between narrow slicing Z and the cohesive function h on a fuzzy graph.

Theorem 3.14 *Let $Z = (C_1, C_2, \ldots, C_m)$ be a narrow slicing of G obtained by successively removing one vertex at a time. Let $G_1 = G \supseteq G_2 \supseteq \ldots \supseteq G_m$ be the sequence of fuzzy subgraphs left after each slicing. Then $h(e) = \wedge\{\lambda(G_i) | e \in G_i, 1 \leq i \leq m\}$.* ■

Example 3.6 *Let G be a fuzzy graph such that*

$M_\rho =$

	a	b	c	d	e
a	0.0	0.8	0.2	0.0	0.0
b	0.8	0.0	0.4	0.0	0.4
c	0.2	0.4	0.0	0.8	03
d	0.0	0.0	0.8	0.0	0.8
e	0.0	0.4	0.3	0.8	0.0

As in the Algorithm 2.6, let G_1 denote the fuzzy graph G. Computing the sum along each row, we have

a	b	c	d	e
1.0	1.6	1.7	1.6	1.5

The minimum value occurs at row a. So we set $C_1 = (\{a\}, \{b, c, d, e\})$ and let G_2 be the fuzzy subgraph induced by the vertex set $\{b, c, d, e\}$. Note that $\lambda(G_1) = 1.0$ and edges (a, b) and (a, c) appear only in G_1. It follows that $h(e) = 1.0$ for $e = (a, b), (a, c)$. Now the matrix associated with the fuzzy subgraph G_2 is given by

	b	c	d	e
b	0.0	0.4	0.0	0.4
c	0.4	0.0	0.8	03
d	0.0	0.8	0.0	0.8
e	0.4	0.3	0.8	0.0

Computing the sum along each row, we have

b	c	d	e
0.8	1.5	1.6	1.5

The minimum value occurs at row b. Hence $C_2 = (\{b\}, \{c, d, e\})$ and G_3 is the fuzzy subgraph generated by the vertex set $\{c, d, e\}$. Note that $\lambda(G_2) = 0.8$ and edges (b, c) and (b, e) appear in G_1 and G_2 and hence $h(e) = 1.0 \wedge 0.8 = 0.8$ for $e = (b, c), (b, e)$. Now the matrix associated with the fuzzy subgraph G_3 is given by

	c	d	e
c	0.0	0.8	03
d	0.8	0.0	0.8
e	0.3	0.8	0.0

Proceeding along these lines, we obtain the following cohesive matrix (where ith row jth column entry denote the cohesiveness of the edge (i, j) if $i \neq j$ and the cohesiveness of the vertex i if $i = j$, for $i, j \in \{a, b, c, d, e\}$)

	a	b	c	d	e
a	0.0	1.0	1.0	1.0	1.0
b	1.0	0.0	1.0	1.0	1.0
c	1.0	1.0	0.0	1.1	1.1
d	1.0	1.0	1.1	0.0	1.1
e	1.0	1.0	1.1	1.1	0.0

and the narrow slicing
$((\{a\}, \{b, c, d, e\}), (\{b\}, \{c, d, e\}), (\{e\}, \{c, d\}), (\{c\}, \{d\}))$.

We are now ready to present an algorithm for the determination of τ-edge components of a fuzzy graph G.

Algorithm 2.7. Determination of τ-edge components of a fuzzy graph G.

1. Obtain the cohesive matrix H of the M_ρ.
2. Obtain the τ-threshold graph of H.
3. Each component of the graph is a maximal τ-edge connected subgraph.

Example 3.7 *Consider the fuzzy graph G in the Example 3.6. The τ-edge components of G for various values τ can be summarized as follows.*

τ	τ-*edge components*
$(1.1, \infty)$	$\langle\{a\}\rangle, \langle\{b\}\rangle, \langle\{c\}\rangle, \langle\{d\}\rangle, \langle\{e\}\rangle$
$(1.0, 1.1]$	$\langle\{a\}\rangle, \langle\{b\}\rangle, \langle\{c, d, e\}\rangle$
$[0, 1.0]$	$\langle\{a, b, c, d, e\}\rangle$

Definition 3.9 *A* disconnection *of a fuzzy graph $G = (V, \mu, \rho)$ is a vertex set D whose removal results in a disconnected or a single vertex graph. The* weight *of D is defined to be $\sum_{v \in D} \wedge\{\rho(v, u) | \rho(v, u) \neq 0, u \in V\}$.*

Definition 3.10 *The vertex connectivity of a fuzzy graph G, denoted by $\Omega(G)$, is defined to be the* minimum weight of disconnection *in G. G is said to be τ-vertex connected if $\Omega(G) \geq \tau$. A* τ-vertex component *is a maximal τ-vertex connected subgraph of G.*

Note that τ-vertex components need not be disjoint as do τ-degree and τ-edge components. The following result is straightforward.

Theorem 3.15 *Let G be a fuzzy graph, then $\Omega(G) \leq \lambda(G) \leq \delta(G)$.* ■

Theorem 3.16 *For any three real numbers a, b, and c such that $0 < a \leq b \leq c$, there exists a fuzzy graph G with $\Omega(G) = a, \lambda(G) = b$, and $\delta(G) = c$.*

Proof. Let n be the smallest integer such that $c/n \leq 1$, and let $a' = a/n, b' = b/n$, and $c' = c/n$. Then $0 < a' \leq b' \leq c' \leq 1$. Let G be the fuzzy graph constructed as follows. The vertex set is the union of three sets $A = \{u_0, u_1, u_2, \ldots, u_n\}, B = \{v_0, v_1, v_2, \ldots, v_n\}$, and $C = \{w_0, w_1, w_2, \ldots, w_n\}$ each containing $n + 1$ vertices. Let $< X >$ denote the fuzzy subgraph induced by the set X, for $X = A, B, C$. In $C, d(w_0) = nc'$ and $d(w_i) = (n-1) + c' + b'$ for $1 \leq i \leq n$. In other words, $< C \setminus \{w_0\} >$ is 1.0-complete and $< C >$ is c'-complete. In $B, d(v_0) = n+1$ and $d(v_i) = n+(n-1)+a'+b'$ for $1 \leq i \leq n$. $< B >$ is 1.0-complete. In $A, d(u_0) = n + 1$ and $d(u_i) = n + (n-1) + a'$ for $1 \leq i \leq n$. $< A >$ is 1.0-complete. Connections between subsets are as follows. Each w_i is connected to v_i with fuzzy value b' for $1 \leq i \leq n$. And each $u_i (i \neq 0)$ is connected to v_i with fuzzy value a' and to v_j's $(j \neq i, 0)$ with fuzzy value 1.0. Finally u_0 is connected to v_0 with fuzzy value 1. All other edges in the fuzzy graph have value 0. Now we will show that G thus constructed satisfies the conditions imposed.

(1) From the process of the construction described above it is clear that $\delta(G) = d(w_0) = nc' = c$.

(2) The number of edges in any cut of the subgraphs $< A >, < B >$ or $< C >$ is greater than or equal to n since $< A >, < B >$ and $< C >$ are c'-

complete. Therefore the weight of a cut is greater than or equal to nc', which means that the weight of any cut which contains a cut of $< A >, < B >$ or $< C >$ is greater than or equal to nc'. Only other cuts which do not contain a cut of $< A >, < B >$ or $< C >$ must contain the cut $(A, B \cup C)$ or $(A \cup B, C)$. The weight of the cut $(A, B \cup C)$ is $1 + n(n-1) + na'$ and that of the cut $(A \cup B, C)$ is nb'. Now $nb' \leq nc'$ and $nb' \leq 1 + n(n-1) + na'$. Hence $\lambda(G) = nb' = b$.

(3) Let us determine the minimum number of vertices in disconnection of G. Since $< A >, < B >$ and $< C >$ are at least c'-complete, they can be disconnected or become a single vertex by removing at least n vertices. Only other possible ways to disconnect G are disconnections between A, B, and C. Since $< (A \setminus \{u_0\}) \cup (B \setminus \{v_0\}) >$ *is* a a'-complete and u_0 and v_0 are connected to each other and to $< (A \setminus \{u_0\}) \cup (B \setminus \{v_0\}) >$, any disconnection must contain at least $n+1$ vertices. On the other hand, since $< B >$ and $< C >$ are connected by n edges, at least n vertices have to be removed to disconnect $< A \cup B >$ and $< C >$. But since vertices on both sides of edges are all different, at least n vertices have to be removed. Therefore, at least n vertices have to be removed to disconnect the graph G. Then since $\wedge\{f(v) | v \in V\} = a'$ and actually $\{v_1, v_2, ..., v_n\}$ is a disconnection of G, the weight of the disconnection $\{v_1, v_2, ..., v_n\}$ specifies the vertex connectivity of the graph G, namely, $\Omega(G) = na' = a$. ■

3.3 Application to Cluster Analysis

The usual graph-theoretical approaches to cluster analysis involve first obtaining a threshold graph from a fuzzy graph and then applying various techniques to obtain clusters as maximal components under different connectivity considerations. These methods have a common weakness, namely, the weight of edges are not treated fairly in that any weight greater (less) than the threshold is treated as 1(0). In this section, we will extend these techniques to fuzzy graphs. It will be shown that the fuzzy graph approach is more powerful.

In Table 2.4, we provide a summary of various graph theoretical techniques for clustering analysis. This table is a modification of table II in Matula [21]. For cluster procedures (1),(2), and (3) the cluster independence can be considered to be disjoint while that of cluster procedure (4) is limited overlap and that of (5) is considerable overlap. The extent of chaining is high, moderate, low, low, and none for cluster procedures (1) – (5), respectively.

In the following definition, clusters will be defined based on various connectivities of a fuzzy graph.

TABLE 3.3 Cluster procedures.

	Cluster procedure	*Graph theoretical interpretation of clusters*
(1)	Single linkage	Maximal connected subgraphs
(2)	k-linkage	Maximal connected subgraphs of minimum degree
(3)	k-edge connectivity	Maximal k-edge connected subgraph
(4)	k-vertex connectivity	Maximal k-vertex connected subgraph and Cliques on k or less vertices
(5)	Complete linkage	Cliques

Definition 3.11 *Let $G = (V, \rho)$ be a fuzzy graph. A* cluster of type k *($k = 1, 2, 3, 4$) is defined by the following conditions (1), (2), (3), and (4) respectively.*

(1) maximal ϵ-connected subgraphs, for some $0 < \epsilon \leq 1$.

(2) maximal τ-degree connected subgraphs.

(3) maximal τ-edge connected subgraphs.

(4) maximal τ-vertex connected subgraphs.

Hierarchial cluster analysis is a method of generating a set of classifications of a finite set of objects based on some measure of similarity between a pair of objects. It follows from the previous definition that clusters of type (1), (2), and (3) are hierarchial with different ϵ and τ, whereas clusters of type (4) are not due to the fact τ-vertex components need not be disjoint.

It is also easily seen that all clusters of type (1) can be obtained by the single-linkage procedure. The difference between the two procedures lies in the fact that ϵ-connected subgraphs can be obtained directly from M_{ρ^∞} by at most $n-1$ matrix multiplications (where n is the rank of M_G), whereas in the single-linkage procedure, it is necessary to obtain as many threshold graphs as the number of distinct fuzzy values in the graph.

Output of hierarchial clustering is called a *dendogram* which is a directed tree that describes the process of generating clusters.

In the following, we will show that not all clusters of types 2, 3 and 4 are obtainable by procedures of k-linkage, k-edge connectivity, and k-vertex connectivity, respectively.

Example 3.8 *Let G be a fuzzy graph given in Figure 3.2(a). The dendrogram in Figure 3.2(b) indicates all the clusters of type 2.*

FIGURE 3.2 A fuzzy graph and its clusters of type 2.

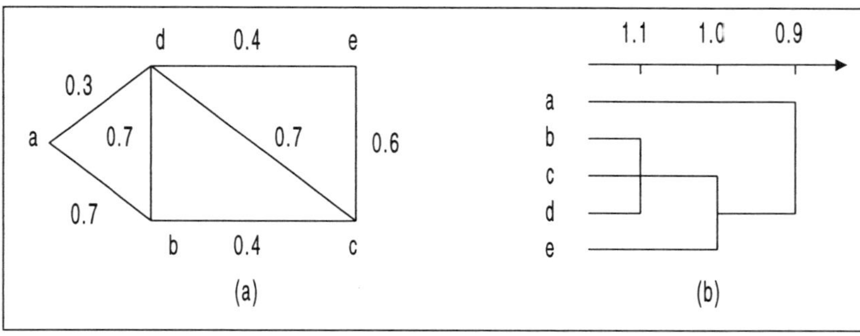

FIGURE 3.3 Dendrograms for clusters obtained by k-linkage method for $k = 1$ and 2.

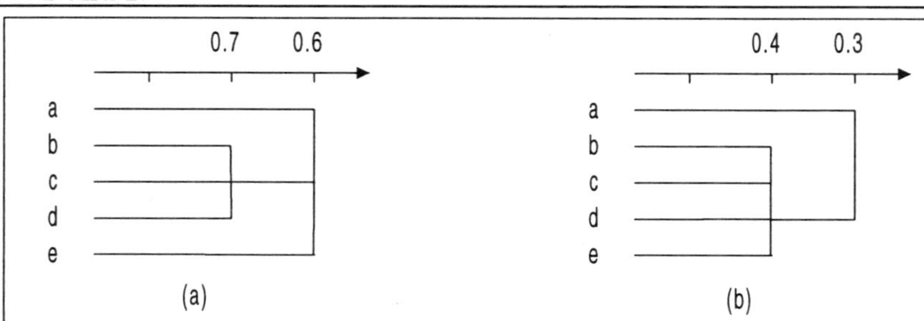

It is easily seen from the threshold graphs of G that the same dendrogram cannot be obtained by the k-linkage procedure. Those for $k = 1$ and 2 are given in Figures 3.3(a) and 3.3(b), respectively.

Theorem 3.17 *The τ-degree connectivity procedure for the construction of clusters is more powerful than the k-linkage procedure.*

Proof. In light of Example 3.8, it is sufficient to show that all clusters obtainable by the k-linkage procedure are also obtainable by the τ-degree connectivity procedure for some τ. Let G be a fuzzy graph. For $0 < \epsilon \leq 1$, let G' be a graph obtained from G by replacing those weights less than ϵ in G by 0. For any k used in the k-linkage procedure, set $\tau = k\epsilon$. It is easily seen that a set is a cluster obtained by applying the k-linkage procedure to G if and only if it is a cluster obtained by applying the τ-degree connectivity procedure to G'. ■

FIGURE 3.4 A fuzzy graph and its clusters of type 3.

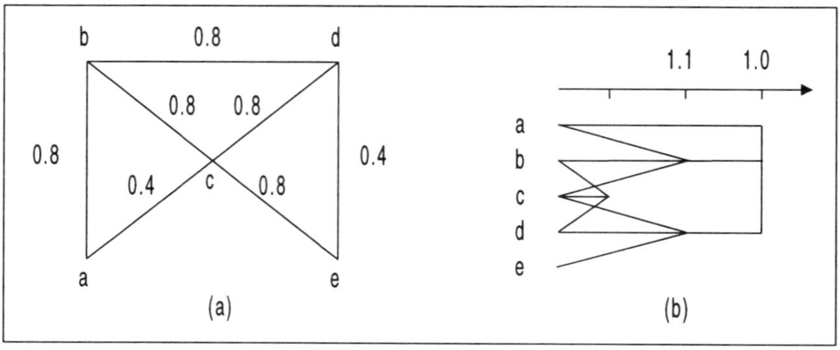

FIGURE 3.5 Dendrograms for clusters obtained from k-edge method for $k = 1$ and 2.

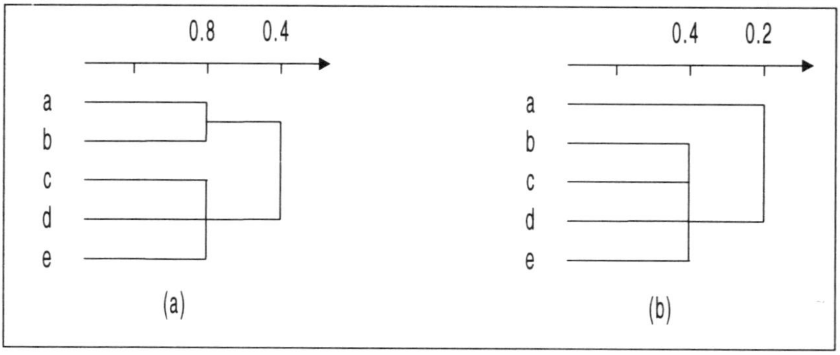

Example 3.9 *Let G be a fuzzy graph given in Figure 3.4(a). The dendrogram in Figure 3.4(b) gives all clusters of type 3.It is clear by examining all the threshold graphs of G that the same dendrogram cannot be obtained by means of the k-edge connectivity technique for any k. Those for $k = 1$ and 2 are given in Figure 3.5.*

By Example 3.9 and following same proof procedure as in Theorem 3.17, we have the following result.

Theorem 3.18 *The τ-edge connectivity procedure for the construction of clusters is more powerful than the k-edge connectivity procedure.* ■

Example 3.10 *Let G be a fuzzy graph given in Figure 3.6(a). The dendrogram in Figure 3.6(b) provides all clusters of type 4.*

FIGURE 3.6 A symmetric graph and its clusters of type 4.

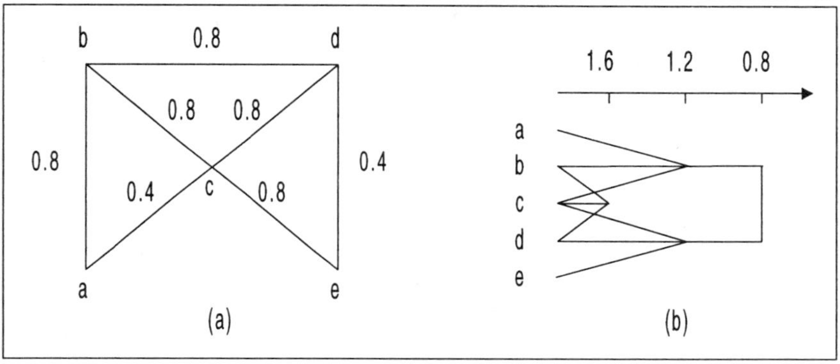

FIGURE 3.7 Dendrograms for clusters obtained from k-vertex method for $k = 1$ and 2.

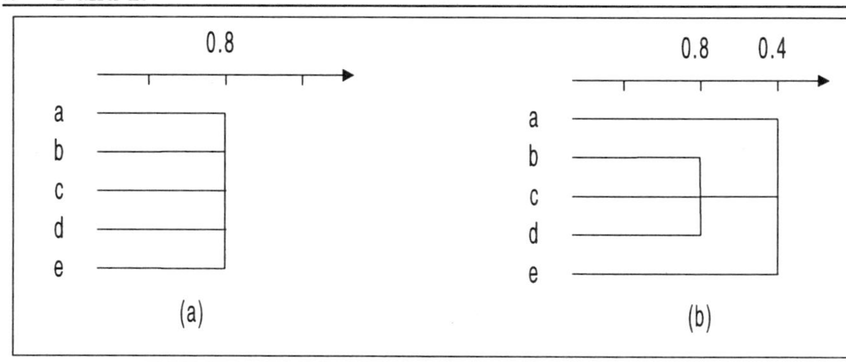

It is easily seen that the same dendrogram cannot be obtained by means of the k-vertex connectivity technique for any k. Those for $k = 1$ and 2 are given in Figure 3.7.

Following the same proof procedure as in Theorems 3.17 and 3.18, we conclude with the result below.

Theorem 3.19 *The $\tau-$vertex connectivity procedure for the construction of clusters is more powerful than the k-vertex connectivity procedure.*

3.4 Fuzzy Intersection Equations

We give necessary and sufficient conditions for the solution of a system of fuzzy intersection equations. We also give an algorithm for the solution of such a system. We apply the results to fuzzy graph theory.

In [20], Liu considered systems of intersection equations of the form

$$\begin{aligned} &e_{11}x_1 \wedge ... \wedge e_{1n}x_n = b_1 \\ &\vdots \\ &e_{m1}x_1 \wedge ... \wedge e_{mn}x_n = b_m \end{aligned} \tag{3.1}$$

where $e_{ij} \in \{0,1\}$ and $b_i, x_j \in L$ where L is a complete distributive lattice, $i = 1,...,m; j = 1,...,n$. In this section, we consider systems of equations of the form (3.1), where L is the closed interval $[0,1]$. Although this case is more restrictive, our approach is entirely different than that in [20]. The specificity of $[0,1]$ yields different types of results than those in [20]. We show that system (3.1) is equivalent to several independent systems of the type where $b_1 = \ldots = b_m$. Also our proofs concerning the existence of solutions are constructive in nature. In fact, we give an algorithm for the solution of a system of intersection equations. We also give two applications. One application is in the area of fuzzy graph theory.

Existence of Solutions

We write the system (3.1) in the matrix form $E\overline{x} = \overline{b}$, where $E = [e_{ij}]$,

$$\overline{x} = \begin{bmatrix} x_1 \\ x_2 \\ \vdots \\ x_n \end{bmatrix} \text{ and } \overline{b} = \begin{bmatrix} b_1 \\ b_2 \\ \vdots \\ b_m \end{bmatrix}.$$

We assume throughout that $\forall j = 1,...,n, \exists i$ such that $e_{ij} = 1$. We also assume that the equations of (3.1) have been ordered so that $b_{q_1+1} = ... = b_{q_2} < b_{q_2+1} = ... = b_{q_3} < ... < b_{q_t+1} = ... = b_{q_{t+1}}$ where $0 = q_1 < q_2 < ... < q_{t+1} = m$. Let $I_r = \{q_r+1, ..., q_{r+1}\}$ for $r = 1,...,t$. For each $j = 1,...,n$, let i_j^* denote the maximum i such that $e_{ij} = 1$. Let $e_{hj}^* = 0 \forall h \in \cup_{s=1}^{r-1} I_s$

and $e^*_{hj} = e_{hj} \forall h \in \cup_{s=r}^{t} I_s$, where $i^*_j \in I_r$. Let

$$E^*_j = \begin{bmatrix} e^*_{1j} \\ e^*_{2j} \\ \vdots \\ e^*_{mj} \end{bmatrix}.$$

Let $E^* = (E^*_1, ..., E^*_n)$.

Theorem 3.20 *$E\bar{x} = \bar{b}$ and $E^*\bar{x} = \bar{b}$ are equivalent systems.*

Proof. Let i be any row of E and j any column. Suppose that $e_{ij} = 1$. Let $i \in I_r$. Suppose $\exists h \in I_s, s < r$, such that $e_{hj} = 1$. Let E' be the matrix $[e'_{ij}]$ where $e'_{uv} = e_{uv}$ if $(u, v) \neq (h, j)$ and $e'_{uv} = 0$ if $(u, v) = (h, j)$. That is, E' is obtained from E by replacing the hj-th component of E with 0. It suffices to show that $E'\bar{x} = \bar{b}$ and $E\bar{x} = \bar{b}$ are equivalent. Now the h-th equations of $E'\bar{x} = \bar{b}$ and $E\bar{x} = \bar{b}$ are

$$e_{h1}x_1 \wedge ... \wedge 0x_j \wedge ... \wedge e_{hn}x_n = b_{q_s+1} \tag{3.2}$$

and

$$e_{h1}x_1 \wedge ... \wedge 1x_j \wedge ... \wedge e_{hn}x_n = b_{q_s+1}, \tag{3.3}$$

respectively. The other equations of $E'\bar{x} = \bar{b}$ and $E\bar{x} = \bar{b}$ are identical to each other. Since $e_{ij} = 1$, we have that $x_j \geq b_{q_r+1} > b_{q_s+1}$. Thus equation (3.2) and the i-th equation are equivalent to equation (3.3) and the i-th equation. Hence the desired result follows. ■

Example 3.11 *Consider the following systems of intersection equations:*

$$x_1 \wedge x_3 \wedge x_4 = 1/2$$

$$x_2 \wedge x_3 = 3/4$$

and

$$x_1 \wedge x_4 = 1/2$$

$$x_2 \wedge x_3 = 3/4$$

and

$$x_1 \wedge x_3 = 1/2$$

$$x_1 \wedge x_2 \wedge x_3 = 3/4$$

and

$$x_1 = 1/2$$

$$x_1 \wedge x_2 \wedge x_3 = 3/4.$$

The first two systems are equivalent while the last two systems are equivalent. The last two systems have no solution. In both pairs of systems, $j = 3$ and $i_j^* = 2$.

Theorem 3.21 *Consider the system $E\bar{x} = \bar{b}$.*

(1) The system has a unique solution if and only if $\forall r = 1, ..., t,$the system

$$E^*_{q_r+1}\bar{x} = b_{q_r+1}, ..., E^*_{q_{r+1}}\bar{x} = b_{q_{r+1}}$$

has a unique solution.

(2) The system is inconsistent if and only if $\exists i \in \{1, ..., m\}$ such that $b_i > 0$ and $e^*_{i1} = ... = e^*_{in} = 0$.

Proof. (1) Suppose that $i \in I_r$ and $h \in I_s$ where $r \neq s$. Then $e^*_{ij} = 1$ implies $e^*_{hj} = 0$. That is, the t systems

$$E^*_{q_r+1}\bar{x} = b_{q_r+1}, ..., E^*_{q_{r+1}}\bar{x} = b_{q_{r+1}}, r = 1, ..., t,$$

pairwise involve distinct unknowns.

(2) Since the t systems in (i) pairwise involve distinct unknowns,

$$E^*\bar{x} = \bar{b}$$

is inconsistent if and only if one of the t systems is inconsistent. The desired result now follows by applying the condition in (ii) individually to the t systems. ■

For the matrix E, let E_i denote the i-th row of $E, i = 1, ..., m$. We write $E_g \leq E_h$ if and only if $\forall k = 1, ..., n, e_{gk} = 1$ implies $e_{hk} = 1$. We write $E_g < E_h$ if and only if $E_g \leq E_h$ and $E_g \neq E_h$. The addition of two rows of E is componentwise with $0 + 0 = 0, 0 + 1 = 1 + 0 = 1 + 1 = 1$.

Corollary 3.22 *Consider the system $E\bar{x} = \bar{b}$. Then $E\bar{x} = \bar{b}$ is inconsistent if and only if $\exists i, h_1, ..., h_k \in \{1, ..., m\}$ such that $b_i > 0, i \in I_r$ and $h_u \in I_{s_u}$ with $r < s_u$ for $u = 1, ..., k$ and $E_i \leq E_{h_1} + ... + E_{h_k}$.*

Proof. There exists $i \in \{1, ..., m\}$ such that $e^*_{i1} = ... = e^*_{in} = 0$ if and only if $\exists i$, h_1, ..., $h_k \in \{1, ..., m\}$ such that $i \in I_r$ and $h_u \in I_{s_u}$ with $r < s_u$ for $u = 1, ..., k$ and $E_i \leq E_{h_1} + ... + E_{h_k}$. ■

We now examine the case where $b_1 = ... = b_m$. Let $i \in \{1, ..., m\}$. Suppose that $\exists E_{i_1}, ..., E_{i_{k_i}} \leq E_i$. If $E_{i_1} + ... + E_{i_{k_i}} < E_i$, then let $c_{ij} = 0$ if $e_{i_r j} = 1$ for some $r = 1, ..., k_i$ and $c_{ij} = e_{ij}$ otherwise, $j = 1, ..., n$. If no such E_{i_r} exist let $c_{ij} = e_{ij}, j = 1, ..., n$. Let $C_i = (c_{i1}, ..., c_{in})$ and $C = (C_1, ..., C_m)^T$, i.e., C is the transpose of the matrix $(C_1, ..., C_m)$. (If $E_{i_1} + ... + E_{i_{k_i}} = E_i$, then the i-th equation may be deleted.)

Theorem 3.23 *Suppose that $b_1 = ... = b_m = b$ in system (3.1). Let C denote the matrix defined above. Then the systems $E\bar{x} = \bar{b}$ and $C\bar{x}R\bar{b}$ are equivalent where R indicates that the relation in the i-th equation is either "$=$" or "$\geq$" depending upon whether $C_i = E_i$ or $C_i \neq E_i$, respectively.*

Proof. Let $S(0)$ denote the system $E\bar{x} = \bar{b}$ and let $S(i)$ denote the system obtained from $S(0)$ by replacing its i-th equation by $C_i\bar{x}R_ib$ where R_i denotes "$=$" or "$\geq$". Let $T(i)$ be the system $C_1\bar{x}R_1b, ..., C_i\bar{x}R_ib, E_{i+1}\bar{x} = b, ..., E_m\bar{x} = b$. It is easily seen that $E_{i_1}\bar{x} = b, ..., E_{i_{k_i}}\bar{x} = b, E_i\bar{x} = b$ and $E_{i_1}\bar{x} = b, ..., E_{i_{k_i}}\bar{x} = b, C_i\bar{x}R_ib$ are equivalent. Thus $S(0)$ and $S(i)$ are equivalent $\forall i = 1, ..., m$. Now $T(1) = S(1)$ and so $S(0)$ and $T(1)$ are equivalent. Assume that $S(0)$ and $T(i)$ are equivalent (the induction hypothesis). We now show that $S(0)$ and $T(i+1)$ are equivalent. Hence the result holds by induction. As noted above, $S(0)$ and $S(i+1)$ are equivalent. Let the $(i+1)$-st equation of $S(0)$ (and thus of $T(i)$) be denoted by $y_1 \wedge ... \wedge y_h \wedge z_1 \wedge ... \wedge z_k = b$ where $y_1, ..., y_h, z_1, ..., z_k \in \{x_1, ..., x_n\}$ and where the $(i+1)$-st inequality of $S(i+1)$ is $z_1 \wedge ... \wedge z_k \geq b$. Now $\{y_1, ..., y_h\} \cap \{z_1, ..., z_k\} = \emptyset$. Also, $y_1 \wedge ... \wedge y_h \wedge z_1 \wedge ... \wedge z_k = b$ is equivalent to $(y_1 \wedge ... \wedge y_h = b$ and $z_1 \wedge ... \wedge z_k \geq b)$ or $(y_1 \wedge ... \wedge y_h \geq b$ and $z_1 \wedge ... \wedge z_k = b)$. Since $T(i)$ and $S(i+1)$ are each equivalent to $S(0)$, $T(i)$ and $S(i+1)$ are equivalent. Hence the system $T(i)$ minus the $(i+1)$-st equation and the system $S(i+1)$ minus the $(i+1)$-st inequality individually imply $y_1 \wedge ... \wedge y_h = b$. Thus we have the equivalence of $T(i+1)$ and $S(i+1)$ and thus the equivalence of $T(i+1)$ and $S(0)$. ■

System (3.1) with $b_1 = ... = b_m$ is consistent if and only if $\forall i, \exists j$ such that $e_{ij} = 1$.

Example 3.12 *Consider the following system $S(0)$:*

$$x_1 \wedge x_2 \wedge x_3 = b$$

$$x_1 \wedge x_2 = b$$

$$x_1 \wedge x_2 \wedge x_3 \wedge x_4 = b.$$

Let $i = 1$. *Then* $E_2 < E_1$. *Applying Theorem 3.23, we obtain* $S(1)$:

$$x_3 \geq b$$

$$x_1 \wedge x_2 = b$$

$$x_1 \wedge x_2 \wedge x_3 \wedge x_4 = b.$$

Let $i = 2$ *in* $S(0)$. *Then* $S(2) = S(0)$. *Let* $i = 3$ *in* $S(0)$. *Then* $E_1 + E_2 < E_3$. *Applying Theorem 3.23, we obtain* $S(3)$:

$$x_1 \wedge x_2 \wedge x_3 = b$$

$$x_1 \wedge x_2 = b$$

$$x_4 \geq b.$$

Thus $T(3)$ *is the system*

$$x_3 \geq b$$

$$x_1 \wedge x_2 = b$$

$$x_4 \geq b.$$

Theorem 3.24 *Suppose that* $b_1 = ... = b_m = b$ *in system (3.1). Let* C *be the matrix as defined above. Suppose that* $c_{hk} = c_{ik} = 1, h \neq i$, *for some* i, k *where* $C_h\bar{x} \geq b$ *and* $C_i\bar{x} = b$. *Suppose that* $C_h \nleq C_i$. *Let* $d_{hk} = 0$ *and* $d_{uv} = c_{uv}$ *if* $(u, v) \neq (h, k)$. *Let* $D = [d_{ij}]$. *Then* $C_h\bar{x} \geq b, C_i\bar{x} = b$ *are equivalent to* $D_h\bar{x} \geq b, D_i\bar{x} = b$.

Proof. Both systems force $x_k > b$.

If $C_h \leq C_i$, then drop the h-th equation. In fact, if $C_h\bar{x} \geq b, C_{h_1}\bar{x} = ... = C_{h_k}\bar{x} = b$ and $C_h \leq C_{h_1} + ... + C_{h_k}$, then drop the h-th equation.

We also note that $x_1 \wedge x_2 \geq b$ is equivalent to $x_1 \geq b$ and $x_2 \geq b$. ■

Example 3.13 *The following systems are equivalent:*

$x_2 \wedge x_3 \geq b$
$x_1 \wedge x_2 = b$
and
$x_3 \geq b$
$x_1 \wedge x_2 = b.$
Here $h = 1$ *and* $i = k = 2$.

Example 3.14 *The following systems are equivalent:*
$x_2 \wedge x_3 \geq b$
$x_1 \wedge x_2 = b$
$x_1 \wedge x_3 = b$
and
$x_1 \wedge x_2 = b$
$x_1 \wedge x_3 = b.$
In the first system, $C_1 \leq C_2 + C_3$.

To solve a general system of intersection equations, we may use the following algorithm. We use the notation E_i' to denote the complement of E_i. We let θ denote the zero vector. We also assume that $b_1 > 0$.

Algorithm 2.8.
1. Sort the E_i so that the b_i's are in nondecreasing order.
2.1. Let Temp and Total each be a row of n zeros
2.2. Let $c = b_m$
2.3. For $i = m$ down to 1 do
    if $b_i = c$ then
        $Temp = Temp + E_i$ and $E_i = E_i'$ NOR $Total$
    if $c > b_i$ then
        $c = b_i, Total = Total + Temp, Temp = E_i$
        and $E_i = E_i'$ NOR $Total$
2.4. If $\exists i, 1 \leq i \leq m, E_i = \theta$, then
    INCONSISTENT and STOP
3. For each distinct b_k
3.1. Let $E_j, i_1 \leq j \leq i_k$, be all rows such that $b_j = b_k$ where $1 \leq i_1, i_k \leq m$
3.2. Let O_j be the number of 1's in E_j
3.3 Sort E_j's such that O_j's are in nondecreasing order
3.4. Let T be a row of n zeros
3.5. For $x = i_1$ to i_k do
3.5.1. For $y = x - 1$ down to i_1 do
    if E_y' NOR $E_x = \theta$ then
        $T = T + E_y$
3.5.2. If $T = E_x$ then
    erase E_x, R_x, b_x
3.5.3 Else if $T \neq \theta$ then
    $C_x = E_x$ XOR T and $R_x =$ '$\geq$'
3.5.4 Else
    $C_x = E_x$
4. For each distinct b_k
4.1. While $\exists C_i$ and C_j such that
    $(1) b_i = b_j = b_k$
    $(2) R_i =$ '$=$'
    $(3) R_j =$ '$\geq$' and
    $(4) C_j'$ NOR $C_i \neq \theta$ do

$C_j = C'_j \ NOR \ C_i$

4.2. Let $T_k = \sum C_i \forall C_i$ such that $b_i = b_k$ and $R_i =$ '$=$'

4.3. If $\exists C_i$ such that $(1) R_i =$ '$\geq$' and $(2) C'_i \ NOR \ T_k = \theta$ then erase C_i and b_i from matrices C and $\bar{b}$, respectively.

The time complexity of the algorithm is easily seen to be $O(m^2n)$. If each row in E and C is denoted as a binary number, then the time complexity becomes $O(m^2)$.

A unique minimal solution can be immediately determined.

The results of this section can be applied to those of Section 2.4 as we now describe. Suppose that G is the Cartesian product of two graphs G_1 and G_2. Let (μ, ρ) be a partial fuzzy subgraph of G. Then (μ, ρ) is a Cartesian product of a partial fuzzy subgraph of G_1 and a partial fuzzy subgraph of G_2 if and only if the system of intersection equations as described in Theorem 2.57 has a solution.

The composition of fuzzy graphs is also defined in Section 2.4. If (μ, ρ) is a partial fuzzy subgraph of the composition $G_1 [G_2]$ of graphs G_1 and G_2, then necessary and sufficient conditions are given in Section 2.4 for (μ, ρ) to be the composition of partial fuzzy subgraphs of G_1 and G_2 in terms of the existence of a solution to a system of fuzzy intersection equations.

3.5 Fuzzy Graphs in Database Theory

We now give an application of fuzzy graphs to database theory as developed in [16]. We examine fuzzy relations which store uncertain relationships between data. In classical relational database theory, design principles are based on functional dependencies. In this section, we generalize this notion for fuzzy relations and fuzzy functional dependencies. Results presented are useful for designing fuzzy relational databases.

Definition 3.12 *Let $U = \{A_1, ..., A_n\}$ be the set of attributes and each A_i is assigned to the set of possible values $DOM(A_i)$. A fuzzy subset ρ of the Cartesian cross product $\times_{i=1}^{n} DOM(A_i)$ is called a* fuzzy relation *on $\times_{i=1}^{n} DOM(A_i)$*

In classical database theory, functional dependencies play important roles. A functional dependency 'X functionally determines Y in R' means for any two tuples of the relation R, if the X values are the same, then the Y values are also same. In other words, $\chi_{X \to Y}$ is equivalent to

$$\forall t_1, t_2((R(t_1) \ .and. \ R(t_2) \ .and. \ t_1 [X] = t_2 [X]) \implies t_1 [Y] = t_2 [Y]).$$

For example, consider the relation R given below:

A	B	C
a	b	c
a	b	d
e	f	g
e	b	d

Note that A functionally determines B since for any two rows (known as tuples in database theory) t_1 and t_2 of R, if their values in column (known as attribute in database theory) A are the same then those tuples have identical values in column B. However, A does not functionally determine C since considering the first two rows observe that while the column A values are identical, the column C values are not identical. It may be noted that C functionally determines B and B does not functionally determine A.

We get a fuzzy version of the formula when we substitute the operators .and., $\forall$ with the operators min ($\wedge$), inf ($\wedge$) and .or., $\exists$ with max ($\vee$), sup ($\vee$), and $\Longrightarrow$ with $\rightarrow$, where the implication $\rightarrow$ is defined as follows:

$$a \rightarrow b = \begin{cases} 1 & \text{if } a \leq b \\ 1 - (a - b), & \text{otherwise} \end{cases}$$

and finally .not. with $\neg$, where $\neg a = 1 - a$. In this way, we get that the truth value of the fuzzy relation ρ satisfies a given functional dependency $X \rightarrow Y$:

$$\mu(X, Y) = 1 - \vee\{\rho(t_1) \wedge \rho(t_2) \mid t_1 [X] = t_2 [X] \text{ but } t_1 [Y] \neq t_2 [Y]\},$$

where t_1 and t_2 are any two tuples of ρ. As in the classical database theory, we denote the union of attributes X and Y by XY.

Example 3.15 *Consider the fuzzy relation ρ on $DOM(A) \times DOM(B) \times DOM(C)$.*

A	B	C	$\rho(t)$
a	b	c	1
a	b	f	0.8
e	d	c	0.7
e	b	f	0.6

The fuzzy relation ρ generates the following truth values.

$$\begin{array}{lll} \mu(A, B) = 0.4, & \mu(B, C) = 0.2, & \mu(C, A) = 0.3, \\ \mu(A, C) = 0.2, & \mu(B, A) = 0.4, & \mu(C, B) = 0.3, \\ \mu(AC, B) = 1, & \mu(BC, A) = 0.4, & \mu(AB, C) = 0.2, \\ \mu(AB, B) = 1, & \mu(AB, A) = 1. & \end{array}$$

Fuzzy functional dependency satisfies the following properties.

A1 If $Y \subseteq X$, then $\mu(X,Y) = 1$,

A2 $\mu(X,Y) \wedge \mu(Y,Z) \leq \mu(X,Z)$,

A3 $\mu(X,Y) \leq \mu(XZ,YZ)$.

From these, other properties can be obtained:

B1 $\mu(X,Y) \wedge \mu(Y,Z) \leq \mu(X,YZ)$,

B2 $\mu(X,Y) \wedge \mu(WY,Z) \leq \mu(XW,Z)$,

B3 if $Z \subseteq Y$, then $\mu(X,Y) \leq \mu(X,Z)$.

An important consequence is that $\mu(X,Y) = \wedge\{\mu(X,A)|A : A \in Y\}$.

Thus a fuzzy relation generates another a fuzzy relation $\mu(X,Y)$ on U^2 with the properties A1 - A3.

Moreover, if there is given an arbitrary fuzzy relation $\tau(X,Y)$ on U^2, then it defines the fuzzy relation $\tau^+(X,Y)$ which is the smallest fuzzy relation on U^2 that contains $\tau(X,Y)$ and has the properties A1 - A3. We call $\tau^+(X,Y)$ the closure of $\tau(X,Y)$. (Recall that $\tau_1(X,Y) \subseteq \tau_2(X,Y)$ if and only if $\tau_1(X,Y) \leq \tau_2(X,Y) \forall X,Y \subseteq U$.)

The closure is well defined because the fuzzy relation $\varsigma(X,Y) \equiv 1$ satisfies A1 - A3 and contains every fuzzy relation on U^2, and if $\tau \subseteq \varsigma_1, \tau \subseteq \varsigma_2$, where ς_1, ς_2 satisfy A1 - A3, then $\tau \subseteq \varsigma_1 \cap \varsigma_2$ and $\varsigma_1 \cap \varsigma_2$ also satisfies A1 - A3. ($\varsigma_1 \cap \varsigma_2(X,Y) := \varsigma_1(X,Y) \wedge \varsigma_2(X,Y)$ for all $X,Y \subseteq U$.)

Proposition 3.25 *$\tau^+(X,Y)$ is a closure, that is*

(1) $\tau(X,Y) \subseteq \tau^+(X,Y)$,

(2) $\tau^{++}(X,Y) = \tau^+(X,Y)$,

(3) if $\tau_1(X,Y) \subseteq \tau_2(X,Y)$, *then* $\tau_1^+(X,Y) \subseteq \tau_2^+(X,Y)$.

Proof. The proof follows from the fact that closure is the smallest with the given properties. ■

Now we extend $\tau^+(X,A)$ for fuzzy subsets σ as follows: Let σ be a fuzzy subset on U and

$$\tau_f^+(\sigma, A) = \vee\{(\tau^+(Z,A) \wedge \lambda) \mid Z \subseteq U, \lambda \in [0,1], Z_\lambda \subseteq \sigma\}$$

where for $\lambda \in [0,1]$ we define

$$Z_\lambda(A) = \begin{cases} \lambda, & \text{if } A \in Z \\ 0, & \text{otherwise.} \end{cases}$$

With the help of $\tau_f^+(\sigma, A)$, we define a closure set on U as follows: Let σ be a fuzzy subset on U. Then σ^+ is also a fuzzy set on U and defined by $\sigma^+(A) = \tau_f^+(\sigma, A)$ for all $A \in U$.

First note that $\tau_f^+(\sigma, A) = \tau^+(X, A)$ if X is a crisp set, that is $X(A) = 1$ or 0 for all $A \in U$. This is true because $\tau^+(X, A)$ is an increasing function in the argument X.

Proposition 3.26 *σ^+ is a closure on U, that is*

(1) $\sigma \subseteq \sigma^+$,

(2) if $\sigma \subseteq \rho$, *then* $\sigma^+ \subseteq \rho^+$,

(3) $\sigma^{++} = \sigma^+$.

Proof.

(1) $\sigma(A) = \tau(A, A) \wedge \sigma(A) \leq \sigma^+(A)$ for all $A \in U$.

(2) $\sigma \subseteq \rho$ implies that for all Z_λ for the definition of $\sigma^+(A)$ are good for the definition of $\rho^+(A)$ as well.

(3) $\sigma^+(A) \leq \sigma^{++}(A)$ holds by (1) and (2). $\sigma^{++}(A) = \tau_f{}^+(\sigma, A)$ and for some $Z = B_1 B_2 ... B_k \subseteq U$, it is equal to $\tau^+(Z, A) \wedge (\wedge_{B \in Z} \sigma^+(B)) = \tau^+(Z, A) \wedge (\wedge_{i=1}^k (\tau^+(\rho_i, B_i) \wedge \lambda_i)$, where $(\rho_i)_{\lambda_i} \subseteq \sigma$. Now let $\rho = \cup_{i=1}^k \rho_i$ and $\lambda = \wedge_{i=1}^k \lambda_i$, then $\rho_\lambda \subseteq \sigma$. We have $\tau^+(\rho, Z) = \wedge_{i=1}^k \tau^+(\rho, B_i)$ and $\tau^+(\rho_i, B_i) \leq \tau^+(\rho, B_i)$ for all $i = 1, ..., k$. Therefore our expression is not greater than $\tau^+(Z, A) \wedge \tau^+(\rho, Z) \wedge \lambda$ and for A2 it is not greater than $\tau^+(\rho, A) \wedge \lambda \leq \vee_{W, \alpha : W_\alpha \subseteq \sigma} (\tau^+(W, A) \wedge \alpha) = \sigma^+(A)$.

■

Representation of Dependency Structure $\tau(X, Y)$ by Fuzzy Graphs

Let $\tau(X, Y)$ be a fuzzy relation on U^2. We correspond to $\tau(X, Y)$ a fuzzy graph $G_T = (\varpi, \rho)$ as follows. The vertices are ordered pairs (X, Y) such that $\tilde{V}(X, Y) = \tau(X, Y)$. Edges are ordered pairs of vertices such that $\rho((X, Y), (X, Z)) = \tau(Y, Z)$.

The following algorithm gives $\tau^+(X, Y)$ by modifying step by step the labels of the graph:

Algorithm 2.9.

1. For all $Y \subseteq X$ let $\varpi((X, Y)) = 1$.
2. while (STAT1 is true or STAT2 is true) do
   (where STAT 1 is true means
   there exists an edge $e = (v_1, v_2)$ so that
   $\varpi(v_2) < \varpi(v_1) \wedge \varpi(e)$,
   and STAT2 is true means
   there are vertices $v_1 = (X, Y)$ and $v_2 = (XZ, YZ)$ so that

$\varpi(v_2) < \varpi(v_1))$
if (STAT1 is true) then
$\varpi(v_2) = \varpi(v_1) \wedge \rho(e)$;
for all edges $d = ((X,Y),(X,Z))$ where $v_2 = (Y,Z)$,
$\rho(d) = \varpi(v_2)$;
if (STAT2 is true) then
$\varpi(v_2) = \varpi(v_1)$;
for all edges $d = ((W,XZ),(W,YZ))$ where $v_2 = (XZ,YZ)$,
$\rho(d) = \varpi(v_2)$;
3. $\tau^+(X,Y) = \varpi(v)$, where $v = (X,Y)$.

Proposition 3.27 *The algorithm is correct.*

Proof. First it stops in finite time because the range of the label function l is finite, and when a label value is modified it is changed with a greater value from the range.

So the algorithm produces a sequence of graphs $G_0, G_1, ..., G_{k-1} = G_k$, where $G_0 = G_T$ and we get G_i when we do all changes of the label values induced by STAT1 or STAT2. Let l_i be the label function of the graph G_i and $\tau^i(X,Y) = l_i(v)$, where $v = (X,Y)$, for all $i = 0,...,k$. Then obviously $\tau \subseteq \tau^k$ and τ^k satisfies A1 - A3, otherwise the graph can still be modified. Thus $\tau^+ \subseteq \tau^k$. If $\tau^+ \subset \tau^k$ were true, then there would be an $i \in [0, k-1]$ such that $\tau^i \subseteq \tau^+ \subset \tau^{i+1}$. If G_{i+1} is obtained by STAT1 with $v_1 = (X,Y), v_2 = (X,Z)$ and $e = (v_1, v_2)$ then $\tau^+(X,Z) < \tau^{i+1}(X,Z) = \tau^i(X,Y) \wedge \tau^i(Y,Z) \leq \tau^+(X,Y) \wedge \tau^+(Y,Z)$, which is a contradiction. If G_{i+1} is obtained from STAT2 with $v_1 = (X,Y)$, $v_2 = (XZ,YZ)$, then $\tau^+(XZ,YZ) < \tau^{i+1}(XZ,YZ) = \tau^i(X,Y) \leq \tau^+(X,Y)$, which is a contradiction.

■

Since $\sigma^+(A)$ is defined by $\tau^+(X,A)$ when X is a crisp set on U, it can be computed by this algorithm as well.

3.6 A Description of Strengthening and Weakening Members of a Group

The results in this section are from [33]. Matrix analysis is used to identify various aspects of group structure. These include redundancies [31], complete cycles [12], liaison persons [32], and cliques [13]. The structural concepts are developed in [9, 10].

There are groups in which some members are seen as exercising a disruptive or divisive influence while other individuals appear to help hold the group together. Although there are a number of ways in which these weakening and strengthening effects of group members may be conceptualized, we shall confine our discussion to one that is suggested by and can be coordinated to various kinds of connectedness of "directed graphs" or "digraphs."

A digraph becomes a mathematical model for the structure of a group of people when its points represent the members of the group and the (directed) lines the relationships between pairs and members. The group structure may be built from many different kinds of relationships, such as sociometric choice, communication, or power. Hence it may be useful to have a way of characterizing, in digraph terms, the way in which individual members contribute to the connectedness of the group.

In order to give precise meanings to these concepts, we coordinate the structure of a group to a digraph. Then the kinds of connectedness a group may possess correspond to the four kinds of connectedness a digraph may have. A *digraph* is a pair (V, P), where V is a finite set and P is a subset of $V \times V$. The elements of V are referred to as *points* or *vertices* or *nodes.* Let $< x, y > \in P$. If $x \neq y$, then $< x, y >$ is called a *(directed) line* or *a directed edge* or an *arc*. Further, $< x, y >$ is sometimes denoted as $x \rightarrow y$. If $x = y$, then $< x, y >$ is called a *loop*. Let $y_0, y_1, \ldots, y_k$ be distinct members of V. The set $\{< y_0, y_1 >, < y_1, y_2 >, < y_2, y_3 >, \ldots, < y_{k-1}, y_k >\}$ is called a directed path from y_0 to y_k of length k. The *distance* between x and y is the length of a shortest path from x to y. Although the digraph model is capable of describing several different kinds of bonds or lines at the same time, we shall confine ourselves to the simplest case in which the existence of lines implies the same characteristic of the relationship between each pair of joined points. This restriction is not serious since the meaning coordinated to the lines may be a summary of many observations. We also note that directed lines may be more useful than symmetric lines since it is not necessary that bonds or lines be in the same direction. Suppose for example, that lines stand for the flow of information in a particular direction. If information flows from x to y and from y to z, then information flows from x to z via y. Such a multistep path indicates a possible channel for communication.

A digraph is said to be *strongly connected* (or *strong*) if for every pair of distinct points, x and y, there exists a directed path from x to y and one from y to x. A digraph is said to be *unilaterally connected* (or *unilateral*) if for every pair of points, x and y, there is a directed path from x to y or one from y to x. A digraph is called *disconnected* if the points can be divided into two sets with no line joining any point in one set with a point in the other set. A digraph is called *weakly connected* (or *weak*) if it is not disconnected. These connectedness definitions are inclusive since every strong digraph is unilateral and every unilateral digraph is weak. We use the term "digraph" and "group" interchangeably.

In order to distinguish between groups on the basis of the kind of connectedness, we require exclusive connectedness categories. These may be obtained as follows. Let U_3 be the collection of all strong digraphs. We define U_2 as the set of all unilateral digraphs. Similarly we define U_1 as the set of all weak digraphs. Finally, U_0 is the collection of all disconnected digraphs. Then $U_3 \subseteq U_2 \subseteq U_1$ and $U_1 \cap U_0 = \emptyset$. Let $C_3 = U_3$, $C_2 = U_2 \setminus U_3, C_1 = U_1 \setminus U_2$, and $C_0 = U_0$. Clearly, any digraph belongs to exactly one of the categories C_3, C_2, C_1, or C_0.

In order to evaluate the effect of a point on the connectedness of its digraph, we require a precise definition of the removal of a point of a digraph. If D is a digraph and x is a point in it, than $D \setminus x$ is the digraph obtained from D by deleting the point x and all lines which are either directed toward x or away from x. We say x is a point of type P_{ij} if the digraph D (with x present) is in class C_i but $D \setminus x$ (with x absent) is in class C_j. Since the four categories C_3, C_2, C_1, and C_0 are numbered in accordance with the convention that the higher the subscript the stronger the kind of connectedness of the digraph, we may utilize this convention to describe and characterize which points are strengthening. Thus a point of a digraph is a *strengthening point* if it is of type P_{ij} such that $i > j$; it is a *weakening point* if $i < j$ and the point is called *neutral* if $i = j$. By *strengthening and weakening group members,* we mean those individuals coordinated with strengthening and weakening points of the digraph that represents the structure of the group. Similarly, if a point is of the type P_{ij}, we speak of the corresponding individual as an (i, j) *member.*

When we contrast D and $D \setminus x$, we assume that the lines between pairs of distinct points exist independently.

Connectedness Criteria

We characterize strengthening and weakening group members. We first show that there are no (1, 3) members in a group. We then introduce the concept of the "reachability matrix" of a group. This matrix, R, is useful for the expression of matrix conditions that characterize the inclusive connectedness categories (i.e., strong, unilateral, weak, and disconnected). A straightforward modification of these theorems leads to a description of the exclusive connectedness categories C_3, C_2, C_1, C_0. We then obtain conditions for each of the (i, j) types. We show that weakening members may be identified from direct examination of the reachability matrix.

To illustrate all the possible types of (i, j) members that can occur, we display the following digraphs in which the ordered pair written at each point indicates its type. Whenever a line between two points of any digraph of Figure 3.8 is drawn without any direction displayed, it stands for two directed lines, one in each direction.

Example 3.16 *All* (i, j) *members, except for* $(1, 3)$, *are possible.*

Theorem 3.28 *There are no* (1, 3) *members in any group.*

Proof. We assume that there exists a (1, 3) member x of group G, and derive a contradiction. Here G is in class C_1 and $G \setminus x$ is strong. But when we join the point x to the group $G \setminus x$ by at least one line, then the original group G must be unilateral (i.e., is in either class C_2 or C_3). This contradicts the hypothesis that G is in class C_1. ■

Throughout the remainder of this section, $D = (V, P)$ is digraph with $x_1, x_2, ..., x_n$ the n members of V or the group. Let the matrix M of the group relationship be defined as follows: If x_i has the relationship to x_j, the i, j entry of M equals 1; otherwise it is 0. We assume that all the diagonal entries are 1. The relationship is not assumed to be symmetric, but may be.

Let R denote the $n \times n$ matrix whose i, j element is 1 if there is a directed path from x_i to x_j and 0 otherwise. Then R is called the *reachability matrix* of the digraph D. The *diameter, d,* of digraph D is the greatest distance between any two points of D. A *loop* is a directed line which begins and ends at the same point. Our convention that all the diagonal elements of M are 1 is equivalent to the existence of a loop at every point.

A *Boolean matrix* is one in which all the elements are either 1 or 0. The usual addition and multiplication operations of matrices are applied to Boolean matrices by means of the rule $1+1 = 1$. Thus in Boolean terms, any positive integer obtained by ordinary matrix operations is replaced by a 1. If the sum or product is 0, it is entered in the matrix as 0. In this section all matrices are Boolean, and all the operations are performed as described above.

A redundant chain from y_1 to y_k is a sequence of lines of the form $y_1 \to y_2, y_2 \to y_3, ..., y_{k-1} \to y_k$ where the points y_1 to y_k are not all distinct. By its constructive definition, the matrix M gives all the (directed) paths of length 1 between distinct pairs of points. The matrix M contains all these paths and also a path of length 1 from each point to itself. Thus each point is reachable from itself. In addition, these loops permit the existence of redundant chains from x to y of any length whenever paths from x to y of shorter length occur. We now form the matrix M^2. In this matrix, the number 1 occurs in the i, j place if and only if there is either a path or redundant chain of length 2 from point i to point j. Continuing in this manner, we eventually obtain the matrix M^d. Since d is the diameter of the given digraph, there are no paths of length greater than d that connect points not already connected. The presence of loops at each point assures us that there will be a redundant chain of length d whenever there is a path of any shorter length. Therefore this last matrix M^d contains all reachability relationships that occur in the digraph. Hence, $R = M^d$. Hence we have

FIGURE 3.8 Possible types of (i, j) members.

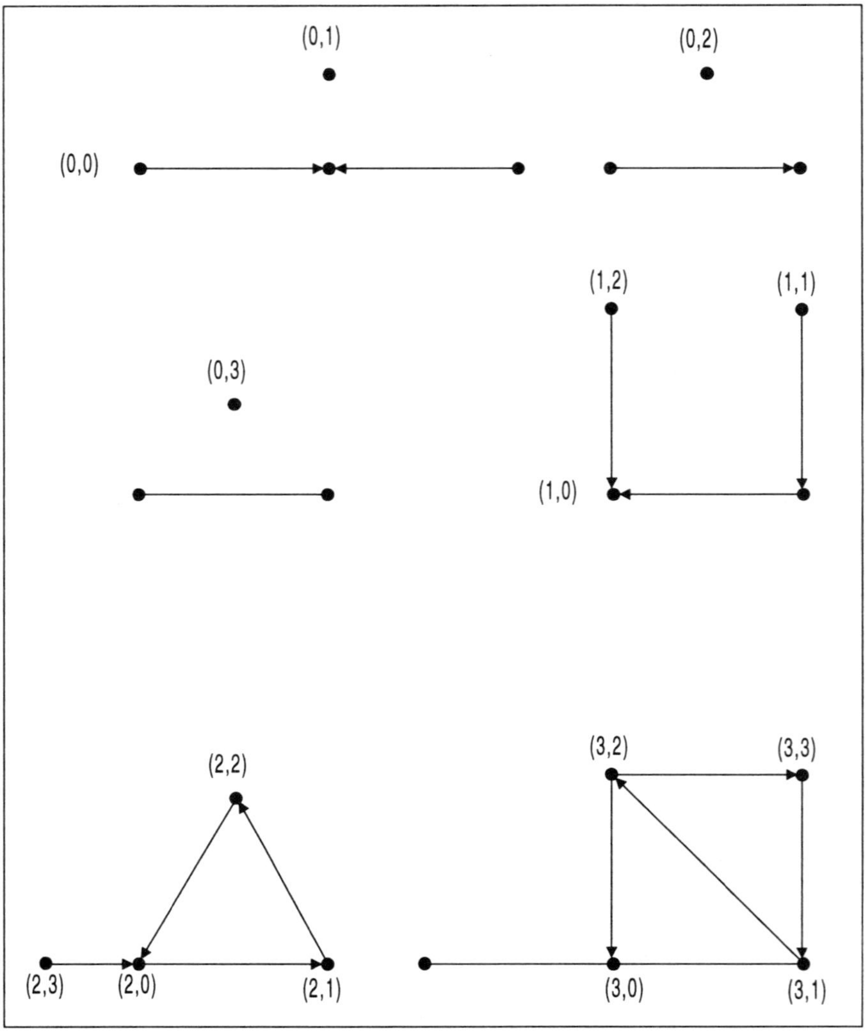

the following theorem which gives a formula for the reachability matrix of a group in terms of its relationship matrix.

Theorem 3.29 $R = M^d$, *where* d *is the diameter of* D.

Corollary 3.30 *If* k *is any integer greater than* d, *then* $M^k = R$.

Proof. The diameter d is the smallest number m such that $M^m = M^{m+1}$. Hence for any integer k greater than d, the matrix M^k is equal to M^d which we have already seen equals R. ■

Corollary 3.31 $R = (I - M)^{-1}$.

Proof. By a well-known matrix identity, $(I - M)^{-1} = I + M + M^2 + \ldots + M^d + M^{d+1} \ldots$ But the terms $M^d + M^{d+1} + \ldots = M^d$ since we are using Boolean addition. Also $I + M + M^2 + \ldots + M^d = M^d$. Hence $(I - M)^{-1} = M^d = R$ by Theorem 3.29. We note that if N is the matrix obtained from M by having 0's on the diagonal, then $R = (-N)^{-1}$. ■

Inclusive Connectedness Categories

The next four theorems serve to characterize the inclusive connectedness categories. These are expressible in terms of the matrix W of the *universal relationship*. Specifically, W is the n by n matrix in which every entry is 1.

Theorem 3.32 *The group* G *is strong if and only if* $R = W$.

Proof. The group G is strong if and only if for each member-pair x_i and x_j there exists a path in both directions. However, this condition holds if and only if for all values of i and j the i, j element of the matrix R is equal to 1. However, i and j are arbitrary positive integers between 1 and n. Therefore all the entries in R must be equal to 1 for the group to be strong, i.e., $R = W$. ■

This result was also found in [29]. Let M' denote the transpose of a matrix M. Then R' is the matrix of "reverse reachability" in the sense that a 1 in the i, j place of R' means that x_i is reachable from x_j. The next theorem uses the matrix $R + R'$. In this matrix, a 1 in the i, j place means x_j is reachable from x_i, or x_i is reachable from x_j, or both (since the addition of R and R' is Boolean).

Theorem 3.33 *The group* G *is unilateral if and only if* $R + R' = W$.

TABLE 3.4 Weakening members.

Exclusive Class	Criterion
C_3	$R = W$
C_2	$R + R' = W,\ R \neq W$
C_1	$R^- = W,\ R + R' \neq W$
C_0	$R^- \neq W$

Proof. Let r_{ij} and r'_{ij} denote the i, j^{th} element of R and R', respectively. Then G is unilateral $\Leftrightarrow \forall i, j, \exists$ a directed path from either x_i to x_j or from x_j to $x_i \Leftrightarrow \forall i, j, r_{ij} = 1$ or $r'_{ij} = 1 \Leftrightarrow R + R' = W$. ■

In order to characterize weak groups, we introduce the digraph D^- obtained from $D = (V, P)$ as follows: Let $D^- = (V, P^-)$, where $P^- = P \cup \{< y, x > \mid < x, y > \in P\}$. This process may be called *symmetrizing* the digraph D. Let $M^-$and R^- be the relationship matrix and the reachability matrix of D^- respectively. Then $M^- = M + M'$.

Theorem 3.34 *A group is weak if and only if* $R^- = W$.

Proof. Clearly, D is weak if and only if D^- is strong. Therefore, by Theorem 3.32, D is weak if and only if $R^- = W$. ■

Theorem 3.35 *A group is disconnected if and only if* $R^- \neq W$.

Proof. By definition, a group is disconnected if and only if it is not weak. The desired result follows by Theorem 3.34. ■

Exclusive Connectedness Categories

We now combine the last four theorems to characterize the exclusive categories C_3, C_2, C_1, and C_0.

Theorem 3.35 gives a criterion for a group G to be in C_0. However, for G to be in C_1 the condition of Theorem 3.34 holds while that of Theorem 3.33 does not. Similarly, G is in C_2 if and only if Theorem 3.33 holds while Theorem 3.32 does not. Finally G is in C_3 if and only if Theorem 3.32 holds. These observations are summarized in Table 3.4 which lists the exclusive connectedness categories in one column and the respective criteria in the other column.

The weakening members of a group can be described further. All the possible kinds of weakening members are:

A. The (0, j) members for $j = 1, 2$, or 3
B. The (1, 2) members
C. The (2, 3) members

This follows from Theorem 3.28 which asserts that there are no (1, 3) members.

Let $x \in V$. Then x is called an *isolate* if $\nexists y \in V \setminus \{x\}$ such that either $< x, y > \in P$ or $< y, x > \in P$.

Theorem 3.36 *Let $x_i \in V$. Then x_i is an isolate if and only if the only nonzero element in the ith row and the ith column of R is the i, i (or diagonal) entry.* ■

A *symmetric matrix* is one which remains unchanged when its rows and columns are interchanged. It is well known that for any square matrix A, the matrix $A + A'$ is symmetric. Hence, in particular $R + R'$ is symmetric.

In the following theorem, we give a characterization of all weakening members.

Theorem 3.37 *(1) For $j = 1, 2$, or 3, x_1 is a (0, j) member if and only if x_i is an isolate and $G \setminus x_i$ is in C_j.*

(2) x_i is a (1, 2) member if and only if $R + R'$ has at least one zero and all its zeros occur in the ith row and in the ith column.

(3) x_i is a (2, 3) member if and only if every element in R is a 1 except in the ith row (or column) of R, where all but the diagonal element are 0.

Proof. (1) If x_i is an isolate and $G \setminus x_i$ is in C_j then x_i is a (0, j) member by definition.

To prove the converse, let x_i be a (0, j) member for $j > 0$ and assume x_i is not an isolate. Then $G \setminus x_i$ is still disconnected. Hence x_i is a (0, 0) member which is a contradiction. Procedurally, (0, j) members can be identified by considering the condition met by the remainder of the matrices R, $R + R'$, and R^- after the row and column belonging to an isolate are deleted.

(2) If $R + R'$ has at least one zero and all its zeros occur in the ith row and in the ith column, then by Table 3.4, G is in $C_1$and $G \setminus x_i$ is in C_2.

To prove the converse, let x_i be a (1, 2) member of G. Then G is in C_1 so that $R + R'$ must have at least one 0. Assume that there is a 0 which is neither in the ith row nor in the ith column or $R + R'$. If x_i is eliminated from $R + R'$, all 0's not in the ith row or column would remain since the elimination of a member and its bonds cannot add connections to members of G. Therefore $G \setminus x_1$ would be in C_1, a contradiction.

(3) Again the criteria of Table 3.4 show that under the stated conditions x_i is a (2, 3) member.

Conversely, let x_i be a given (2, 3) member. Then either x_i can reach all other members but no other members can reach x_i or vice versa. In the first case all the elements in the ith row are 1 and all the nondiagonal elements in the ith column are 0, while every element of R not in the ith row or column is a 1. For the other case, we interchange the words "row" and "column." ■

Corollary 3.38 *The group consisting of exactly two isolates, i.e ,* $(\{x, y\}, \emptyset)$, *has two* (0, 3) *members. Any other disconnected group has at most one weakening member.* ■

Corollary 3.39 *A* C_1 *group has at most two* (1, 2) *members.* ■

The members x and z are $(1,2)$ members in the following digraphs: $(\{x,y,z\},\{(x,y\},\{z,y\}\})$ and $(\{x,y,z\},\{(y,x\},\{y,z\}\})$.

Corollary 3.40 *A* C_2 *group has at most two* (2, 3) *members.* ■

The members x and y are $(2,3)$ members in the digraphs $(\{x,y\},\{(x,y\}\})$.

The following is probably the most startling result in this section. It must be kept in mind that weakening members are defined in terms of the kinds of connectedness introduced here.

Theorem 3.41 *Any group has at most two weakening members.*

Proof. This is an immediate consequence of three corollaries since the possibilities there listed are exhaustive and mutually exclusive. ■

Unfortunately there do not appear to be analogous theorems describing the strengthening members of a group. However, we can still identify these strengthening members by using the results of Table 3.4 on the given group G and on $G \setminus x$. In particular, the strengthening members in classes $(i, 0)$ for $p = 1, 2$, or 3 correspond to the liaison persons [32] of the symmetrized group.

3.7 An Application to the Problem Concerning Group Structure

The results of this section are from [37]. A fuzzy directed graph is utilized to characterize the role played by an individual member in such a group that a class of group members having relationship with any given member has no sharply defined boundary. The concepts of weakening and strengthening points of an ordinary graph presented in the previous section and by Ross and Harary, [33], are generalized to those of a fuzzy directed graph.

The theory of graphs is one of the most important tools in the study of the group structure. Recall that a strengthening member of the group is one whose presence causes the graph corresponding to the group to be more highly connected than that obtained when he is absent, while a weakening member is one whose presence causes the graph to belong to a weaker category of connectedness. Besides this, the graph has been widely utilized to study the problems concerning redundancies, liaison persons, cliques, structural balance and so forth.

In many cases, however, the mere presence or absence of a relation is not adequate to represent a given group structure. There may be different strengths of the relations between individuals. There may even be situations in which it is fuzzy rather than well-defined whether or not an arbitrary individual has relationship with a given member, that is, a class of group members being in relationship with any given member does not have a sharply defined boundary. In such cases, an ordinary graph may not fully represent the group structure. Instead, the fuzzy graph seems to be a more relevant mathematical model.

Connectedness of a Fuzzy Graph

Definition 3.13 *Let V be a finite set of points and let Γ be a function of V into the set of all fuzzy subsets of V, $\mathfrak{F}\wp(V)$. Then $G(V,\Gamma)$ is called a* fuzzy directed graph.

Let Γ_x denote $\Gamma(x)$, for all $x \in V$. For $x, y \in V$, we can think of $\Gamma_x(y)$ as the strength of the directed line from x to y. If for all x, $y \in V$, $\Gamma_x(y)$ is either 0 or 1, then G reduces to an ordinary directed graph.

In order to evaluate the effect of the removal of a point on the connectedness of its fuzzy directed graph, we introduce the following definition.

Definition 3.14 *A fuzzy directed subgraph of $G = (V, \Gamma)$ is defined to be a fuzzy directed graph of the form (Y, Γ'), where Y is a subset of V and the function Γ' is defined as*

$$\Gamma'_y = \Gamma_y \mid_Y \text{ for any } y \in Y.$$

Definition 3.15 *For a fuzzy subset μ of V, two fuzzy subsets Γ_μ and Γ_μ^{-1} are defined by $\forall x \in V$,*

$$\begin{cases} \Gamma_\mu(x) = \vee\{\mu(y) \wedge \Gamma_y(x) \mid y \in V\} \\ \Gamma_\mu^{-1}(x) = \vee\{\mu(y) \wedge \Gamma_x(y) \mid y \in V\} \end{cases}$$

respectively.

Proposition 3.42 *Let μ and ν be two fuzzy subsets of V. Then,*

(1) $\Gamma_\mu \subseteq \Gamma_\nu$ *if* $\mu \subseteq \nu$,

(2) $\Gamma_\mu^{-1} \subseteq \Gamma_\nu^{-1}$ *if* $\mu \subseteq \nu$,

(3) $\Gamma_{\mu\cap\nu} \subseteq \Gamma_\mu \cap \Gamma_\nu$,

(4) $\Gamma_{\mu\cap\nu}^{-1} \subseteq \Gamma_\mu^{-1} \cap \Gamma_\nu^{-1}$,

(5) $\Gamma_{\mu\cup\nu} = \Gamma_\mu \cup \Gamma_\nu$,

(6) $\Gamma_{\mu\cup\nu}^{-1} = \Gamma_\mu^{-1} \cup \Gamma_\nu^{-1}$,

Proof. Properties (1) and (2) are obvious from definition of Γ_μ and Γ_μ^{-1}. Properties (3) and (4) directly follow from (1) and (2), respectively. For property (5), we have $\Gamma_{\mu\cup\nu}(x_i) = \vee\{(\mu(x)\vee \nu(x)) \wedge \Gamma_y(x) \mid y \in V\} = (\vee\{\mu(x) \wedge \Gamma_y(x) \mid y \in V\}) \vee (\vee\{\nu(x) \wedge \Gamma_y(x) \mid y \in V\}) = \Gamma_\mu(x) \vee \Gamma_\nu(x) = (\Gamma_\mu \cup \Gamma_\nu)(x)$.

The property (6) is shown in the same way as (5). ■

Let $G = (V, \Gamma)$ be a fuzzy directed graph. If we were to write $\Gamma(x, y)$ for $\Gamma_x(y)$ for all $x, y \in V$, then we could consider Γ as a fuzzy subset of $V \times V$, that is, a fuzzy relation on V. With this interpretation we have the following definition.

Definition 3.16 *Let $G = (V, \Gamma)$ be a fuzzy directed graph. Let Γ^{-1} be the fuzzy subset of $V \times V$ defined by $\Gamma^{-1}(y, x) = \Gamma(x, y)$ $\forall (x, y) \in V \times V$. Let Δ be the fuzzy subset of $V \times V$ defined by $\Delta = \Gamma \cup \Gamma^{-1}$. Let $\hat{\Gamma}$, $\hat{\Gamma}^{-1}$ and $\hat{\Delta}$ denote the transitive closures of Γ, Γ^{-1} and Δ, respectively.*

If A and B are functions of V into $\mathfrak{F}\wp(V)$, then we define the function $A \cup B$ of V into $\mathfrak{F}\wp(V)$ by $\forall x \in V, (A \cup B)(x) = A(x) \cup B(x)$. Then $\forall x, y \in V, ((A \cup B)(x))(y) = A(x)(y) \vee B(x)(y)$. It is now easy to see that $\forall x \in V$,

$$\hat{\Gamma}_x = \Gamma_x^0 \cup \Gamma_x \cup \Gamma_x^2 \cup \ldots \cup \Gamma_x^{n-1} \tag{3.4}$$

and

$$\hat{\Gamma}_x^{-1} = \Gamma_x^0 \cup \Gamma_x^{-1} \cup \Gamma_x^{-2} \cup \ldots \cup \Gamma_x^{-n+1} \quad \text{for } x \in V,$$

where $\forall x, y \in V, \Gamma_x^0(x, y) = 0$ if $x \neq y$ and $\Gamma_x^0(x, y) = 1$ if $x = y$.

The grades of membership $\hat{\Gamma}_x(y)$ and $\hat{\Gamma}_y^{-1}(x)$ may be interpreted as the degree of the existence of a directed path from x to y and that from y to x, respectively. Similarly, we have $\forall x \in V$,

$$\hat{\Delta}_x = \Delta_x^0 \cup \Delta_x \cup \Delta_x^2 \cup \ldots \cup \Delta_x^{n-1} \tag{3.5}$$

The value of $\hat{\Delta}_x(y)$ may be interpreted as the degree for two points x and y to be joined by a semipath, that is, an alternating sequence $v_0, e_0, v_1, \ldots, e_n, v_n$ of vertices v_{i-1} and directed edges e_i, where each edge e_i may be either $< v_{i-1}, v_i >$ or $< v_i, v_{i-1} >$.

Definition 3.17 *The grades of membership of a fuzzy directed graph $G = (V, \Gamma)$ in U_3, U_2, U_1, and U_0 are defined by*

$$\begin{cases} \mu_{U_3}(G) = & \wedge \{ \hat{\Gamma}_{x_i}(x_j) \mid i,j = 1,\ldots,n\}, \\ \mu_{U_2}(G) = & \wedge \{\hat{\Gamma}_{x_i}(x_j) \vee \hat{\Gamma}_{x_j}(x_i) \mid i,j = 1,\ldots,n\}, \\ \mu_{U_1}(G) = & \wedge \{ \hat{\Delta}_{x_i}(x_j) \mid i,j = 1,\ldots,n\}, \\ \mu_{U_0}(G) = 1 - & \wedge \{ \hat{\Delta}_{x_i}(x_j) \mid i,j = 1,\ldots,n\}, \end{cases}$$

respectively.

It follows that

$$\mu_{U_3}(G) \le \mu_{U_2}(G) \le \mu_{U_1}(G) \quad \text{for any } G = (V, \Gamma). \tag{3.6}$$

Specifically, we can see that for any digraph G in C_i, $\mu_{U_j}(G) = 0$ for $3 \ge j > i$; $\mu_{U_j}(G) = 1$ for $i \ge j \ge 1$, where we recall that $C_3 = U_3, C_2 = U_2 \setminus U_3, C_1 = U_1 \setminus U_2$ and $C_0 = U_0$.

We now present two simple examples. Let $V = \{x, y, z\}$ and $\Gamma_x(y) = 0.5$, $\Gamma_y(x) = 0$, $\Gamma_x(z) = 0$, $\Gamma_z(x) = 0$, $\Gamma_y(z) = 0$ and $\Gamma_z(y) = 0.25$. Then $\hat{\Gamma}_x(y) = 0.5$, $\hat{\Gamma}_y(x) = 0$, $\hat{\Gamma}_x(z) = 0$, $\hat{\Gamma}_z(x) = 0$, $\hat{\Gamma}_y(z) = 0$ and $\hat{\Gamma}_z(y) = 0.25$. Hence $\mu_{U_3}(G) = 0, \mu_{U_2}(G) = 0$, and $\mu_{U_1}(G) = 0.25$. Note that $G = (V, \Gamma)$ is in C_1.

Now let $V = \{x, y, z\}$ and Γ be defined as before except that $\Gamma_z(z) = 0.25$ and $\Gamma_z(y) = 0$. Then $\hat{\Gamma}$ is as above except that $\hat{\Gamma}_x(z) = 0.25$, $\hat{\Gamma}_y(z) = 0.25$, and $\hat{\Gamma}_z(y) = 0$. We have that $G = (V, \Gamma)$ is in C_2 and that $\mu_{U_3}(G) = 0, \mu_{U_2}(G) = 0.25$, and $\mu_{U_1}(G) = 0.25$.

Weakening and Strengthening Points of a Fuzzy Directed Graph

In this section, we define weakening and strengthening points of a fuzzy directed graph as a natural extension of those of an ordinary digraph. We then investigate their fundamental properties.

Definition 3.18 *For a fuzzy directed graph $G = (V, \Gamma)$, let G_k be the fuzzy directed subgraph $(V \setminus \{x_k\}, \Gamma')$ obtained from G by the removal of a point*

FIGURE 3.9 A point of the type (W_1, S_2, N_3).

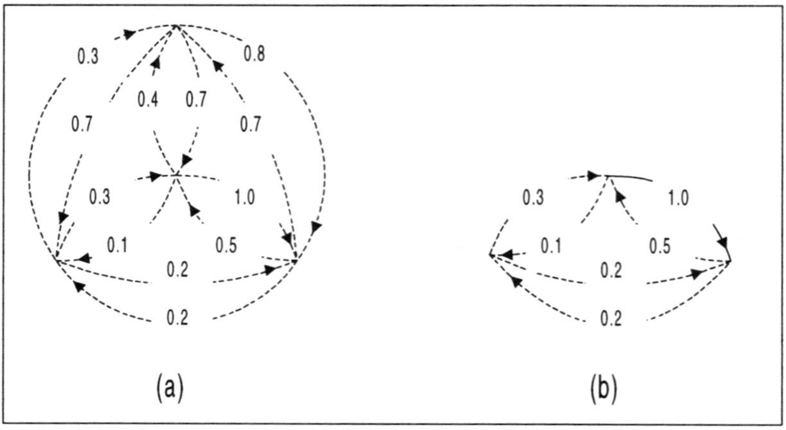

x_k. Then, the point x_k is a weakening point for U_i (a W_i point, for short) if $\mu_{U_i}(G) < \mu_{U_i}(G_k)$; it is a neutral point for U_i (an N_i point) if $\mu_{U_i}(G) = \mu_{U_i}(G_k)$; and it is a strengthening point for U_i (an S_i point) if $\mu_{U_i}(G) > \mu_{U_i}(G_k)$, where $i = 1, 2, 3$.

For instance, a point x_k, as shown in Figure 3.9, is a weakening point for U_1 because the grade of membership in U_1 of the fuzzy directed subgraph G_k is greater than that of G. In the similar way, it is also an S_2 point and an N_3 point, so we say x_i is a point of the type (W_1, S_2, N_3).

In what follows, for brevity of notation, let

$$p_{ij} = \hat{\Gamma}_{x_i}(x_j), \quad i,\ j = 1, 2, \ldots, n, \tag{3.7}$$

$$q_{ij} = \hat{\Delta}_{x_i}(x_j), \quad i,\ j = 1, 2, \ldots, n, \tag{3.8}$$

and

$$r_{ij} = \hat{\Gamma}'_{x_i}(x_j), \quad i,\ j \neq k; \quad i,\ j = 1, 2, \ldots, n, \tag{3.9}$$

where $\hat{\Gamma}$ and $\hat{\Delta}$ are respectively as defined in (3.4) and (3.5), and $\hat{\Gamma}'$ is the transitive closure of Γ'.

Let P and Q denote respectively $n \times n$ matrices with elements p_{ij} and q_{ij} and let R be an $n \times n$ matrix, whose elements in the k-th row and in the k-th column are zeros and each $(i,\ j)$ element is r_{ij}, where $i,\ j \neq k;\quad i,\ j = 1, 2, \ldots, n$.

The next lemma serves to characterize weakening points for each connectedness category.

Lemma 3.43 *(1) A point x_k is a W_3 point if and only if the elements of P which are equal to $\mu_{U_3}(G)$ are all in the k-th row or in the k-th column of P.*

(2) A point x_k is a W_2 point if and only if any $(i,\ j)$ elements of P such that $p_{ij} \vee p_{ji} = \mu_{U_2}(G)$ are in the k-th row and in the k-th column of P.

(3) A point x_k is a W_1 point if and only if all the elements of Q which are equal to $\mu_{U_1}(G)$ are in the k-th row and in the k-th column of Q.

Proof. (1) Let x_k be a W_3 point. Suppose that there exists an element, say an (l, m) element, $l, m \neq k$, which is equal to $\mu_{U_3}(G)$. Since

$$r_{ij} \leq p_{ij}, \quad i,\ j \neq k;\ i,\ j = 1, 2, ..., n,$$

we have

$$\mu_{U_3}(G_k) = \wedge\{r_{ij} \mid i,\ j = 1, 2, \ldots, n, i \neq k \neq j\} \leq p_{lm} = \mu_{U_3}(G),$$

which contradicts the assumption that x_k is a W_3 point. Therefore, every element of P which is equal to $\mu_{U_3}(G)$ is in the k-th row or in the k-th column of P.

Conversely, assume that the elements of P which are equal to $\mu_{U_3}(G)$ are all in the k-th row or in the k-th column of P. First, notice that if an element which is equal to $\mu_{U_3}(G)$ is in the k-th row (column) of P, then every non-diagonal element in the k-th row (column) of P is equal to $\mu_{U_3}(G)$. Hence

$$p_{ik} \wedge p_{kj} = \mu_{U_3}(G) < p_{ij}, \quad i,\ j \neq k; i,\ j = 1, 2, ..., n,$$

which yields

$$r_{ij} = p_{ij} > \mu_{U_3}(G), \quad i,\ j \neq k;\ i,\ j = 1, 2, ..., n.$$

Therefore

$$\mu_{U_3}(G_k) = \ \wedge\{r_{ij} \mid i,\ j = 1, 2, \ldots, n, i \neq k \neq j\} > \mu_{U_3}(G),$$

so that x_k is a W_3 point, which completes the proof of (i).

The proofs of (ii) and (iii) are similar to that of (i). ■

The following theorem is an immediate consequence of Lemma 3.43.

Theorem 3.44 *There exist at most two W_i points in any fuzzy directed graph, where $i = 1, 2, 3$. Further, any fuzzy directed graph with n $(n \geq 3)$ points has at most one W_1 (W_3) point.* ■

Before proving the following lemma, we recall that an *oriented graph* is a digraph having no symmetric pair of directed lines and a *tournament* is an oriented complete graph. A *Hamiltonian path* in a digraph D is a spanning path in D.

Lemma 3.45 *For any fuzzy directed graph $G = (V, \Gamma)$, there exists a path $\{x_{i_1}, x_{i_2}, ..., x_{i_s}\}$ $(s \geq n)$ such that:*

(1) every point of V appears in the path;

(2) $\Gamma_{x_{i_l}}(x_{i_{l+1}}) \geq \mu_{U_2}(G)$, $\quad l = 1, 2, ..., s-1$.

Proof. The result is trivial if $\mu_{U_2}(G) = 0$. Hence we assume $\mu_{U_2}(G) > 0$. Let us construct an ordinary digraph $G' = (V, \Gamma'')$ from G as follows: For $i,\ j = 1, 2, ..., n$,

$$\Gamma''_{x_i}(x_j) = \begin{cases} 1 & \text{if } p_{ij} \geq \mu_{U_2}(G), \\ 0 & \text{if } p_{ij} < \mu_{U_2}(G). \end{cases}$$

Since $p_{ij} \vee p_{ji} \geq \mu_{U_2}(G)$, G includes a tournament as a partial graph of G. Since every tournament has a Hamiltonian path, G has a Hamiltonian path. On the other hand, we can easily see from Definition 3.16 that if $p_{ij} \geq \mu_{U_2}(G)$, then there exists at least a path $\{x_i, x_u, ..., x_v, x_j\}$ such that

$$\begin{aligned} \Gamma_{x_i}(x_u) &\geq \mu_{U_2}(G), \\ &\vdots \\ \Gamma_{x_v}(x_j) &\geq \mu_{U_2}(G). \end{aligned}$$

■

The following theorem shows that in any fuzzy graph with n $(n \geq 2)$ points, it is impossible for all points to be strengthening ones for U_2 (U_1).

Theorem 3.46 *In any fuzzy directed graph G with n $(n \geq 2)$ points, there exist at least two points which are either weakening or neutral ones for U_2 (U_1).*

Proof. Let a path $\{x_{i_1}, x_{i_2}, ..., x_{i_s}\}$ satisfy (1) and (2) of Lemma 3.45. Without loss of generality, we can assume that the initial and final points x_{i_1} and x_{i_s} appear exactly once in the path. For, if the initial point (the final point) appears more than once in the path, we can delete the first point (the last point) of the path, so that the remaining path also meets the requirements (1) and (2).

Now, according to the above assumption, a path $\{x_{i_2}, x_{i_3}, ..., x_{i_s}\}$ and a path $\{x_{i_1}, x_{i_2}, ..., x_{i_{s-1}}\}$ contain respectively all points in $V \setminus \{x_{i_1}\}$ and all points in $V \setminus \{x_{i_s}\}$. Therefore

$$\mu_{U_2}(G_{i_1}) \geq \mu_{U_2}(G),$$

and

$$\mu_{U_2}(G_{i_s}) \geq \mu_{U_2}(G).$$

Thus, each of x_{i_1} and x_{i_s} is either a W_2 or N_2 point, which completes the proof for U_2. The proof for U_1 is similar. ■

Corollary 3.47 *Any fuzzy directed graph with n $(n \geq 3)$ points has at least one N_1 point.*

Proof. It is an immediate consequence of Theorems 3.44 and 3.46. ■

Theorem 3.48 *If a fuzzy directed graph G with n $(n \geq 3)$ points has two W_2 points, then*

$$\mu_{U_2}(G) < \mu_{U_1}(G).$$

Proof. Let x_k and x_l be W_2 points, i.e.,

$$\mu_{U_2}(G_k) > \mu_{U_2}(G), \tag{3.10}$$

and

$$\mu_{U_2}(G_l) > \mu_{U_2}(G).$$

Suppose that

$$\mu_{U_2}(G) = \mu_{U_1}(G). \tag{3.11}$$

From (3.6) and (3.10) through (3.11), we find that $\mu_{U_1}(G) < \mu_{U_2}(G_k) \leq \mu_{U_1}(G_k)$ and $\mu_{U_1}(G) < \mu_{U_2}(G_l) \leq \mu_{U_1}(G_l)$. Thus x_k and x_l must be W_1 points, which contradicts Theorem 3.44. Hence $\mu_{U_2}(G) < \mu_{U_1}(G)$. ■

Theorem 3.49 *Any W_3 point is either a W_2 one or an N_2 one.*

Proof. Let x_k be a W_3 point. From the proof of Lemma 3.43, we get

$$r_{ij} = p_{ij}, \quad i, \ j = 1, 2, ..., n.$$

Therefore,

$$\mu_{U_2}(G_k) \geq \mu_{U_2}(G).$$

■

The following theorem directly follows from Definition 3.18 and (3.6).

Theorem 3.50 *If $\mu_{U_i}(G) = \mu_{U_j}(G)$ for some $i < j$, then an S_i point is also an S_j point.* ■

Theorem 3.51 *If $\mu_{U_i}(G) = \mu_{U_j}(G)$ for some $i > j$, then a W_i point is also a W_l point, where $1 \leq l \leq i$.*

Proof. Let $i = 3$ and $j = 1, i.e.,$

$$\mu_{U_3}(G) = \mu_{U_1}(G). \tag{3.12}$$

Let x_k be a W_3 point. From (3.6) and (3.12) we obtain, $\mu_{U_3}(G_k) \leq \mu_{U_2}(G_k)$ and $\mu_{U_1}(G) = \mu_{U_2}(G) = \mu_{U_3}(G_k)$. Thus

$$\mu_{U_2}(G_k) > \mu_{U_2}(G),$$

and

$$\mu_{U_1}(G_k) > \mu_{U_1}(G).$$

Thus, x_k is a W_l point, where $1 \leq l \leq 3$.

Next, assume that x_k is a W_2 point and that $\mu_{U_2}(G) = \mu_{U_1}(G)$. It follows that

$$\mu_{U_1}(G_k) > \mu_{U_1}(G).$$

Thus, x_k is a W_l point, where $1 \leq l \leq 2$.

Finally, we shall prove that if x_k is a W_3 point and $\mu_{U_3}(G) = \mu_{U_2}(G)$ then it is a W_l point, where $1 \leq l \leq 3$. Since it is obvious that x_k is a W_2 point, it suffices to show that x_k is a W_l point. Using Lemma 3.43, it follows that both in the k-th and in the k-th column of p there exists an element which is equal to $\mu_{U_3}(G)$. Hence we get from the proof of Lemma 3.43

$$p_{kj} = p_{jk} = \mu_{U_3}(G), \quad j \neq k;\ j = 1, 2, ..., n.$$

Thus we have

$$\Gamma_{x_k} \cup \Gamma_{x_k}^{-1}(x_j) \leq \mu_{U_3}(G), \quad j \neq k;\ j = 1, 2, ..., n,$$

which yields

$$q_{kj} = q_{jk} \leq \mu_{U_3}(G), \quad j \neq k;\ j = 1, 2, ..., n.$$

Therefore we get

$$\mu_{U_1}(G) = \mu_{U_3}(G),$$

so that

$$\mu_{U_1}(G_k) > \mu_{U_1}(G).$$

■

Theorem 3.52 *Let x_k be a W_i point. If $\mu_{U_i}(G_k) = \mu_{U_j}(G_k)$ for some $i < j$, then x_k is also a W_l point, where $1 \leq l \leq j$.*

Proof. The proof of this theorem is similar to that of Theorem 3.51. ■

In closing, we shall show how results of Ross and Harary can be obtained from our results as the special cases. First, note that, in the case of the ordinary digraph G^d, $\mu_{U_i}(G_k^d) > \mu_{U_i}(G^d)$ if and only if $\mu_{U_i}(G_k^d) = 1$ and $\mu_{U_i}(G^d) = 0$, that is, $G_k^d \in U_i$ and $G^d \notin U_i$. With the understanding that a weakening point for U_0 is one whose presence makes its fuzzy graph more highly disconnected than it would be without the point, the W_0 point is defined to be the W_1 point. We can easily see from Theorem 3.44 that any digraph has at most two weakening points. And, from Theorem 3.51, we can find that there are no $(1, 3)$ points in any digraph.

3.8 References

1. Bellman, R.E., and Zadeh, L.A., *Mgmt Sci.,* Vol. 17, No. 4, 1970.
2. Bezdek, J.C. and Harris, J.D., Fuzzy partitions and relations an axiomatic basis for clustering, *Fuzzy Sets and Systems* **1:**111–127, 1978.
3. Bhattacharya, P., Some remarks on fuzzy graphs, *Pattern Recognition Letters* **6:**297–302, 1987.
4. Bhattacharya, P., and Suraweera, F, An algorithm to compute the supremum of max-min powers and a property of fuzzy graphs, *Pattern Recognition Letters* **12:**413–420, 1991.

5. Delgado, M. and Verdegay, J.L., On valuation and optimization problems in fuzzy graphs: A general approach and some particular cases, *ORSA J. on Computing* **2:**74–83, 1990.

6. Ding, B., A clustering dynamic state method for maximal trees in fuzzy graph theory, *J. Numer. Methods Comput. Appl.* **13:**157–160, 1992.

7. Dunn, J.C., A graph theoretic analysis of pattern classification via Tamura's fuzzy relation, *IEEE Trans. on Systems, Man, and Cybernetics* 310–313, 1974.

8. Harary, F., Graph Theoretic Methods in the Management Sciences, *Management Science,* **5:**387-403, 1959.

9. Harary, F., and R.Z. Norman, *Graph Theory as a Mathematical Model in Social Science*, Ann Arbor, Mich.: Institute for Social Research, 1953.

10. Harary, F., R.Z. Norman and Cartwright, D., *Structural Models: An Introduction to the Theory of Directed Graphs,* John Wiley & Sons, Inc., New York, 1965.

11. Harary, F., *Graph Theory*, Addison Wesley, Third printing, October 1972.

12. Harary, F., and Ross, I.C., The Number of Complete Cycles in a Communication Network, *Journal of Social Psychology*, **40**:329-332, 1953.

13. Harary, F., and Ross, I.C., A Procedure for Clique Detention using the Group Matrix, *Sociometry*, **20**:205-215, 1957.

14. Kaufmann, A., Introduction a la Theorie des sons-ensembles flous, Vol. 1, Masson Paris, 41–189, 1973.

15. Kaufmann, A., *Introduction to the Theory of Fuzzy Subsets, Vol. 1,* Academic Press, New York, 1975.

16. Kiss, A., An application of fuzzy graphs in database theory, Automata, languages and programming systems (Salgotarjan 1990) *Pure Math, Appl. Ser. A,* **1**:337 - 342, 1991.

17. Kóczy, L.T., Fuzzy graphs in the evaluation and optimization of networks, *Fuzzy Sets and Systems* **46:**307–319, 1992.

18. Leenders, J.H., Some remarks on an article by Raymond T. Yeh and S.Y. Bang dealing with fuzzy relations: Fuzzy relations, fuzzy graphs, and their applications to clustering analysis, Fuzzy sets and their applications to cognitive and decision processes (Proc. U.S.-Japan Sem.,

Univ. Calif., Berkeley, Calif., 1974), 125–149, *Simon Stevin* **51:**93–100, 1977/78.

19. Ling, R.F., On the theory and construction of k-cluster, *The Computer J.* **15:**326–332, 1972.

20. Liu, W-J., On some systems of simultaneous equations in a completely distributive lattice, *Inform. Sci.* **50**:185–196, 1990.

21. Matula, D.W., Cluster analysis via graph theoretic techniques, *Proc. of Lousiana Conf. on Combinatrics, Graph Theory, and Computing,* 199–212, March 1970.

22. Matula, D.W., k-components, clusters, and slicings in graphs, *SIAM J. Appl. Math.* **22:**459–480, 1972.

23. Mordeson, J.N. and Peng, C-S, Fuzzy intersection equations, *Fuzzy Sets and Systems* **60:**77–81, 1993.

24. Mordeson, J.N. and Peng, C-S, Operations on fuzzy graphs, *Inform. Sci.* **79:**159–170, 1994.

25. Mori, M. and Kawahara, Y., Fuzzy graph rewritings, *Theory of rewriting systems and its applications (Japanese)* **918:**65–71, 1995.

26. Morioka, M., Yamashita, H., and Takizawa, T., Extraction method of the difference between fuzzy graphs, *Fuzzy information, knowledge representation and decision analysis* (Marseille, 1983), 439–444, *IFAC Proc. Ser.,* **6,** IFAC, Lexenburg, 1984.

27. Nance, R.E., Korfhage, R.R., and Bhat, U.N., Information networks: Definitions and message transfer models, *Tech. Report CP-710011,* Computer Science/Operations Research Center, SMU, Dallas, Texas, July 1971.

28. Ramamoorthy, C.V., Analysis of graphs by connectivity considerations, *JACM,* **13:**211–222, 1966.

29. Rosenblatt, D., On Linear Models and the Graphs of Minkowski - Leontief Matrices, *Econometrica,* **25**:325-338, 1957.

30. Rosenfeld, A., Fuzzy graphs, In: L. A. Zadeh, K. S. Fu, M. Shimura, Eds., *Fuzzy Sets and Their Applications*, 77–95, Academic Press, 1975.

31. Ross, I.C., and Harary, F., On the Determination of Redundancies in Sociometric Chains, *Psychometrika,* **17**:195-208, 1952.

32. Ross, I.C., and Harary, F., Identification of the Liaison Persons of an Organization using the Structure Matrix, *Management Science,* **1**:251-258, 1955.

33. Ross, I.C., and Harary, F., A Description of Strengthening and Weakening Members of a Group, *Sociometry,* **22**:139-147, 1959.

34. Sibson, R., Some observation on a paper by Lance and Williams, *The Computer J.* **14:**156–157, 1971.

35. Sunouchi, H. and Morioka, M., Some properties on the connectivity of a fuzzy graph (Japanese), *Bull. Sci. Engrg. Res. lab. Waseda Univ.* no. 132, 70–78, 1991.

36. Takeda, E., Connectvity in fuzzy graphs, *Tech. Rep. Osaka Univ.* **23:**343–352, 1973.

37. Takeda, E. and Nishida, T., An application of fuzzy graph to the problem concerning group structure, *J. Operations Res. Soc. Japan* **19:**217–227, 1976.

38. Tong, Z. and Zheng, D., An algorithm for finding the connectedness matrix of a fuzzy graph, *Congr. Numer.* **120:**189–192, 1996.

39. Ullman, J. D., *Principles of Database and Knowledge-base Systems,* Vol 1–2, Computer Science Press, Rockville, MD., 1989.

40. Wu, L. G. and Chen, T.P., Some problems concerning fuzzy graphs (Chinese), *J. Huazhong Inst. Tech.* no 2, Special issue on fuzzy math, iv, 58–60, 1980.

41. Xu, J., The use of fuzzy graphs in chemical structure research, In: D.H. Rouvry, Ed., *Fuzzy Logic in Chemistry,* 249–282, Academic Press, 1997.

42. Yamashita, H., Approximation algorithm for a fuzzy graph (Japanese), *Bull. Centre Inform.* **2:**59–60, 1985.

43. Yamashita, H., Structure analysis of fuzzy graph and its application (Japanese), *Bull. Sci. Engrg. Res. Lab. Waseda Univ.* no. 132, 61–69, 1991.

44. Yamashita, H. and Morioka, M., On the global structure of a fuzzy graph, *Analysis of Fuzzy Information,* **1:**167–176, CRC, Boca Raton, Fla., 1987.

45. Yeh, R.T. and Bang, S.Y., Fuzzy relations, fuzzy graphs, and their applications to clustering analysis, In: L. A. Zadeh, K. S. Fu, M. Shimura, Eds., *Fuzzy Sets and Their Applications*, 125–149, Academic Press, 1975.

46. Zadeh, L.A., Fuzzy Sets, *Information and Control,* **8**:338-353, 1965.

47. Zadeh, L.A., Similarity relations and fuzzy orderings, *Information Sciences,* **3**:177-200, 1971.

48. Zhu, R.Y., The critical number of the connectivity degree of a fuzzy graph (Chineses), *Fuzzy Math.* **2:**113–116, 1982.

49. Zykov, A.A., On some properties of linear complexes (Russian), *Mat. Sbornik* **24**:163–188, 1949, *Amer. Math. Soc. Translations* N. **79**, 1952.

4 FUZZY HYPERGRAPHS

Graph theory has found many application areà in science, engineering, and mathematics. In order to expand the application base, the notion of a graph was generalized to that of a hypergraph, that is, a set X of vertices together with a collection of subsets of X. In this chapter, we fuzzify the notion of a hypergraph and state some possible applications. The results are taken from [9,10,11,12,22].

The degrees of membership in an edge may vary; this feature is essential in the latter part of this section where fuzzy hypergraphs are used to interpret ideas of Hebb on cortical development.

4.1 Fuzzy Hypergraphs

A (crisp) *hypergraph* on a set X is a pair $H = (X, \mathbf{E})$ where X is a finite set and $\mathbf{E}$ is a finite family of nonempty subsets of X such that $\forall x \in X, \exists E \in \mathbf{E}$ such that $x \in E$. We call X the *vertex set* and $\mathbf{E}$ the *edge set* of H. Repeated or multiple edges are allowed. We use $\mathbf{E}$ and H interchangeably to designate a hypergraph with the understanding that if only the edge set $\mathbf{E}$ is specified, then $X = \cup\{E \mid E \in \mathbf{E}\}$. A hypergraph $H = (X, \mathbf{E})$ is called *simple* if $\mathbf{E}$ contains no repeated edges and whenever $E, F \in \mathbf{E}$ and $E \subseteq F$, then $E = F$. We sometimes denote the vertex set of H by $\mathbf{V}(H)$.

Definition 4.1 *Let X be a finite set and let $\mathcal{E}$ be a finite family of non-trivial fuzzy subsets of X such that*

$$X = \underset{\mu\in\mathcal{E}}{\cup} \operatorname{supp}(\mu). \tag{4.1}$$

The pair $\mathcal{H} = (X, \mathcal{E})$ is called a fuzzy hypergraph *(on X) and $\mathcal{E}$ is called the* edge set *of $\mathcal{H}$ which is sometimes denoted $\mathcal{E}(\mathcal{H})$. The members of $\mathcal{E}$ are called the* fuzzy edges *of $\mathcal{H}$. The* height *of $\mathcal{H}$, $h(\mathcal{H})$, is defined by $h(\mathcal{H}) = \vee \{h(\mu) \mid \mu \in \mathcal{E}\}$, where we recall that $h(\mu)$ denotes the height of μ.*

We use $\mathcal{E}$ and $\mathcal{H}$ interchangeably to designate a fuzzy hypergraph with the understanding that the vertex set X always satisfies (4.1).

Definition 4.2 *A fuzzy hypergraph $\mathcal{H} = (X, \mathcal{E})$ is called* simple *if $\mathcal{E}$ has no repeated fuzzy edges and whenever $\mu, \nu \in \mathcal{E}$ and $\mu \subseteq \nu$, then $\mu = \nu$.*

Definition 4.3 *A fuzzy hypergraph $\mathcal{H} = (X, \mathcal{E})$ is called* support simple, *if, whenever $\mu, \nu \in \mathcal{E}$, $\mu \subseteq \nu$ and $supp(\mu) = supp(\nu)$, then $\mu = \nu$.*

Definition 4.4 *Let $\sigma \in \mathfrak{F}\wp(X)$. If $|\sigma(supp(\sigma))| = 1$, then σ is called* elementary *on X. If σ is elementary on X, we sometimes write $\sigma(A, r)$ where $A = supp(\sigma)$ and $r = h(\sigma)$ is the constant value assumed by σ on A. If $|supp(\sigma)| = 1$, σ is called a* spike. *An* elementary fuzzy hypergraph *$\mathcal{H} = (X, \mathcal{E})$ is a fuzzy hypergraph where all fuzzy edges are elementary.*

Definition 4.5 *Let $\mathcal{H} = (X, \mathcal{E})$ be a fuzzy hypergraph. Suppose that $t \in [0, 1]$. Let*

$$\mathbf{E}^t = \{\mu^t \neq \emptyset \mid \mu \in \mathcal{E}\} \text{ and } X^t = \underset{\mu\in\mathcal{E}}{\cup} \mu^t. \tag{4.2}$$

If $\mathbf{E}^t \neq \emptyset$, then the (crisp) hypergraph

$$H^t = (X^t, \mathbf{E}^t) \tag{4.3}$$

is the t-level hypergraph *of $\mathcal{H}$.*

Clearly, it is possible that $\mu^t = \nu^t$ for $\mu \neq \nu$; by using distinct markers to identify the various members of $\mathcal{E}$ a distinction between μ^t and ν^t to represent multiple edges in H^t. However, we do not take this approach; unless otherwise stated, we will always regard H^t as having no repeated edges.

The families of crisp sets (hypergraphs) produced by the t-cuts of a fuzzy hypergraph share an important relationship with each other, as expressed below. Suppose $\mathcal{A}$ and $\mathcal{B}$ are two families of sets such that for each set A belonging to $\mathcal{A}$ there is at least one set B belonging to $\mathcal{B}$ which contains A.

In this case we say $\mathcal{B}$ *absorbs* $\mathcal{A}$ and symbolically write $\mathcal{A} \sqsubseteq \mathcal{B}$ to express this relationship between $\mathcal{A}$ and $\mathcal{B}$. Since it is possible for $\mathcal{A} \sqsubseteq \mathcal{B}$ while $\mathcal{A} \cap \mathcal{B} = \emptyset$, we have that $\mathcal{A} \subseteq \mathcal{B} \Rightarrow \mathcal{A} \sqsubseteq \mathcal{B}$, whereas the converse is generally false. If $\mathcal{A} \sqsubseteq \mathcal{B}$ and $\mathcal{A} \neq \mathcal{B}$, then we write $\mathcal{A} \sqsubset \mathcal{B}$.

Definition 4.6 *Let $\mathcal{H} = (X, \mathcal{E})$ be a fuzzy hypergraph, and for $0 < t \leq h(\mathcal{H})$, let $H^t = (X^t, \mathbf{E}^t)$ be the t-level hypergraph of $\mathcal{H}$. The sequence of real numbers $\{r_1, r_2, \ldots, r_n\}$, $0 < r_n < \cdots < r_1 = h(\mathcal{H})$, which satisfies the properties*

(1) if $r_{i+1} < s \leq r_i$, then $\mathbf{E}^s = \mathbf{E}^{r_i}$, and

(2) $\mathbf{E}^{r_i} \sqsubset \mathbf{E}^{r_{i+1}}$,

is called the fundamental sequence *of $\mathcal{H}$, and is denoted by $\mathbf{F}(\mathcal{H})$ and the set of r_i-level hypergraphs $\{H^{r_1}, H^{r_2}, \ldots, H^{r_n}\}$ is called the set of* core hypergraphs *of $\mathcal{H}$ or, simply, the* core set *of $\mathcal{H}$, and is denoted by $\mathbf{C}(\mathcal{H})$.*

If $r_1 < s \leq 1$ in Definition 4.6, then $\mathbf{E}^s = \{\emptyset\}$ and H^s does not exist. For simplicity, whenever there is no confusion, we shall use H_i to denote the r_i-level hypergraph H^{r_i}. Further, X_i and $\mathbf{E}_i$ shall usually denote the vertex and edge set of the core hypergraph H^{r_i}. Thus, $H^{r_i} = (X^{r_i}, \mathbf{E}^{r_i}) = H_i = (X_i, \mathbf{E}_i), i = 1, \ldots, n$.

Definition 4.7 *Suppose $\mathcal{H} = (X, \mathcal{E})$ is a fuzzy hypergraph and $\mathbf{F}(\mathcal{H}) = \{r_1, r_2, \ldots, r_n\}$. Then $\mathcal{H}$ is called* sectionally elementary *if for each $\mu \in \mathcal{E}$ and each $r_i \in \mathbf{F}(\mathcal{H})$, $\mu^t = \mu^{r_i}$ for all $t \in (r_{i+1}, r_i]$. (We assume $r_{n+1} = 0$.)*

Definition 4.8 *Suppose $H = (X, \mathbf{E})$ and $H' = (X', \mathbf{E}')$ are (crisp) hypergraphs. H is called a* partial hypergraph *of H' if $\mathbf{E} \subseteq \mathbf{E}'$. If H is a partial hypergraph of H', we write $H \subseteq H'$. If $H \subseteq H'$ and $\mathbf{E} \subset \mathbf{E}'$, we write $H \subset H'$.*

Definition 4.9 *Suppose $\mathcal{H} = (X, \mathcal{E})$ and $\mathcal{H}' = (X', \mathcal{E}')$ are fuzzy hypergraphs. $\mathcal{H}$ is called a* partial fuzzy hypergraph *of $\mathcal{H}'$ if $\mathcal{E} \subseteq \mathcal{E}'$. If $\mathcal{H}$ is a partial hypergraph of $\mathcal{H}'$, we write $\mathcal{H} \subseteq \mathcal{H}'$. If $\mathcal{H} \subseteq \mathcal{H}'$ and $\mathcal{E} \subset \mathcal{E}'$, we write $\mathcal{H} \subset \mathcal{H}'$.*

We now illustrate some of the above definitions.

Example 4.1 *Consider the fuzzy hypergraph $\mathcal{H} = (X, \mathcal{E})$, where $X = \{a, b, c, d\}$ and $\mathcal{E} = \{\mu_1, \mu_2, \mu_3, \mu_4, \mu_5\}$, which is represented by the following incidence matrix:*

$$\begin{array}{c} \\ a \\ b \\ c \\ d \end{array} \begin{array}{c} \begin{array}{ccccc} \mu_1 & \mu_2 & \mu_3 & \mu_4 & \mu_5 \end{array} \\ \left(\begin{array}{ccccc} 0.7 & 0.9 & 0 & 0 & 0.4 \\ 0.7 & 0.9 & 0.9 & 0.7 & 0 \\ 0 & 0 & 0.9 & 0.7 & 0.4 \\ 0 & 0.4 & 0 & 0.4 & 0.4 \end{array} \right) \end{array}.$$

From this matrix we understand, for example, that $\mu_2 : X \to [0,1]$ *satisfies* $\mu_2(a) = 0.9,\ \mu_2(b) = 0.9,\ \ \mu_2(c) = 0,\ \ \mu_2(d) = 0.4$. *Clearly* $h(\mathcal{H}) = 0.9$ *and so* $r_1 = 0.9$. *Now*

$\mathbf{E}^{0.9} = \{\{a,b\},\ \{b,c\}\} = \mathbf{E}^{0.7}$

and

$\mathbf{E}^{0.4} = \{\{a,b\},\ \{a,b,d\},\ \{b,c\},\ \{b,c,d\},\ \{a,c,d\}\}$.

Thus for $0.4 < t \le 0.9, \mathbf{E}^t = \{\{a,b\},\ \{b,c\}\}$ *and for* $0 < t \le 0.4, \mathbf{E}^t = \{\{a,b\},\ \{a,b,d\},\ \{b,c\},\ \{b,c,d\},\ \{a,c,d\}\}$. *Further note that* $\mathbf{E}^{0.9} \sqsubseteq \mathbf{E}^{0.4}$ *and* $\mathbf{E}^{0.9} \ne \mathbf{E}^{0.4}$. *Therefore the fundamental sequence is* $\mathbf{F}(\mathcal{H}) = \{r_1 = 0.9, r_2 = 0.4\}$ *and the set of core hypergrahs is* $\mathbf{C}(\mathcal{H}) = \{H_1 = (X_1, \mathbf{E}_1) = H^{0.9}, H_2 = (X_2, \mathbf{E}_2) = H^{0.4}\}$ *where*

$X_1 = \{a,b,c\}$,

$\mathbf{E}_1 = \{\{a,b\},\ \{b,c\}\}$,

$X_2 = \{a,b,c,d\}$

and

$\mathbf{E}_2 = \{\{a,b\},\ \{a,b,d\},\ \{b,c\},\ \{b,c,d\},\ \{a,c,d\}\}$.

$\mathcal{H}$ *is support simple, but not simple. Since* $(\mu_1)^t \ne (\mu_1)^{0.9}$, *for* $t = 0.7$, $\mathcal{H}$ *is not sectionally elementary. The following partial fuzzy hypergraphs of* $\mathcal{H}$ *illustrate several of the above definitions:* $\mathcal{E}' = \{\mu_2,\ \mu_3,\ \mu_4,\ \mu_5\}$ *is simple;* $\mathcal{E}'' = \{\mu_2,\ \mu_3,\ \mu_5\}$ *is sectionally elementary, but not elementary and* $\mathcal{E}''' = \{\mu_1,\ \mu_3,\ \mu_5\}$ *is elementary. For* $\mathcal{E}'''$, *we have* $(E''')^{0.9} = \{\{b,c\}\}, (E''')^{0.7} = \{\{a,b\},\{b,c\}\}$ *and* $(E''')^{0.4} = \{\{a,b\},\{b,c\},\{a,c,d\}\}$. *Hence the corresponding fundamental sequence is* 0.9, 0.7, 0.4.

Definition 4.10 *A sequence of crisp hypergraphs* $H_i = (X_i, \mathbf{E}_i)$, $1 \le i \le n$, *is said to be* ordered *if* $H_1 \subset H_2 \subset \cdots \subset H_n$. *The sequence* $\{H_i \mid 1 \le i \le n\}$ *is said to be* simply ordered *if it is ordered, and if whenever* $E \in \mathbf{E}_{i+1} \setminus \mathbf{E}_i$, *then* $E \not\subseteq X_i$.

Definition 4.11 *A fuzzy hypergraph* $\mathcal{H}$ *is said to be* ordered *if* $\mathbf{C}(\mathcal{H})$ *is ordered. That is, if* $\mathbf{C}(\mathcal{H}) = \{H^{r_1}, H^{r_2}, \ldots, H^{r_n}\}$, *then* $H^{r_1} \subset H^{r_2} \subset \cdots \subset H^{r_n}$. *The fuzzy hypergraph* $\mathcal{H}$ *is said to be* simply ordered *if* $\mathbf{C}(\mathcal{H})$ *is simply ordered.*

We note that the fuzzy hypergraph $\mathcal{H}$ given in Example 4.1 is simply ordered.

Proposition 4.1 *If $\mathcal{H} = (X, \mathcal{E})$ is an elementary fuzzy hypergraph, then $\mathcal{H}$ is ordered. Also, if $\mathcal{H} = (X, \mathcal{E})$ is an ordered fuzzy hypergraph with $\mathbf{C}(\mathcal{H}) = \{H^{r_1}, H^{r_2}, \ldots, H^{r_n}\}$ and if H^{r_n} is simple, then $\mathcal{H}$ is elementary.* ■

Consider the situation where the vertex set of a crisp hypergraph is fuzzified. Suppose that each edge is given a uniform degree of membership consistent with the weakest vertex of the edge. Such constructions describe the following subclass of fuzzy hypergraphs.

Definition 4.12 *A fuzzy hypergraph $\mathcal{H} = (X, \mathcal{E})$ is called a μ* tempered fuzzy hypergraph *of H if there is a crisp hypergraph $H = (X, \mathbf{E})$ and a fuzzy subset $\mu : X \to (0, 1]$ such that $\mathcal{E} = \{\nu_E \mid E \in \mathbf{E}\}$, where*

$$\nu_E(x) = \begin{cases} \wedge \{\mu(y) \mid y \in E\} & \text{if } x \in E, \\ 0 & \text{otherwise.} \end{cases}$$

We let $\mu \otimes H$ denote the μ tempered fuzzy hypergraph of H determined by the crisp hypergraph $H = (X, \mathbf{E})$ and the fuzzy subset μ: $X \to (0, 1]$.

Example 4.2 *Consider the fuzzy hypergraph $\mathcal{H} = (X, \mathcal{E})$, where $X = \{a, b, c, d\}$ and $\mathcal{E} = \{\mu_1, \mu_2, \mu_3, \mu_4\}$, which is represented by the following incidence matrix:*

$$\begin{array}{c} \\ a \\ b \\ c \\ d \end{array} \begin{array}{c} \begin{array}{cccc} \mu_1 & \mu_2 & \mu_3 & \mu_4 \end{array} \\ \begin{pmatrix} 0.7 & 0 & 0 & 0.7 \\ 0.7 & 0.4 & 0.9 & 0 \\ 0 & 0 & 0.9 & 0.7 \\ 0 & 0.4 & 0 & 0 \end{pmatrix} \end{array}$$

Then

$\mathbf{E}^{0.9} = \{\{b, c\}\}$,

$\mathbf{E}^{0.7} = \{\{a, b\}, \{a, c\}, \{b, c\}\}$

and

$\mathbf{E}^{0.4} = \{\{a, b\}, \{a, c\}, \{b, c\}, \{b, d\}\}$.

Define $\mu : X \to (0, 1]$ by $\mu(a) = 0.7, \mu(b) = \mu(c) = 0.9$ and $\mu(d) = 0.4$. Note that $\nu_{\{a,b\}}(a) = \mu(a) \wedge \mu(b) = 0.7, \nu_{\{a,b\}}(b) = \mu(a) \wedge \mu(b) = 0.7, \nu_{\{a,b\}}(c) = 0$ and $\nu_{\{a,b\}}(d) = 0$. Thus $\mu_1 = \nu_{\{a,b\}}$. Also $\mu_2 = \nu_{\{b,d\}}$, $\mu_3 = \nu_{\{b,c\}}, \mu_4 = \nu_{\{a,c\}}$. Thus $\mathcal{H}$ is μ tempered.

The fuzzy hypergraphs $\mu \otimes H$ can be characterized as follows (under the restriction that H has no repeated edges).

Theorem 4.2 *A fuzzy hypergraph $\mathcal{H} = (X, \mathcal{E})$ is a μ tempered fuzzy hypergraph of some crisp hypergraph H if and only if $\mathcal{H}$ is elementary, support simple and simply ordered.*

Proof. Suppose $\mathcal{H} = (X, \mathcal{E})$ is a μ tempered fuzzy hypergraph of some crisp hypergraph H. Clearly, $\mathcal{H}$ is elementary and support simple. We show that $\mathcal{H}$ is simply ordered. Let $\mathbf{C}(\mathcal{H}) = \{H^{r_1} = (X_1, \mathbf{E}_1),\ H^{r_2} = (X_2, \mathbf{E}_2), \ldots, H^{r_n} = (X_n, \mathbf{E}_n)\}$. Since $\mathcal{H}$ is elementary, it follows from Proposition 4.1 that $\mathcal{H}$ is ordered. To show that $\mathcal{H}$ is simply ordered, suppose there exists $E \in \mathbf{E}_{i+1} \setminus \mathbf{E}_i$. Then there exists $x^* \in E$ such that $\mu(x^*) = r_{i+1}$. Since $\mu(x^*) = r_{i+1} < r_i$, it follows that $x^* \notin X_i$ and $E \nsubseteq X_i$; hence, $\mathcal{H}$ is simply ordered.

Conversely, suppose $\mathcal{H} = (X, \mathcal{E})$ is elementary, support simple and simply ordered. Given $\mathbf{C}(\mathcal{H})$ as above where it is understood that $\mathbf{F}(\mathcal{H}) = \{r_1, \ldots, r_n\}$ with $0 < r_n < \cdots r_1$. Recall that $H^{r_n} = H_n = (X_n, \mathbf{E}_n)$ and define $\mu : X_n \to (0, 1]$ by

$$\mu(x) = \begin{cases} r_1 & \text{if } x \in X_1 \\ r_i & \text{if } x \in X_i \setminus X_{i-1},\ i = 2, 3, \ldots, n. \end{cases}$$

We show that $\mathcal{E} = \{\nu_E \mid E \in \mathbf{E}_n\}$, where

$$\nu_E(x) = \begin{cases} \wedge\{\mu(e) \mid e \in E\} & \text{if } x \in E, \\ 0 & \text{otherwise.} \end{cases}$$

Let $E \in \mathbf{E}_n$. Since $\mathcal{H}$ is elementary and support simple there is a unique fuzzy edge ω_E in $\mathcal{E}$ having support E. Indeed, distinct edges in $\mathcal{E}$ must have distinct supports that lie in $\mathbf{E}_n$. Thus, to show that $\mathcal{E} = \{\nu_E \mid E \in \mathbf{E}_n\}$, it suffices to show that for each $E \in \mathbf{E}_n, \nu_E = \omega_E$.

As all edges are elementary and different edges have different supports, it follows from the definition of the fundamental sequence that $h(\omega_E)$ is equal to some member r_i of $\mathbf{F}(\mathcal{H})$. Consequently, $E \subseteq X_i$. Moreover, if $i > 1$, then $E \in \mathbf{E}_i \setminus \mathbf{E}_{i-1}$.

Since $E \subseteq X_i$, it follows from the definition of μ that for each $x \in E$, $\mu(x) \geq r_i$.

We claim that $\mu(x) = r_i$ for some $x \in E$. For if not, then, by definition of $\mu, \mu(x) \geq r_{i-1}$ for all $x \in E$ which implies that $E \subseteq X_{i-1}$and so $E \in \mathbf{E}_i \setminus \mathbf{E}_{i-1}$ and since $\mathcal{H}$ is simply ordered $E \nsubseteq X_{i-1}$, a contradiction. Thus it follows from the definition of ν_E that $\nu_E = \omega_E$. ■

Corollary 4.3 *Suppose $\mathcal{H} = (X, \mathcal{E})$ is a simply ordered fuzzy hypergraph and $\mathbf{F}(\mathcal{H}) = \{r_1,\ r_2, \ldots,\ r_n\}$. If H^{r_n} is a simple hypergraph, then there is a partial fuzzy hypergraph $\mathcal{H}' = (X, \mathcal{E}')$ of $\mathcal{H}$ such that the following assertions hold.*

(1) $\mathcal{H}'$ is a μ tempered fuzzy hypergraph of H_n.

(2) $\mathcal{E} \sqsubseteq \mathcal{E}'$; that is, $\forall \nu \in \mathcal{E}, \exists \nu' \in \mathcal{E}'$ such that $\nu \subseteq \nu'$.

(3) $\mathbf{F}(\mathcal{H}') = \mathbf{F}(\mathcal{H})$ and $\mathbf{C}(\mathcal{H}') = \mathbf{C}(\mathcal{H})$.

Proof. By Proposition 4.1, it follows that $\mathcal{H}$ is an elementary fuzzy hypergraph. By removing all edges of $\mathcal{H}$ that are properly contained in another edge of $\mathcal{H}$, we obtain the partial fuzzy hypergraph $\mathcal{H}' = (X, \mathcal{E}')$ of $\mathcal{H}$, where $\mathcal{E}' = \{\mu \in \mathcal{E} \mid \text{if } \mu \subseteq \nu \text{ and } \nu \in \mathcal{E}, \text{ then } \nu = \mu\}$.

Since H^{τ_n} is simple and all edges are elementary, one edge cannot contain another edge in $\mathcal{H}$ unless both have the same support. Hence (3) holds. In addition, $\mathcal{H}'$ is support simple. Thus $\mathcal{H}'$satisfies all conditions in Theorem 4.2 and (1) is seen to hold. ■

4.2 Fuzzy Transversals of Fuzzy Hypergraphs

Let $H = (X, \mathbf{E})$ represent a (crisp) hypergraph on X. A *transversal* of H is any subset T of X with the property that $T \cap E \neq \emptyset$ for each $E \in \mathbf{E}$. A transversal T of the hypergraph H is called a *minimal transversal* of H if no proper subset of T is a transversal of H. Clearly every transversal contains a minimal transversal. The collection of minimal transversals of $H = (X, \mathbf{E})$ forms the edge set of a hypergraph, denoted by $Tr(H)$, in which the vertex set is a (possibly proper) subset of X.

Conceivably, a fuzzy transversal of a fuzzy hypergraph $\mathcal{H}$ over X could simply be defined as a fuzzy subset of X that "intersects" every fuzzy edge of $\mathcal{H}$. However, such a definition would not permit the existence of minimal fuzzy transversals (unless the interval of fuzzy membership values was discretized), nor (more importantly) would there necessarily be a useful association with the core set of crisp hypergraphs of $\mathcal{H}$. Both concerns are answered in the following definition.

Definition 4.13 *Let $\mathcal{H} = (X, \mathcal{E})$ be a fuzzy hypergraph. A* fuzzy transversal *τ of $\mathcal{H}$ is a fuzzy subset of X with the property that $\tau^{h(\mu)} \cap \mu^{h(\mu)} \neq \emptyset$ for each $\mu \in \mathcal{E}$, where $h(\mu)$ is the height of μ. A* minimal fuzzy transversal *τ for $\mathcal{H}$ is a transversal of $\mathcal{H}$ with the property that if $\rho \subset \tau$, then ρ is not a fuzzy transversal of $\mathcal{H}$.*

We let $Tr(\mathcal{H})$ denote the family of minimal fuzzy transversals of $\mathcal{H}$.

Proposition 4.4 *If τ is a fuzzy transversal of a fuzzy hypergraph $\mathcal{H} = (X, \mathcal{E})$, then $h(\tau) \geq h(\mu)$ for each $\mu \in \mathcal{E}$. Moreover, if τ is a minimal fuzzy transversal of $\mathcal{H}$, then $h(\tau) = \vee\{h(\mu) \mid \mu \in \mathcal{E}\} = h(\mathcal{H})$.*

The proof of the next result follows from Definition 4.13.

Theorem 4.5 *If $\mathcal{H}$ is a fuzzy hypergraph then $Tr(\mathcal{H}) \neq \emptyset$.* ■

Proposition 4.6 *Let $\mathcal{H} = (X, \mathcal{E})$ be a fuzzy hypergraph. The following two statements are equivalent:*

(1) τ is a fuzzy transversal of $\mathcal{H}$.

(2) For each $\mu \in \mathcal{E}$ and each t, $0 < t \leq h(\mu)$, $\tau^t \cap \mu^t \neq \emptyset$.

If the t-cut τ^t is a subset of the vertex set of H^t for each t, $0 < t \leq h(\mathcal{H})$, then the following statement (3) is equivalent to (1) or (2).

(3) *For each t, $0 < t \leq h(\mathcal{H})$, τ^t is a transversal of H^t.*

(4) *Every fuzzy transversal of $\mathcal{H}$ contains a fuzzy transversal that satisfies statement (3).* ■

Clearly, property (3) in Proposition 4.6 is valid for each $\tau \in Tr(\mathcal{H})$. But, it does not necessarily follow that if τ is a minimal fuzzy transversal of $\mathcal{H}$, then τ^t must belong to $Tr(H^t)$ for every t satisfying $0 < t \leq h(\mathcal{H})$. However, it is interesting to identify those cases where all t-cuts of τ do belong to $Tr(H^t)$. Let $Tr^*(\mathcal{H})$ represent the collection of those minimal fuzzy transversals, τ, of $\mathcal{H}$ where τ^t is a minimal transversal of H^t for every t satisfying $0 < t \leq h(\mathcal{H})$. In other words,

$Tr^*(\mathcal{H}) = \{\tau \in Tr(\mathcal{H}) \mid h(\tau) = h(\mathcal{H}) \text{ and } \tau^t \in Tr(H^t) \text{ for every } t \text{ satisfying } 0 < t \leq h(\mathcal{H})\}$. The members of $Tr^*(\mathcal{H})$ are called *locally minimal fuzzy transversals* of $\mathcal{H}$. Clearly, $Tr^*(\mathcal{H}) \subseteq Tr(\mathcal{H})$; however, the converse of this statement is not generally valid as later examples will show.

The members of $Tr(\mathcal{H})$ (or $Tr^*(\mathcal{H})$) can be determined by Algorithm 4.1 (or Algorithm 4.2) below. Justification of the procedure depends upon several properties stated in Lemma 4.10 which are shared by all members of $Tr(\mathcal{H})$ (or $Tr^*(\mathcal{H})$).

To determine $Tr(\mathcal{H})$, Algorithm 4.1 utilizes a procedure that appears in [2, pp.52 - 53], which can be used to determine $Tr(H)$ for any crisp hypergraph H. In order to execute the following algorithm, we use the following notation: Let $H_i = (X_i, \mathbf{E}_i)$, $i = 1, 2$, be a pair of (crisp) hypergraphs. Then $H_1 \cup H_2$ denotes the hypergraph with edge set $\mathbf{E}_1 \cup \mathbf{E}_2$.

In the following algorithms assume $\mathcal{H} = (X, \mathcal{E})$ is a fuzzy hypergraph where $\mathbf{F}(\mathcal{H}) = \{r_1, r_2, \ldots, r_n\}$ with $r_1 > r_2 > \cdots > r_n > 0$ and $\mathbf{C}(\mathcal{H}) = \{H^{r_i} \mid r_i \in \mathbf{F}(\mathcal{H})\}$.

Algorithm 4.1.

Step 1: Determine $Tr(H^{r_1})$.

Assume

$Tr(H^{r_1}) = \{T^1_{i^1} \mid i^1 = 1, \ldots, m_1\}$.

For each $i^1 = 1, 2, \ldots, m_1$ proceed to

Step 2(i^1): Determine $Tr(H^{r_2} \cup \{\{x_j\} \mid x_j \in T^1_{i^1}\})$.

Note that $\{\{x_j\} \mid x_j \in T^1_{i^1}\}$ is interpreted in Step 2(i^1) as a crisp hypergraph. Assume

$Tr(H^{r_2} \cup \{\{x_j\} \mid x_j \in T^1_{i^1}\}) = \{T^2_{i^1,i^2} \mid i^2 = 1, \ldots, m_2^{i^1}\}$.

For each $i^2 = 1, \ldots, m_2^{i^1}$ proceed to

Step 3(i^1, i^2): Execute Step 3 in a manner consistent with the process described in Step 2. Then continue recursively terminating with *Step* $n(i^1, \ldots, i^{n-1})$: Determine $Tr(H^{r_n} \cup \{\{x_j\} \mid x_j \in T^{n-1}_{i^1,\ldots,i^{n-1}}\})$. Assume

$Tr(H^{r_n} \cup \{\{x_j\} \mid x_j \in T^{n-1}_{i^1,\ldots,i^{n-1}}\}) = \{T^n_{i^1,\ldots,i^{n-1},i^n} \mid i^n = 1,\ldots,m_n^{i^1,\ldots,i^{n-1}}\}$.

Each member $T^n_{i^1,\ldots,i^{n-1},i^n}$ of $Tr(H^{r_n} \cup \{\{x_j\} \mid x_j \in T^{n-1}_{i^1,\ldots,i^{n-1}}\})$ corresponds uniquely to a nested (recursively defined) sequence,

$S_{i^1,\ldots,i^n} = \{T^1_{i^1}, \{T^k_{i^1,\ldots,i^k} \in Tr(H^{r_k} \cup \{\{x_j\} \mid x_j \in T^{k-1}_{i^1,\ldots,i^{k-1}}\}) \mid k = 2,\ldots,n\}\}$

from which we form a unique fuzzy subset

$\tau^{i^1,i^2,\ldots,i^n} = \cup\{\sigma(T^k_{i^1,\ldots,i^k}, r_k) \mid k = 1,\ldots,n\}$.

Recall that $\sigma(A,r)$ denotes an elementary fuzzy subset with support A and height r. Then,

$Tr(\mathcal{H}) = \{\tau^{i^1,i^2,\ldots,i^n} \mid i^1 \in \{1,\ldots,m_1\},\ldots,i^n \in \{1,\ldots,m_n^{i^1,\ldots,i^{n-1}}\}\}$.

Algorithm 4.2.

Step 1: Compute $\{Tr(H^{r_k}) \mid k = 1,\ldots,n\}$.

Step 2: Determine the family $\mathcal{L}$ of all possible nested sequences:

$\mathcal{L} = \{s = \{T^s_1,\ldots,T^s_k,\ldots,T^s_n\} \mid T^s_k \in Tr(H^{r_k})$, and $T^s_k \subseteq T^s_{k+1}$, for $k = 1,\ldots,n-1\}$.

Step 3: To every $s \in \mathcal{L}$ correspond the unique fuzzy subset:

$\tau^s = \cup\{\sigma(T^s_k, r_k) \mid k = 1,\ldots,n\}$.

Then

$Tr^*(\mathcal{H}) = \{\tau^s \mid s \in \mathcal{L}\}$.

It appears that Algorithm 4.2 is easier to execute than Algorithm 4.1; but, as latter examples will show, sometimes $Tr^*(\mathcal{H}) \subset Tr(\mathcal{H})$ and sometimes $Tr^*(\mathcal{H}) = \emptyset$.

If $\mathcal{H}$ is replaced with $\mathcal{H}^s$ in Algorithm 4.1, it is possible to ease the execution of the algorithm since $\mathcal{H}^s$ is simpler. Recall that $\mathcal{H}^s$ is developed in Construction 4.2 and that $Tr(\mathcal{H}^s) = Tr(\mathcal{H})$.

Algorithm 4.1 clearly affirms the existence of $Tr(\mathcal{H})$ for every fuzzy hypergraph $\mathcal{H}$. The same is not always true for $Tr^*(\mathcal{H})$ as the next example demonstrates.

Example 4.3 *Define $\mathcal{H}$ by the following incidence matrix.*

$$\begin{array}{c} \\ a \\ b \\ c \end{array} \begin{array}{c} \begin{array}{ccc} \mu_1 & \mu_2 & \mu_3 \end{array} \\ \begin{pmatrix} 0.9 & 0 & 0.4 \\ 0.4 & 0.4 & 0.4 \\ 0 & 0 & 0.4 \end{pmatrix} \end{array}.$$

Clearly τ given by

$$\begin{array}{c} a \\ b \end{array} \begin{pmatrix} 0.9 \\ 0.4 \end{pmatrix} \text{ is the only element of } Tr(\mathcal{H}).$$

Since $\{b\}$ is the only minimal transversal of the .4-cut, $\{\{a,b\},\{b\},\{a,b,c\}\}$, of $\mathcal{H}$ it follows that the minimal fuzzy transversal, τ, is not a member of $Tr^*(\mathcal{H})$. Hence $Tr^*(\mathcal{H}) = \emptyset$.

We now consider two basic questions:

(i) When is $Tr^*(\mathcal{H}) \neq \emptyset$?
(ii) When is $Tr^*(\mathcal{H}) = Tr(\mathcal{H})$?

We give partial answers to both questions and begin our investigation with a few preliminary results about crisp hypergraphs.

Lemma 4.7 *Suppose $\hat{H} = (\hat{X}, \hat{\mathbf{E}})$ is a (crisp) partial hypergraph of a (crisp) hypergraph $H = (X, \mathbf{E})$.*

(1) If T is a minimal transversal of H, then there exists a minimal transversal $\hat{T}$ of $\hat{H}$ such that $\hat{T} \subseteq T$.

(2) If $\hat{H}$ and H are simply ordered, and if $\hat{T}$ is minimal transversal of $\hat{H}$, then there exists a minimal transversal T of H such that $\hat{T} \subseteq T$.

Proof. (1) Let $S = T \cap \hat{X}$. Then S is a transversal of $\hat{H}$. Clearly, there exists a minimal transversal $\hat{T}$ of $\hat{H}$ such that $\hat{T} \subseteq S \subseteq T$.

(2) Let $\mathbf{E}^* = \{E \mid E \in E \setminus \hat{\mathbf{E}} \text{ and } E \cap \hat{T} = \emptyset\}$. If $\mathbf{E}^* = \emptyset$, then $\hat{T}$ is also a minimal transversal of H, and we are done. Suppose that $\mathbf{E}^* = \{E_1, E_2, \ldots, E_k\}$. Since $\hat{H}$ and H are simply ordered, there exists a vertex $x_i \in E_i \setminus \hat{X}$, for each integer $i, 1 \leq i \leq k$. Let $S = \hat{T} \cup \{x_1, x_2, \ldots, x_k\}$. Then S is a transversal of H and contains a minimal transversal T, which must contain $\hat{T}$. ■

There are examples where the pair $(\hat{H}, H)$ is ordered and the conclusion in Lemma 4.7(2) is false, so a condition stronger than "ordering" was needed in (2).

Theorem 4.8 *Suppose $\mathcal{H} = (X, \mathcal{E})$ is an ordered fuzzy hypergraph with $\mathbf{F}(\mathcal{H}) = \{r_1, r_2, \ldots, r_n\}$ and $\mathbf{C}(\mathcal{H}) = \{H^{r_i} \mid r_i \in \mathbf{F}(\mathcal{H})\}$. Then $Tr^*(\mathcal{H}) \neq \emptyset$. Moreover, if T_n is a minimal transversal of H^{r_n}, then there exists $\tau \in Tr^*(\mathcal{H})$ such that $supp(\tau) = T_n$.*

Proof. Let T_n be a minimal transversal of H^{r_n}. Since $\mathcal{H}$ is ordered, $H^{r_{n-1}}$ is a partial hypergraph of H^{r_n}, and hence, by (1) of Lemma 4.7 there is a minimal transversal T_{n-1} of $H^{r_{n-1}}$ such that $T_{n-1} \subseteq T_n$. Continuing this argument we construct a nested sequence, $T_1 \subseteq T_2 \subseteq \cdots \subseteq T_{n-1} \subseteq T_n$, of minimal transversals where each T_i is a minimal transversal of H^{r_i}. For $1 \leq i \leq n$, let $\sigma_i = \sigma(T_i, r_i)$ be the elementary fuzzy subset with support T_i and height r_i. Then $\tau = \cup\{\sigma_i \mid 1 \leq i \leq n\}$ is a locally minimal fuzzy transversal of $\mathcal{H}$ with support T_n. ■

Theorem 4.9 *Suppose $\mathcal{H} = (X, \mathcal{E})$ is a simply ordered fuzzy hypergraph with $\mathbf{F}(\mathcal{H}) = \{r_1, r_2, \ldots, r_n\}$ and $\mathbf{C}(\mathcal{H}) = \{H^{r_i} \mid r_i \in \mathbf{F}(\mathcal{H})\}$. Then if T_i is a minimal transversal of H^{r_i}, there exists $\tau \in Tr^*(\mathcal{H})$ such that $\tau^{r_i} = T$.*

Proof. Suppose T_i is a minimal transversal of H^{r_i}. Since $\mathcal{H}$ is simply ordered, (1) and (2) of Lemma 4.7 imply that a nested sequence, $T_i \supseteq \cdots \supseteq T_k \supseteq \cdots \supseteq T_1$, from T_i to T_1 and a nested sequence, $T_i \subseteq \cdots \subseteq T_k \subseteq \cdots \subseteq T_n$, from T_i to T_n can be constructed, which together produce a nested sequence

$T_1 \subseteq \cdots \subseteq T_k \subseteq \cdots \subseteq T_i \subseteq \cdots \subseteq T_k \subseteq \cdots \subseteq T_n$,

where $T_k \in Tr(H^{r_k})$, for all integers k from 1 to n. As before, for each k, let σ_k be the elementary fuzzy subset with support T_k and height r_k. Then $\tau = \cup\{\sigma_k \mid 1 \leq k \leq n\}$ has the desired properties. ■

Lemmas 4.10 and 4.12 below provide essential information about the members of $Tr(\mathcal{H})$. We first need a few definitions.

Definition 4.14 *A fuzzy hypergraph $\mathcal{H} = (X, \mathcal{E})$ with fundamental sequence $0 < r_n < \cdots < r_1$ is* sectionally elementary *if for each fuzzy edge $\mu \in \mathcal{E}$ and for each $i \in \{1, \ldots, n\}, \mu^t = \mu^{r_i}$, for all $t \in (r_{i+1}, r_i]$, where it is understood that $r_{n+1} = 0$.*

Definition 4.15 *Let μ be a fuzzy subset. The* lower truncation *of μ at level t, $0 < t \leq 1$, is the fuzzy subset $\mu_{(t)}$ defined by*

$$\mu_{(t)}(x) = \begin{cases} \mu(x) & \text{if } x \in \mu^t, \\ 0 & \text{otherwise.} \end{cases}$$

The upper truncation *of μ at level t, $0 < t \leq 1$, is the fuzzy subset $\mu^{(t)}$ defined by*

$$\mu^{(t)}(x) = \begin{cases} t & \text{if } x \in \mu^t, \\ \mu(x) & \text{otherwise.} \end{cases}$$

Definition 4.16 *Let $\mathcal{E}$ be a collection of fuzzy subsets of X and let*

$\mathcal{E}^{(t)} = \{\nu^{(t)} \mid \nu \in \mathcal{E}\}$,

$\mathcal{E}_{(t)} = \{\nu_{(t)} \mid \nu \in \mathcal{E}\}$.

Then the upper and lower truncations of a fuzzy hypergraph $\mathcal{H} = (X, \mathcal{E})$ at level t are the fuzzy hypergraphs, $\mathcal{H}^{(t)}$ and $\mathcal{H}_{(t)}$, defined by

$\mathcal{H}^{(t)} = (X, \mathcal{E}^{(t)})$ and $\mathcal{H}_{(t)} = (X', \mathcal{E}_{(t)})$, respectively.

We note that $X' = \bigcup_{\nu \in \mathcal{E}} \text{supp}(\nu_{(t)})$ may be a proper subset of X.

Definition 4.17 *Let X be a finite set and $\mu \in \mathfrak{F}\wp(X)$. Then each $t \in (0, h(\mu))$ for which $\mu^c \subset \mu^t, t < c \leq h(\mu)$, is called a* transition level *of μ.*

We emphasize that the height $h(\mu)$ of a fuzzy subset μ is not considered a transition level of μ. Thus, an elementary fuzzy subset has no transition levels.

Definition 4.18 *Let X be a finite set and suppose $\mu \in \mathfrak{F}\wp(X) \setminus \{\emptyset\}$. Then*

(1) *the* basic sequence *of* μ, *denoted by* $\mathbf{S}(\mu)$, *is the sequence* $\mathbf{S}(\mu) = \{t_1^\mu, t_2^\mu, \ldots, t_{n^\mu}^\mu\}$ *determined by* μ *where it is understood that*

 (a) $t_1^\mu > t_2^\mu > \cdots > t_{n^\mu}^\mu > 0$,

 (b) $t_1^\mu = h(\mu)$,

 (c) $\{t_2^\mu, \ldots, t_{n^\mu}^\mu\}$ *represents the set of transition levels of* μ;

(2) *the* set of basic cuts *of* μ, *denoted by* $\mathbf{C}(\mu)$, *is the set of cuts of* μ *defined by* $\mathbf{C}(\mu) = \{\mu^t \mid t \in \mathbf{S}(\mu)\}$;

(3) *the* basic elementary join *(or simply* basic join*) of* μ *is the synthesized join,* $\mu = \bigcup_{t \in \mathbf{S}(\mu)} \sigma(\mu^t, t)$, *of the* basic elementary fuzzy subsets, $\mathbf{E}(\mu) = \{\sigma(\mu^t, t) \mid t \in \mathbf{S}(\mu)\}$ *of* μ.

Lemma 4.10 *Let* $\mathcal{H}$ *be a fuzzy hypergraph with* $\mathbf{F}(\mathcal{H}) = \{r_1, \ldots, r_n\}$, *where* $0 < r_n < \cdots < r_1$. *Then*

(1) *If* t *is a transition level of* $\tau \in Tr(\mathcal{H})$, *then there exists an* $\varepsilon > 0$ *such that, for all* $c \in (t, t+\varepsilon]$, τ^t, *is a minimal* H^t*-transversal extension of* τ^c *(i.e., if* $\tau^c \subseteq A \subset \tau^t$, *then* A *is not a transversal of* H^t*).*

(2) $Tr(\mathcal{H})$ *is sectionally elementary.*

(3) $\mathbf{F}(Tr(\mathcal{H})) \subseteq \mathbf{F}(\mathcal{H})$.

(4) *For each* $\tau \in Tr(\mathcal{H})$, τ^c *is a minimal transversal of* H^c *for* $r_2 < c \leq r_1$.

Proof. (1) Let $\hat{t}$ be a transition level of a minimal fuzzy transversal $\tau \in Tr(\mathcal{H})$. Then $\tau^c \subset \tau^{\hat{t}}$ for all c satisfying $\hat{t} < c \leq h(\mathcal{H})$. Since the support of τ is finite, there exists an $\varepsilon > 0$ such that τ^c is constant on $(\hat{t}, \hat{t}+\varepsilon]$. Let $c' \in (\hat{t}, \hat{t}+\varepsilon]$ and assume there exists a transversal T of $H^{\hat{t}}$ satisfying

$$\tau^{c'} \subseteq T \subset \tau^{\hat{t}}. \tag{4.4}$$

We now show that this assumption is false. That is, that (4.4) is false. Let $\mathbf{E}(\tau) = \{\sigma(\tau^t, t) \mid t \in \mathbf{S}(\tau)\}$ denote the collection of basic elementary fuzzy subsets of τ (see Definition 4.18). Note, in view of (4.4), that $\mathbf{C}(\tau) \cup T$ forms a nested sequence of subsets of X, where $\mathbf{C}(\tau)$ denotes the set of basic cuts of τ (see Definition 4.18). In addition, since $\mathcal{H} = (X, \mathcal{E})$ is defined on a finite set X and $\mathcal{E}$ is a finite subset of $\mathfrak{F}\wp(x)$, for each $r \in (0, h(\mathcal{H}))$ there is a corresponding number $\varepsilon_r > 0$ such that
(i) H^c is constant on $(r, r+\varepsilon_r]$, and
(ii) H^c is constant on $(r - \varepsilon_r, r]$.
From these considerations it follows that the level cuts $\hat{\tau}^c$ of the join $\hat{\tau} = \cup\{\{\mathbf{E}(\tau) \setminus \sigma(\tau^{\hat{t}}, \hat{t})\} \cup \sigma(\tau^{\hat{t}}, \hat{t} - \varepsilon_{\hat{t}}) \cup \sigma(T, \hat{t})\}$

satisfy

$$\hat{\tau}^c = \begin{cases} T & \text{if } c \in (\hat{t} - \varepsilon_{\hat{t}}, \hat{t}], \\ \tau^c & \text{if } c \in (0, h(\mathcal{H})] \setminus (\hat{t} - \varepsilon_{\hat{t}}, \hat{t}]. \end{cases} \tag{4.5}$$

Eq. (4.5) is derived under the supposition that $\varepsilon_{\hat{t}}$ is sufficiently small so that no transition level of τ is contained in the open interval $(\hat{t} - \varepsilon_{\hat{t}}, \hat{t})$.

Notice that, since T is presumed to be a transversal of $H^{\hat{t}}$ and H^c is constant on $(\hat{t}-\varepsilon_{\hat{t}}, \hat{t}]$, T is a transversal of H^c for all c belonging to $(\hat{t}-\varepsilon_{\hat{t}}, \hat{t}]$. In addition, τ^c is a transversal of H^c for all $c \in (0, h(\mathcal{H})]$ (see Proposition 4.6). Therefore, it follows from (4.5) and Proposition 4.6 that $\hat{t}$ is a fuzzy transversal of $\mathcal{H}$. Now $\hat{\tau} \subset \tau$ by (4.4) and (4.5) and so we conclude that $\tau \notin Tr(\mathcal{H})$, which is a contradiction. Hence (4.4) is false.

(2) Let $\tau \in Tr(\mathcal{H})$. Then, in view of Proposition 4.6,

$$\tau^c \text{ is a transversal of } H^c \text{ for } 0 < c < h(\mathcal{H}). \tag{4.6}$$

Suppose t is a transition level of τ. Associate with t an interval $(t, t+\varepsilon]$, $\varepsilon > 0$, on which τ^c is constant. Then select a value $c' \in (t, t+\varepsilon]$. From part (1), $\tau^{c'}$ is not a transversal of H^t. Thus by (4.6),

$$\tau^{c'} \notin (Tr(\mathcal{H}))^t, \tag{4.7}$$

where $(Tr(\mathcal{H}))^t$ is the t-cut of $Tr(\mathcal{H})$. However, according to the definition of $\mathbf{F}(Tr(\mathcal{H}))$ (see Definition 4.6), (4.7) implies $t \in \mathbf{F}(Tr(\mathcal{H}))$.

(3) First, we consider the following claim.

We claim that if t is a transition level of a member τ of $Tr(\mathcal{H})$, then $t \in \mathbf{F}(\mathcal{H})$.

To prove this claim, assume there is a $\tau \in Tr(\mathcal{H})$ with a transition level $t \notin \mathbf{F}(\mathcal{H})$. Then for some $r_j \in \mathbf{F}(\mathcal{H})$, $r_{j+1} < t < r_j$, where it is understood that $r_{n+1} = 0$. Therefore, as $H^c = H^{r_j}$ for all $c \in (r_{j+1}, r_j]$, it follows that τ^t is a transversal of $H^t = H^{r_j}$. Moreover, there is an $\varepsilon > 0$ such that τ^c is constant on $(t, t+\varepsilon]$. Without loss of generality, assume $t + \varepsilon \leq r_j$ and select $c' \in (t, t+\varepsilon]$. Then, since t is a transition level of τ, $\tau^{c'} \subset \tau^t$ and from part (1) $\tau^{c'}$ is not a transversal of H^t. But this is impossible, since $\tau^{c'}$ is a transversal of $H^{c'}$ and $H^{c'} = H^{r_j} = H^t$. This establishes the claim and proves that

$\{t \mid t \text{ is a transition level of some } \tau \in Tr(\mathcal{H})\} \subseteq \mathbf{F}(\mathcal{H})$.

With this latter result and the fact that $h(\tau) = r_1 \in \mathbf{F}(\mathcal{H})$, for all $\tau \in Tr(\mathcal{H})$, it follows at once that $\mathbf{F}(Tr(\mathcal{H})) \subseteq \mathbf{F}(\mathcal{H})$.

(4) First we show that τ^{r_1} is a minimal transversal of H^{r_1}. Suppose this were not the case. Then, since τ^{r_1} is a transversal of H^{r_1}, there is a minimal transversal T of H^{r_1} properly contained in τ^{r_1}. Thus let

$\hat{\tau} = \tau^{(r_2)} \cup \sigma_1$,

where σ_1 is the elementary fuzzy subset with support T and height r_1, and $\tau^{(r_2)}$ is the upper truncation of τ at level r_2. Clearly, $\hat{\tau}$ is a fuzzy transversal of $\mathcal{H}$ such that $\hat{\tau} \subset \tau$. This is impossible. Therefore, $\tau^{r_1} \in Tr(H^{r_1})$. From parts (2) and (3) of this lemma, it follows at once that $\tau^c \in Tr(H^c)$ for $r_2 < c \leq r_1$. ■

Properties (1)-(4) also hold for $Tr^*(\mathcal{H})$. This is easily established from the fact that if t is a transition level of some $\tau \in Tr^*(\mathcal{H})$, then τ^t cannot appear as a higher level t-cut for any member of $Tr^*(\mathcal{H})$ and from the fact that $Tr^*(\mathcal{H}) \subseteq Tr(\mathcal{H})$.

Before continuing, we illustrate Lemma 4.10. Let $\mathcal{H} = (X, \mathcal{E})$ be a fuzzy hypergraph of Example 4.3. Recall that $Tr(\mathcal{H}) = \{\tau\}$, where $\tau(a) = 0.9$, $\tau(b) = 0.4$, and $\tau(c) = 0$. Now $H^{0.4} = (\{a,b,c\}, \{\{a,b\}, \{b\}, \{a,b,c\}\})$ and $H^{0.9} = (\{a\}, \{\{a\}\})$. The transition level of τ is 0.4. Let $t = 0.5$. Then $\forall c \in (0.4, 0.4+\epsilon]$, $\tau^c = \{a\}$. Clearly, then $\tau^{0.4} = \{a,b\}$ is a minimal $H^{0.4}$–transversal extension of τ^c. Note however that $\tau^{0.4}$ is not a minimal transversal of $H^{0.4}$ since $\tau^{0.4} \supset \{b\}$.

Proposition 4.11 *Let $\mathcal{H} = (X, \mathcal{E})$ be a fuzzy hypergraph, and let $r_1 = h(\mathcal{H})$. Then the r_1-cut, $(Tr(\mathcal{H}))^{r_1}$ of $Tr(\mathcal{H})$ satisfies $(Tr(\mathcal{H}))^{r_1} = Tr(H^{r_1})$.*

Proof. The proof follows from part (4) of Lemma 4.10 and Algorithm 4.1. ■

The following lemma expresses a fundamental property shared by all fuzzy transversals.

Lemma 4.12 *Every fuzzy transversal of a fuzzy hypergraph $\mathcal{H}$ contains at least one minimal fuzzy transversal of $\mathcal{H}$.*

Proof. Let $\mathbf{F}(\mathcal{H}) = \{r_1, \ldots, r_n\}$, where $0 < r_n < \cdots < r_1$, and assume ν is a nonminimal fuzzy transversal of $\mathcal{H}$. A $\tau \in Tr(\mathcal{H})$ such that $\tau \subseteq \nu$ is constructed through a series of reductions, $\{\rho_i \in \mathfrak{F}\wp(X) \mid i = 0, 1, \ldots, n\}$ satisfying

$$\tau = \rho_n \subseteq \cdots \subseteq \rho_1 \subseteq \rho_0 \subseteq \nu. \tag{4.8}$$

From Propositions 4.4 and 4.6 it is clear that $h(\nu) \geq h(\mathcal{H}) = r_1$ and ν^c is (or contains) a crisp transversal of H^c for $0 < c \leq r_1$. Therefore, we begin our reduction process by setting

$\rho_0 = \nu^{(r_1)}$,

where $\nu^{(r_1)}$ signifies the upper truncation of ν at level r_1 (see Definition 4.15). As the top cut of ρ_0, namely ν^{r_1}, contains a crisp minimal transversal T_1 of H^{r_1}, we define

$\rho_1 = \nu^{(r_2)} \cup \sigma_{T_1}$,

where σ_{T_1} is the elementary fuzzy subset with support T_1 and height r_1. Clearly, $\rho_1 \subseteq \rho_0 \subseteq \nu$. The remaining members in (4.8) can be determined in a similar manner. For example, we set

$\rho_2 = \nu^{(r_3)} \cup \sigma_{T_1} \cup \sigma_{T_2}$,
where σ_{T_2} is an elementary fuzzy subset of height r_2 and support T_2:

$$T_2 = \begin{cases} T_1 & \text{if } T_1 \text{ is a transversal of } H^{r_2}, \\ A_2 & \text{otherwise,} \end{cases} \tag{4.9}$$

where A_2 is a minimal H^{r_2}-transversal extension of T_1 (that is, if $T_1 \subseteq B \subset A_2$, then B is not a transversal of H^{r_2}) which is contained in the r_2-level cut of ν, namely ν^{r_2}. The properties of A_2 are possible since ν^{r_2} contains a transversal H^{r_2}. Moreover, since $T_2 \subseteq \nu^{r_2}$, it is clear that $\rho_2 \subseteq \rho_1$. When the reduction process is completed, $\tau = \rho_n$ is certainly a fuzzy transversal of $\mathcal{H}$ (according to Proposition 4.6) and is contained in ν.

We now show that $\tau \in Tr(\mathcal{H})$.

Suppose μ is a fuzzy transversal of $\mathcal{H}$ such that $\mu \subset \tau$. Then

(1) $\mu^c \subseteq \tau^c$ for all $c \in (0, h(\mathcal{H})]$, and

(2) $\mu^{\hat{c}} \subset \tau^{\hat{c}}$ for some $\hat{c} \in (0, h(\mathcal{H})]$.

However, by a simple inductive argument, no such $\hat{c}$ can exist. To begin, suppose $c \in (r_2, r_1]$. Then as $\mu^c \subseteq \tau^c$, μ^c is a transversal of $H^c = H^{r_1}$, and $\tau^c \in Tr(H^{r_1})$, it follows that

$$\mu^c = \tau^c \quad \text{on } (r_2, r_1]. \tag{4.10}$$

Now suppose $c \in (r_3, r_2]$. Then, by (4.10),

$$\mu^c \supseteq \tau^{r_1} \quad \text{for all } c \in (r_3, r_2]. \tag{4.11}$$

If $T_2 = T_1 = \tau^{r_1}$, then by (4.11), $\mu^c = \tau^c$ on $(r_3, r_2]$. On the other hand, if $T_1 \subset T_2$ and $T_1 \subseteq \mu^c \subset T_2$, then by (4.9), μ^c would not be a transversal of $H^c = H^{r_2}$, which would contradict the fact that μ is a transversal of $\mathcal{H}$. Thus we conclude that

$$\mu^c = \tau^c \quad \text{on } (r_3, r_2]. \tag{4.12}$$

The argument establishing (4.12) is typical, and we are forced to conclude that

$$\mu^c = \tau^c \quad \text{on } (0, h(\mathcal{H})]. \tag{4.13}$$

Hence the strict inclusion $\mu \subset \tau$ is impossible. ■

We now examine the question, "when is $Tr^*(\mathcal{H}) = Tr(\mathcal{H})$?" A partial answer appears below in Theorem 4.13. The result is connected to the structural property presented in Definition 4.20.

Definition 4.19 *An ordered pair (H, K) of crisp hypergraphs is T*-related *if whenever Q_H is a minimal transversal of H, P_K is a transversal of K, and $Q_H \subseteq P_K$, then there is a minimal transversal $\hat{P}_K$ of K such that $Q_H \subseteq \hat{P}_K \subseteq P_K$.*

Definition 4.20 *A fuzzy hypergraph $\mathcal{H}$ with $\mathbf{F}(\mathcal{H}) = \{r_1, \ldots, r_n\}$, where $0 < r_n < \cdots < r_1$, is T*-related *if every successive ordered pair $(H^{r_i}, H^{r_{i-1}})$ of members from the core set, $\mathbf{C}(\mathcal{H}) = \{H^{r_1}, H^{r_2}, \ldots, H^{r_n}\}$, is T-related. In the particular case where the fundamental sequence of $\mathcal{H}$ has but one member, $\mathcal{H}$ is considered (vacuously) to be T-related.*

Theorem 4.13 *Let $\mathcal{H} = (X, \mathcal{E})$ be a T-related fuzzy hypergraph. Then $Tr^*(H) = Tr(H)$.*

Proof. Suppose that $\mathcal{H}$ is T-related, and let $\mathbf{F}(\mathcal{H}) = \{r_1, \ldots, r_n\}$, where $0 < r_n < \cdots < r_1$. First consider the case where $\mathbf{F}(\mathcal{H}) = \{r_1\}$. In view of part (4) of Lemma 4.10, it follows that for every $\tau \in Tr(\mathcal{H})$, the c-cut τ^c belongs to $Tr(H^c)$ for all c satisfying $0 < c \leq h(\mathcal{H})$. Thus $Tr^*(\mathcal{H}) = Tr(\mathcal{H})$.

Next, assume $|\mathbf{F}(\mathcal{H})| \geq 2$. Since $Tr^*(\mathcal{H}) \subseteq Tr(\mathcal{H})$, it remains to show that $Tr(\mathcal{H}) \subseteq Tr^*(\mathcal{H})$. Let $\tau \in Tr(\mathcal{H})$, and suppose that $\tau^{r_1} \subset \tau^{r_2}$. We know that τ^{r_2} is a transversal of H^{r_2}, τ^{r_1} is a minimal transversal of H^{r_1}, and (H^{r_1}, H^{r_2}) is a T-related ordered pair. Therefore, if τ^{r_2} were not a minimal transversal of H^{r_2}, then there would exist a minimal transversal T_2 of H^{r_2} such that $\tau^{r_1} \subseteq T_2 \subset \tau^{r_2}$. Hence we could define a fuzzy transversal $\hat{\tau}$ of $\mathcal{H}$ with the property that $\hat{\tau} \subset \tau$ (which is a contradiction) as follows: Let $\tau^{r_1} = T_1$ and set $\hat{\tau} = \tau^{(r_3)} \cup \sigma_2 \cup \sigma_1$, where σ_i is an elementary fuzzy subset with support T_i and height r_i $(i = 1, 2)$.

This impossibility shows that τ^{r_2} is a minimal transversal of H^{r_2} and therefore, in view of parts (2) and (3) of Lemma 4.10, τ^c is a minimal transversal of H^c for $c \in (r_3, r_1]$.

We can continue recursively in this fashion and show for each $c \in (0, r_1]$ that $\tau^c \in Tr(H^c)$. ■

The following example shows that the converse of Theorem 4.13 is not true.

Example 4.4 *Suppose $\mathcal{H} = (X, \mathcal{E})$ is a fuzzy hypergraph given by the incidence matrix*

$$\mathcal{H} = (X, \mathcal{E}) = \begin{array}{c} \\ a \\ b \\ c \\ d \\ e \end{array} \begin{array}{c} \begin{array}{ccc} \mu_1 & \mu_2 & \mu_3 \end{array} \\ \left(\begin{array}{ccc} 0.9 & 0.9 & 0 \\ 0.9 & 0 & 0.9 \\ 0 & 0.9 & 0.9 \\ 0.6 & 0.6 & 0.3 \\ 0.3 & 0.3 & 0.6 \end{array} \right) \end{array}$$

Then $\mathbf{E}^{0.9} = \{\{a,b\},\{a,c\},\{b,c\}\}, \mathbf{E}^{0.6} = \{\{a,b,d\},\{a,c,d\},\{b,c,e\}\}$, $\mathbf{E}^{0.3} = \{\{a,b,d,e\},\{a,c,d,e\},\{b,c,d,e\}\}$.

Clearly, $\mathbf{F}(\mathcal{H}) = \{r_1 = 0.9, r_2 = 0.6, r_3 = 0.3\}$, *and*

$$Tr(\mathcal{H}) = Tr^*(\mathcal{H}) = \begin{matrix} \\ a \\ b \\ c \\ d \\ e \end{matrix} \begin{matrix} \tau_1 & \tau_1 & \tau_3 \\ \end{matrix} \begin{pmatrix} 0.9 & 0.9 & 0 \\ 0.9 & 0 & 0.9 \\ 0 & 0.9 & 0.9 \\ 0 & 0 & 0 \\ 0 & 0 & 0 \end{pmatrix}$$

Since $\{d,e\} \in Tr(H^{r_2})$ *and* $\{d\} \in Tr(H^{r_3})$, *no minimal transversal of* $Tr(H^{r_3})$ *contains* $\{d,e\}$. *Thus,* (H^{r_2}, H^{r_3}) *and in turn* $\mathcal{H}$ *is not* T*-related.*

The next theorem is a partial converse of Theorem 4.13. Later an unrestricted necessary condition for $Tr^*(\mathcal{H}) = Tr\ (\mathcal{H})$ is developed.

Theorem 4.14 *Let* $\mathcal{H} = (X, \mathcal{E})$ *be an ordered fuzzy hypergraph. Then* $Tr(\mathcal{H}) = Tr^*(\mathcal{H})$ *if and only if* $\mathcal{H}$ *is* T*-related.*

Proof. In view of Theorem 4.13, it suffices to show that $Tr(\mathcal{H}) = Tr^*(\mathcal{H})$ implies $\mathcal{H}$ is T-related. Assume $\mathbf{F}(\mathcal{H}) = \{r_1, r_2, \ldots, r_n\}$, where $0 < r_n < \cdots < r_1$, and suppose that $\mathcal{H}$ is not T-related. We construct a minimal fuzzy transversal of $\mathcal{H}$ that fails to belong to $Tr^*(\mathcal{H})$. Towards this goal we begin with a pair $(H^{r_i}, H^{r_{i+1}})$ that is not T-related, where, $\mathbf{C}(\mathcal{H}) = \{H^{r_i} \mid r_i \in \mathbf{F}(\mathcal{H})\}$. Then there exists a minimal transversal T_i of H^{r_i} contained in a transversal T of $H^{r_{i+1}}$ with the property that S is not a minimal transversal of $H^{r_{i+1}}$ for every S satisfying $T_i \subseteq S \subseteq T$. Since $\mathcal{H}$ is ordered, the edges of H^{r_i} form a subset of the edges of $H^{r_{i+1}}$. Therefore, T_i is not a transversal of $H^{r_{i+1}}$, for otherwise T_i would be a minimal transversal of $H^{r_{i+1}}$, contradicting the above assumption that this is not the case. Let $\hat{T}$ be any transversal $H^{r_{i+1}}$ such that

$$T_i \subseteq \hat{T} \subseteq T \tag{4.14}$$

and

$$\text{if } T_i \subseteq B \subset \hat{T}, \text{ then } B \text{ is not a transversal of } H^{r_{i+1}}. \tag{4.15}$$

Then, as noted earlier,

$$\hat{T} \text{ is not a minimal transversal of } H^{r_{i+1}} \text{and } T_i \subset \hat{T}. \tag{4.16}$$

Thus we are now able to construct a minimal fuzzy transversal τ which does not belong to $Tr^*(\mathcal{H})$. First, using the process described in Algorithm 4.1, we find a minimal fuzzy transversal $\hat{\tau}$ of $\mathcal{H}^{(r_i)}$ where T_i is the top cut of $\hat{\tau}$ at level r_i and , in addition, satisfies the property that $\hat{\tau}^{r_{i+1}} \subseteq T$. In conjunction with part (1) of Lemma 4.10, it is clear that the r_{i+1}-cut, $\hat{\tau}^{r_{i+1}}$, of $\hat{\tau}$ must equal some $\hat{T}$ satisfying both (4.14) and (4.15) and, therefore, (4.16) as well. As a result, $\hat{\tau} \in Tr(\mathcal{H}^{(r_i)}) \setminus Tr^*(\mathcal{H}^{(r_i)})$.

Second, we assume for the sake of completeness that $r_i < r_1$. As $\mathcal{H}$ is ordered, part (1) of Lemma 4.7 implies that there exists a nested sequence of crisp minimal transversals $T_i \supseteq T_{i-1} \supseteq \cdots \supseteq T_1$ of $H^{r_i}, H^{r_{i-1}}, \ldots, H^{r_1}$, respectively. Let σ_j denote the elementary fuzzy subset with support T_j and height r_j. Then $\tau = \sigma_1 \cup \ldots \cup \sigma_{i-1} \cup \hat{\tau}$ belongs to $Tr(\mathcal{H})$ but not to $Tr^*(\mathcal{H})$. ■

In the proof of Theorem 4.14 above, we were given a pair $(H^{r_i},\ H^{r_{i+1}})$ which is not T-related and then constructed a minimal fuzzy transversal τ which does not belong $Tr^*(\mathcal{H})$. In the process we showed that $t = r_{i+1}$ is a transition level for τ. This implies (by part (2) of Lemma 4.10) that $r_{i+1} \in \mathbf{F}(Tr(\mathcal{H}))$. Thus, we have the following result.

Corollary 4.15 *Let $\mathcal{H}$ be an ordered fuzzy hypergraph with $\mathbf{F}(\mathcal{H}) = \{r_1, \ldots, r_n\}$, where $0 < r_n < \cdots < r_1$, and $\mathbf{C}(\mathcal{H}) = \{H^{r_i} \mid r_i \in \mathbf{F}(\mathcal{H})\}$. If an ordered pair $(H^{r_i},\ H^{r_{i+1}})$ is not T-related, then*

(1) $r_{i+1} \in \mathbf{F}(Tr(\mathcal{H}))$;

(2) r_{i+1} is a transition level for some $\tau \in Tr(\mathcal{H}) \setminus Tr^*(\mathcal{H})$.

If $\mathcal{H}$ is a μ tempered fuzzy hypergraph, then $\mathcal{H}$ is simply ordered by Theorem 4.2. However, for such an $\mathcal{H}$ it is not necessarily the case that $Tr^*(\mathcal{H}) = Tr(\mathcal{H})$. This is shown in the following example by displaying such an $\mathcal{H}$ which is not T-related.

Example 4.5 *Let $\mu\colon X \to (0,1]$ be a fuzzy subset of $X = \{a,b,c,d,e,f,g\}$ such that $\mu(g) = 0.4$ and $\mu(x) = 0.9$ if $x \in X \setminus \{g\}$. Let $H = (X, \mathbf{E})$ be the crisp hypergraph on X with edges $E_1 = \{a,b,d\}$, $E_2 = \{a,c,d\}$, $E_3 = \{d,e,f\}$, $E_4 = \{a,e\}$ and $E_5 = \{e,g\}$. Then*

$$\mathcal{H} = \mu \otimes H = \begin{array}{c} \\ a \\ b \\ c \\ d \\ e \\ f \\ g \end{array} \begin{array}{c} \begin{array}{ccccc} \nu_1 & \nu_2 & \nu_3 & \nu_4 & \nu_5 \end{array} \\ \left(\begin{array}{ccccc} 0.9 & 0.9 & 0 & 0.9 & 0 \\ 0.9 & 0 & 0 & 0 & 0 \\ 0 & 0.9 & 0 & 0 & 0 \\ 0.9 & 0.9 & 0.9 & 0 & 0 \\ 0 & 0 & 0.9 & 0.9 & 0.4 \\ 0 & 0 & 0.9 & 0 & 0 \\ 0 & 0 & 0 & 0 & 0.4 \end{array} \right) \end{array}.$$

Clearly, $\mathbf{F}(\mathcal{H}) = \{r_1 = 0.9,\ r_2 = 0.4\}$ and $\mathbf{C}(\mathcal{H}) = \{H^{r_1},\ H^{r_2}\}$, where $H^{r_1} = (X \setminus \{g\},\ \mathbf{E}^{r_1})$ with $\mathbf{E}^{r_1} = \{E_i \mid i = 1, \ldots, 4\}$, and $H^{r_2} = H$. We have $\{a,d\} \in Tr(H^{r_1})$, but $\{a,d\} \notin Tr(H^{r_2})$, and that $\{a,d,e\}$ is a transversal, but not a minimal transversal of H^{r_2}. Therefore, the ordered pair $(H^{r_1},\ H^{r_2})$ is not T-related as well.

Remark 1.

(1) We have by Example 4.5 that some simply ordered fuzzy hypergraphs are not T-related.

(2) According to Theorem 4.9, every simply ordered fuzzy hypergraph $\mathcal{H}$ satisfies $(Tr^*(\mathcal{H}))^t = Tr(H^t)$ for all $t \in (0, h(\mathcal{H})]$.

Nevertheless, in view of Theorem 4.14 and part (1) of Remark 1, there exist simply ordered fuzzy hypergraphs for which
$Tr^*(\mathcal{H}) \subset Tr(\mathcal{H})$.

Properties of $Tr(H)$

In this section necessary and sufficient conditions are determined which characterize the set of points that belong to the support of some minimal fuzzy transversal. This information is used later to construct a structurally rich elementary fuzzy hypergraph $\mathcal{H}^s$, where $Tr(\mathcal{H}^s) = Tr(\mathcal{H})$.

Lemma 4.16 *A vertex x of a crisp hypergraph $H = (X, \mathbf{E})$ is a member of some minimal transversal of H if and only if x belongs to an edge which does not properly contain another edge of H.*

Proof. Without loss of generality, we assume H has no repeated edges. Let $\hat{\mathbf{E}} = \{E \in \mathbf{E} \mid E' \in \mathbf{E} \text{ and } E' \subseteq E \Rightarrow E' = E\}$ and let $\hat{X} = \cup\hat{\mathbf{E}}$. (Note that $\hat{\mathbf{E}} \neq \emptyset$.) Suppose $E_0 \in \hat{\mathbf{E}}$ and let $\{E_1, \ldots, E_n\} = \mathbf{E} \setminus E_0$.

For each $j = 1, \ldots, n$, pick some $x_j \in E_j \setminus E_0$ and let $T' = \{x_1, \ldots, x_n\}$. Clearly T' is not a transversal of H, since $T' \cap E_0 = \emptyset$. However, for each $x \in E_0$, $T' \cup \{x\}$ is a transversal of H and, therefore, must contain a minimal transversal T of H. Obviously, T contains x; otherwise T would be contained in T', which is impossible, since T' is not a transversal of H. Therefore, $x \in \mathbf{V}(Tr(H))$ and consequently, $\hat{X} \subseteq \mathbf{V}(Tr(H))$.

Conversely, suppose $x_0 \in X \setminus \hat{X}$. Then x_0 belongs only to edges which properly contain another edge. Thus, whenever x_0 belongs to an edge E there must exist another edge E' such that $E' \subseteq E \setminus \{x_0\}$. This implies that if T is a transversal then $T \setminus \{x_0\}$ is also a transversal of H. Hence $x_0 \notin \mathbf{V}(Tr(H))$. Thus $(X \setminus \hat{X}) \cap \mathbf{V}(Tr(H)) = \emptyset$. Hence $\hat{X} = \mathbf{V}(Tr(H))$. ■

Proposition 4.17 *Let $\hat{H} = (\hat{X}, \hat{\mathbf{E}})$ be the partial hypergraph obtained by deleting from the edge set $\mathbf{E}$ of H all edges which properly contain another edge. Then*

(1) $Tr(\hat{H}) = Tr(H)$,

(2) $\cup Tr(H) = \hat{X}$. ■

To establish a proof for a fuzzy version of Lemma 4.16, we need the following definition and lemma.

Definition 4.21 *The* join of a fuzzy hypergraph $\mathcal{H}$, *denoted* $\mathbf{J}(\mathcal{H})$, *is defined by*

$\mathbf{J}(\mathcal{H}) = \bigcup_{\mu \in \mathcal{E}} \mu,$

where $\mathcal{E}$ is the fuzzy edge set of $\mathcal{H}$.

For each $t \in (0, h(\mathcal{H})]$, the t-level cut, $(\mathbf{J}(\mathcal{H}))^t$, of $\mathbf{J}(\mathcal{H})$ is the vertex set, $X^t = \mathbf{V}(H^t)$, of the t-level hypergraph H^t of $\mathcal{H}$.

Lemma 4.18 *Let $\mathcal{H} = (X, \mathcal{E})$ be a fuzzy hypergraph and suppose $\tau \in Tr(\mathcal{H})$. If $x \in supp(\tau)$, then there exists a fuzzy edge μ of $\mathcal{H}$ for which*

(1) $\mu(x) = h(\mu) = \tau(x) > 0$

(2) $\tau^{h(\mu)} \cap \mu^{h(\mu)} = \{x\}$.

Proof. Assume x_0 belongs to the support of some fuzzy minimal transversal τ of $\mathcal{H}$ and let $t_0 = \tau(x_0)$. Since every transversal of $\mathcal{H}$ contains a transversal which is contained in $\mathbf{J}(\mathcal{H})$, we have

$$\tau \subseteq \mathbf{J}(\mathcal{H}). \tag{4.17}$$

Now (4.17) implies $x_0 \in X^{t_0}$ $(= \mathbf{V}(H^{t_0}))$. Therefore, there is at least one fuzzy edge μ of $\mathcal{H}$ which satisfies $\mu(x_0) \geq t_0$. Let $M = \{\mu_1, \ldots, \mu_m\}$ denote the set of edges in $\mathcal{H}$ where the degree of membership for x_0 equals or exceeds t_0.

We now show that at least one member of M has height t_0.

For suppose otherwise, then

$$h(\mu_j) = t_j > t_0, \quad j = 1, \ldots, m, \tag{4.18}$$

for each member μ_j of M. This would imply that for each member μ_j of M there is a member $x_j \in \text{supp}(\tau)$ for which

$$x_j \in (\mu_j)^{t_j} \cap \tau^{t_j} \tag{4.19}$$

at a level t_j greater than t_0. However, since $\tau(x_0) = t_0$, (4.18) and (4.19) would imply

$$x_j \neq x_0, \quad j = 1, \ldots, m, \tag{4.20}$$

for each member μ_j of M. But if (4.18)-(4.20) were true, it could then be shown that $\tau \notin Tr(\mathcal{H})$ by constructing a fuzzy transversal $\hat{\tau}$ of $\mathcal{H}$ which satisfies $\hat{\tau} \subset \tau$. This contention follows by the fact that, since both X and $\mathcal{E}$ are finite, there is an ε-interval $(t_0 - \varepsilon, t_0]$ such that

$$H^t = H^{t_0} \text{ on } (t_0 - \varepsilon, t_0]. \tag{4.21}$$

Define

$$\hat{\tau}(x) = \begin{cases} \tau(x) & \text{if } x \neq x_0, \\ t_0 - \varepsilon & \text{if } x = x_0. \end{cases}$$

Clearly, $\hat{\tau} \subset \tau$. Moreover, $\hat{\tau}$ is a transversal of $\mathcal{H}$. For, in view of (4.18)-(4.20), $\tau^{t_0} \setminus \{x_0\}$ contains $\{x_j \mid j = 1, \ldots, m\}$ and, therefore, in view of (4.18), (4.19) and the definition of M, $\tau^{t_0} \setminus \{x_0\}$ is a transversal of H^{t_0}. Also in view of (4.21), the same is true for every H^t where $t \in (t_0 - \varepsilon, t_0]$. Thus, since $\hat{\tau}^t = \tau^t$ for all $t \in (0, h(\mathcal{H})] \setminus (t_0 - \varepsilon, t_0]$, it is evident that $\hat{\tau}$ is a transversal of $\mathcal{H}$. This establishes the claim and, consequently, there exists a $\mu \in \mathcal{H}$ for which $\mu(x_0) = h(\mu) = \tau(x_0) > 0$.

Finally, suppose all edges in $M = \{\mu_1, \ldots, \mu_m\}$ of height $\tau(x_0)$ contain more than one member of τ^{t_0}. Then, by basically repeating the above argument, it can be established that τ is not a fuzzy minimal transversal of $\mathcal{H}$, which is a contradiction. ■

In the following example, we illustrate some of the concepts recently introduced.

Example 4.6 *Consider the fuzzy hypergraph defined in Example 4.1. Then μ_1, μ_3 and μ_5 have no transition levels, while μ_2 and μ_4 have transition levels 0.4. The basic sequences (see Definition 4.18) are as follows: $S(\mu_1) = \{0.7\}$, $S(\mu_2) = \{0.9, 0.4\}, S(\mu_3) = \{0.9\}, S(\mu_4) = \{0.7, 0.4\}$, and $S(\mu_5) = \{0.4\}$. Thus $\mathbf{C}(\mu_1) = \{\mu_1^{0.7}\}$, $\mathbf{C}(\mu_2) = \{\mu_2^{0.9}, \mu_2^{0.4}\}, \mathbf{C}(\mu_3) = \{\mu_3^{0.9}\}, \mathbf{C}(\mu_4) = \{\mu_4^{0.7}, \mu_4^{0.4}\}$, and $\mathbf{C}(\mu_5) = \{\mu_5^{0.4}\}$. Now $\mu_2^{0.9} = \{a, b\}$ and $\mu_2^{0.4} = \{a, b, d\}$. We see that $\mu_2 = \sigma(\mu_2^{0.9}, 0.9) \cup \sigma(\mu_2^{0.4}, 0.4)$ and $\mathbf{E}(\mu_2) = \{\sigma(\mu_2^{0.9}, 0.9), \sigma(\mu_2^{0.4}, 0.4)\} = \{(\mu_2)_{(0.9)} = (\mu_2)_{(0.7)}, (\mu_2)^{(0.4)}\}$. Recall that $\mathbf{F}(\mathcal{H}) = \{0.9, 0.4\}$ and $\mathbf{C}(\mathcal{H}) = \{H^{r_i} \mid r_i \in \mathbf{F}(\mathcal{H})\} = \{H^{0.9} = (\{a, b, c\}, \{\{a, b\}, \{b, c\}\})$, $H^{0.4} = (\{a, b, c, d\}, \{\{a, b\}, \{a, b, d\}, \{b, c\}, \{b, c, d\}, \{a, c, d\}\})\}$. We now determine $Tr(\mathcal{H})$ and $Tr^*(\mathcal{H})$. If $\tau \in Tr(\mathcal{H})$, then $\tau^{h(\mu_i)} \cap \mu_i^{h(\mu_i)} \neq \emptyset$ for $i = 1, 2, 3, 4, 5$. Thus $\tau^{0.7} \cap \{a, b\} \neq \emptyset$, and $\tau^{0.9} \cap \{a, b\} \neq \emptyset, \tau^{0.9} \cap \{b, c\} \neq \emptyset, \tau^{0.7} \cap \{b, c\} \neq \emptyset$, and $\tau^{0.4} \cap \{a, c, d\} \neq \emptyset$. We see that $Tr(\mathcal{H}) = \{\tau_1, \tau_2, \tau_3, \tau_4\}$, where $\{\tau_1, \tau_2, \tau_3, \tau_4\}$ is represented by the following incidence matrix.*

$$\begin{array}{c} \\ a \\ b \\ c \\ d \end{array} \begin{array}{c} \begin{array}{cccc} \tau_1 & \tau_2 & \tau_3 & \tau_4 \end{array} \\ \begin{pmatrix} 0.9 & 0 & 0 & 0.4 \\ 0 & 0.9 & 0.9 & 0.9 \\ 0.9 & 0.4 & 0 & 0 \\ 0 & 0 & 0.4 & 0 \end{pmatrix} \end{array}$$

Now $Tr(H^{0.9}) = \{\{b\}, \{a, c\}\}$ and $Tr(H^{0.4}) = \{\{a, c\}, \{b, c\}, \{b, d\}, \{\{a, b\}, \{a, c, d\}\}$. Hence $\tau_i^t \in Tr(H^t)$ for all t such that $0 < t \leq h(\mathcal{H}) = 0.9$ only for $i = 1$. Thus $Tr^*(\mathcal{H}) = \{\tau_1\}$.

We now illustrate Lemma 4.18.

$\mu_2(a) = h(\mu_2) = \tau_1(a) = 0.9, \mu_3(c) = h(\mu_3) = \tau_1(c) = 0.9,\ \tau_1^{0.9} \cap \mu_2^{0.9} = \{a\}, \tau_1^{0.9} \cap \mu_3^{0.9} = \{c\}$.

$\mu_2(b) = h(\mu_2) = \tau_2(b) = 0.9, \mu_5(c) = h(\mu_5) = \tau_2(c) = 0.4,\ \tau_2^{0.9} \cap \mu_2^{0.9} = \{b\}, \tau_2^{0.4} \cap \mu_5^{0.4} = \{c\}$.

$\mu_3(b) = h(\mu_3) = \tau_3(b) = 0.9, \mu_5(d) = h(\mu_5) = \tau_3(d) = 0.4,\ \tau_3^{0.9} \cap \mu_3^{0.9} = \{b\}, \tau_3^{0.4} \cap \mu_5^{0.4} = \{d\}$.

$\mu_5(a) = h(\mu_5) = \tau_4(b) = 0.4, \mu_3(b) = h(\mu_3) = \tau_4(b) = 0.9,\ \tau_4^{0.4} \cap \mu_5^{0.4} = \{a\}, \tau_4^{0.9} \cap \mu_3^{0.9} = \{b\}$.

Note that $0.9 = h(\mathcal{H})$ *and* $(Tr(\mathcal{H}))^{0.9} = \{\tau_1^{0.9}, \tau_2^{0.9}, \tau_3^{0.9}, \tau_4^{0.9}\} = \{\{a,c\}, \{b\}, \{b\}, \{b\}\} = \{\{a,c\},\{b\}\} = Tr(H^{0.9})$. *(See Proposition 4.11).*

Theorem 4.19 *Let* $\mathcal{H} = (X, \mathcal{E})$ *be a fuzzy hypergraph and suppose* $x \in X$. *If there exists a* $\tau \in Tr(\mathcal{H})$ *with* $x \in supp(\tau)$, *then there is a* $\mu \in \mathcal{E}$ *such that:*

(1) $\mu(x) = h(\mu)$.

(2) For every $\nu \in \mathcal{E}$ *in which* $h(\nu) > h(\mu)$, *the* $h(\nu)$*-level cut of* ν *is not a (proper or improper) subset of the* $h(\mu)$*-level cut of* μ.

(3) The $h(\mu)$*-level cut of* μ *does not properly contain another edge of* $H^{h(\mu)}$.

(4) $\tau(x) = \mu(x)$.

Conversely, if there exists a pair $\{x, \mu\}$, *where* $x \in X$ *and* $\mu \in \mathcal{E}$, *which satisfies (1), (2), and (3), then* x *belongs to the support of some minimal fuzzy transversal* τ *of* $\mathcal{H}$ *which satisfies (4).*

Proof. Suppose x belongs to the support of a minimal fuzzy transversal τ of $\mathcal{H}$. By Lemma 4.18, there exists an edge $\mu_1 \in \mathcal{E}$ which satisfies properties (1) and (4) and the condition

$$\tau^{h(\mu_1)} \cap (\mu_1)^{h(\mu_1)} = \{x\}. \tag{4.22}$$

Clearly, μ_1 satisfies property (2) also. For otherwise, there would exist a $\nu \in \mathcal{E}$ such that $h(\nu) > h(\mu_1)$ and

$$\nu^{h(\nu)} \subseteq (\mu_1)^{h(\mu_1)}. \tag{4.23}$$

However, since $\tau^{h(\nu)} \cap \nu^{h(\nu)}$ is nonempty, there exists y in this intersection, and $y \neq x$ since $\tau(y) \geq h(\nu) > \tau(x)$. Therefore, in view of (4.23), condition (4.22) would not hold.

Finally, suppose μ_1 does not satisfy (3). Then there exists a $\mu_2 \in \mathcal{E}$ such that

$$(\mu_2)^{h(\mu_1)} \subset (\mu_1)^{h(\mu_1)} \tag{4.24}$$

Clearly,

$$h(\mu_2) = h(\mu_1); \tag{4.25}$$

for otherwise μ_1 would not satisfy (2).

Furthermore, in view of (4.22), (4.24) and (4.25) and the fact that τ is a transversal of $\mathcal{H}$,

$\tau^{h(\mu_2)} \cap (\mu_2)^{h(\mu_2)} = \{x\}$.

Thus, μ_2 satisfies (1), (2), and (4). If, in addition, μ_2 satisfies (3) we are done. If not, the above process can continue until eventually a $\mu_k \in \mathcal{E}$ is determined which satisfies all four conditions. That this is true follows from the finiteness of X along with the fact that

$(\mu_k)^{h(\mu_1)} \subset \cdots \subset (\mu_2)^{h(\mu_1)} \subset (\mu_1)^{h(\mu_1)}$,

together with the fact that each μ_j, $j = 1, \ldots, k$, has height $h(\mu_1)$ and satisfies conditions (1), (2), and (4).

Conversely, suppose the pair $\{x, \mu\}$, where $x \in \mathbf{V}(\mathcal{H})$ and $\mu \in \mathcal{E}(\mathcal{H})$ satisfy conditions (1), (2), and (3).

Then $h(\mu) \in \mathbf{F}(\mathcal{H})$. For otherwise there would exist a pair $\{r_j, r_{j+1}\} \subseteq \mathbf{F}(\mathcal{H})$ such that

$$r_{j+1} < h(\mu) < r_j. \tag{4.26}$$

However, as H^t is constant on $(r_{j+1}, r_j]$, (4.26) would imply that there exists a $\nu \in \mathcal{E}$ such that

$\mu^{h(\mu)} = \nu^{r_j}$;

this, however, indicates that μ does not satisfy (2), which is a contradiction.

Let $\mathbf{F}(\mathcal{H}) = \{r_1, r_2, \ldots, r_n\}$ and $r_1 > r_2 > \cdots > r_n$. There are two cases to consider.

First, suppose $h(\mu) = r_1$. By condition (3) and Lemma 4.16, there exists a minimal transversal T of H^{r_1} containing x. Since there is a $\tau \in Tr(\mathcal{H})$ for which $\tau^{r_1} = T$ by Proposition 4.11, it follows that x belongs to the support of a minimal fuzzy transversal τ and $\tau(x) = \mu(x)$.

Secondly, suppose $h(\mu) = r_k < r_1$ and let $\mathbf{C}(\mathcal{H}) = \{H^{r_j} = (X^{r_j}, \mathbf{E}^{r_j}) \mid r_j \in \mathbf{F}(\mathcal{H})\}$. Then, for $j \in \{1, 2, \ldots, k-1\}$, condition (2) implies that

$$E \setminus \mu^{h(\mu)} \neq \emptyset, \quad \text{for every } E \in \mathbf{E}^{r_j}. \tag{4.27}$$

We now use property (4.27) in the construction of a nested sequence of sets

$T_1 \subseteq T_2 \subseteq \cdots \subseteq T_{k-1}$

which satisfies

(a) $T_{k-1} \cap \mu^{h(\mu)} = \emptyset$,

(b) $T_1 \in Tr(H^{r_1})$,

(c) T_j is a transversal of H^{r_j} with the property that any S which satisfies $T_{j-1} \subseteq S \subset T_j$ is not a transversal of H^{r_j}, for $j = 2, \ldots, k-1$.

The construction of the sequence $T_1, \ldots, T_{k-1}$ begins with the determination of some $T_1 \in Tr(H^{r_1})$ such that $T_1 \cap \mu^{h(\mu)} = \emptyset$; which is possible in view of (4.27). By property (4.27), it follows that T_1 is contained in a transversal $\hat{T}_2$ of H^{r_2} where $\hat{T}_2 \cap \mu^{h(\mu)} = \emptyset$. T_2 is then determined as a minimal H^{r_2}-transversal extension of T_1 which is contained in $\hat{T}_2$, or T_2 is T_1, the latter situation occurring whenever T_1 is a transversal of H^{r_2}. By the above recursive technique the sequence is constructed.

Moreover, as μ satisfies condition (3), every edge of H^{r_k} not equal to $\mu^{h(\mu)}$ contains a point not in $\mu^{h(\mu)}$. Thus, there exists a transversal T of H^{r_k} such that

$$T_{k-1} \subset T \quad \text{and} \quad T \cap \mu^{h(\mu)} = \{x\}.$$

Therefore, there must exist a minimal H^{r_k}-transversal extension T_k of T_{k-1} that contains x and is contained in T.

If $r_k \neq r_n$ then the established sequence $T_1, \ldots, T_{k-1}, T_k$ is continued by computing recursively $T_{k+1}, \ldots, T_n$, so that the completed sequence

$$T_1 \subseteq \cdots \subseteq T_{k-1} \subseteq T_k \subseteq T_{k+1} \subseteq \cdots \subseteq T_n$$

satisfies the above mentioned property (c) for $j = 2, \ldots, n$. Let

$$\tau = \sigma_1 \cup \cdots \cup \sigma_{k-1} \cup \sigma_k \cup \sigma_{k+1} \cup \cdots \cup \sigma_n,$$

where $\sigma_j = \sigma(T_j, r_j)$ is the elementary fuzzy subset with support T_j and height r_j. Hence $\tau \in Tr(\mathcal{H})$, and the constructive process yields

$$\tau(x) = r_k = \mu(x),$$

which establishes property (4). ■

Corollary 4.20 *Let $\mathcal{H} = (X, \mathbf{E})$. If $\mu \in \mathbf{E}$ satisfies condition (2) in Theorem 4.19, then $h(\mu) \in \mathbf{F}(\mathcal{H})$.* ■

This observation sheds light upon the generality of the constructive process used to determine $\hat{\mathbf{C}}(\mathcal{H})$ in Construction 4.1 below.

Construction of $\mathcal{H}^s$

The characterization given in Theorem 4.19 suggests that the procedure described in Proposition 4.17 is extendable, after modifications, to fuzzy hypergraphs. In particular, given $\mathcal{H}$, the process (described below) is applied to the members of $\mathbf{C}(\mathcal{H})$. Once this process is completed, an elementary fuzzy hypergraph, designated $\mathcal{H}^s$, is (uniquely) constructed with the property that $Tr(\mathcal{H}^s) = Tr(\mathcal{H})$. The usefulness of $\mathcal{H}^s$ will be made apparent in what follows. In addition, some efficiency can be obtained in determining $Tr(\mathcal{H})$ if Algorithm 4.1 is applied directly to $\mathcal{H}^s$.

We now consider the construction of $\hat{\mathbf{C}}(\mathcal{H})$.

Construction 4.1 Let $\mathcal{H}$ be a fuzzy hypergraph with fundamental sequence $\mathbf{F}(\mathcal{H}) = \{r_1, \ldots, r_n\}$, where $0 < r_n < \cdots < r_1$, and core set $\mathbf{C}(\mathcal{H}) = \{H^{r_1}, \ldots, H^{r_n}\}$. The construction of $\hat{\mathbf{C}}(\mathcal{H})$ from $\mathbf{C}(\mathcal{H})$ is a recursive process:

Step 1: Determine a partial hypergraph $\hat{H}^{r_1}$ of H^{r_1} by eliminating all edges in H^{r_1} that properly contain another edge of H^{r_1}.

Step 2: Eliminate all edges of H^{r_2} that either properly contain another edge of H^{r_2} or contain (properly or improperly) an edge of $\hat{H}^{r_1}$. Then, either all edges of H^{r_2} are eliminated (and $\hat{H}^{r_2}$ does not exist) or the remaining edges form a partial hypergraph $\hat{H}^{r_2}$ of H^{r_2}.

Assume Step i has been carried out for $i = 1, 2, \ldots, k; 1 \leq k \leq n-1, n \geq 2$.

Step k+1: Eliminate all edges of $H^{r_{k+1}}$ that either properly contain another edge of $H^{r_{k+1}}$ or contain (properly or improperly) an edge of $\hat{H}^{r_i}$ for $i = 1, 2, \ldots, k$ (if they exist). Then, either all edges of $H^{r_{k+1}}$ are eliminated (and $\hat{H}^{r_{k+1}}$ does not exist) or the remaining edges form a partial hypergraph $\hat{H}^{r_{k+1}}$ of $H^{r_{k+1}}$. Continuing recursively in this manner through Step n, we obtain a subsequence

$$\hat{\mathbf{F}}(\mathcal{H}) = \{r_1^s, \ldots, r_m^s\} \tag{4.28}$$

of $\mathbf{F}(\mathcal{H})$, where $0 < r_m^s < \cdots < r_1^s (= r_1)$, and a corresponding set

$$\hat{\mathbf{C}}(\mathcal{H}) = \{\hat{H}^{r_1^s}, \ldots, \hat{H}^{r_m^s}\} \tag{4.29}$$

of partial hypergraphs from $\mathbf{C}(\mathcal{H})$. (We emphasize that each member of $\hat{\mathbf{C}}(\mathcal{H})$ has a nonempty edge set and that for every $r_i \in \mathbf{F}(\mathcal{H}) \setminus \{r_1^s, \ldots, r_m^s\}$ the entire core hypergraph H^{r_i} was eliminated in the recursive process.)

Next, let us consider the construction of $\mathcal{H}^s$.

Construction 4.2. Let $\mathcal{H} = (X, \mathcal{E})$ be a fuzzy hypergraph. Assume $\hat{\mathbf{F}}(\mathcal{H})$ and $\hat{\mathbf{C}}(\mathcal{H})$ have been determined as described in Construction 4.1 (see (4.28) and (4.29)). Define $\mathcal{H}^s = (X^s, \mathcal{E}^s)$ to be the elementary fuzzy hypergraph satisfying:

(1) $\mathbf{F}(\mathcal{H}^s) = \hat{\mathbf{F}}(\mathcal{H}) = \{r_1^s, \ldots, r_m^s\}$,

(2) if $\mu \in \mathcal{E}^s$, then $h(\mu) \in \hat{\mathbf{F}}(\mathcal{H})$, and

(3) for each j, $1 \leq j \leq m$, the family of edges in $\mathcal{E}^s$ of height r_j^s is the set of elementary fuzzy subsets $\{\sigma(E, r_j^s) \mid E \in \hat{H}^{r_j^s}\}$.

$\mathcal{H}^s$ is called the *skeleton* of $\mathcal{H}$. For convenience the notation $(\mathcal{H})^s$ is sometimes used in place of $\mathcal{H}^s$.

The following two lemmas describe structural properties of $\hat{\mathbf{C}}(\mathcal{H})$.

Lemma 4.21 *Let $\mathcal{H} = (X, \mathcal{E})$ be a fuzzy hypergraph and let*
$\hat{C}(H) = \{\hat{H}^{r_j^s} = (\hat{X}^{r_j^s}, \hat{E}^{r_j^s}) \mid r_j^s \in \hat{F}(H)\}$
represent the set of crisp hypergraphs determined from the core set of $\mathbf{C}(\mathcal{H})$ by the process described in Construction 4.1. Then

(1) $\hat{\mathbf{E}}^{r_j^s} = \{\mu^{h(\mu)} \mid \mu \in \mathcal{E},\ h(\mu) = r_j^s,$ *and* μ *satisfies conditions (2) and (3) of Theorem 4.19*$\}$.

(2) $\hat{\mathbf{E}} = \cup\{\hat{\mathbf{E}}^{r_j^s} \mid r_j^s \in \hat{\mathbf{F}}(\mathcal{H})\} = \{\mu^{h(\mu)} \mid \mu \in \mathcal{E}$ *and satisfies conditions (2) and (3) of Theorem 4.19*$\}$.

(3) $\hat{\mathbf{F}}(\mathcal{H}) = \{h(\mu) \mid \mu \in \mathcal{E}$ *and satisfies conditions (2) and (3) of Theorem 4.19*$\}$.

Proof. (1) Suppose μ is an edge of $\mathcal{H}$ such that $h(\mu) = r_j^s$ and which satisfies conditions (2) and (3) of Theorem 4.19. Then, $E = \mu^{h(\mu)}$ contains no edge of $H^{r_1} \in \mathbf{C}(\mathcal{H})$ if $r_i > r_j^s$. (For if otherwise then there is an edge E_i of H^{r_i}, where $r_i > r_j^s$, such that $E_i \subseteq E$. Consequently, there is a $\nu \in \mathcal{E}$ such that $\nu^{r_i} = E_i$. This implies $\nu^{h(\nu)} \subseteq E$ and $h(\mu) < r_i \leq h(\nu)$, which contradicts the assumption that μ satisfies (2) in Theorem 4.19.) Moreover, since μ satisfies condition (3) of Theorem 4.19, it follows from Construction 4.1 that $E = \mu^{h(\mu)}$ is a member of $\hat{\mathbf{E}}^{r_j^s}$, and therefore the right side of (1) contains $\hat{\mathbf{E}}^{r_j^s}$.

Conversely, suppose $E \in \hat{\mathbf{E}}^{r_j^s}$ and let $M = \{\mu \in \mathcal{E} \mid \mu^{r_j^s} = E\}$. Note that M is not empty. Suppose $h(\mu) > r_j^s$ for some $\mu \in M$. Then, according to the definition of $\mathbf{F}(\mathcal{H})$, $\mu^{h(\mu)}$ must be an edge of some $H^{r_i} \in \mathbf{C}(\mathcal{H})$ where $r_i > r_j^s$. This implies E contains an edge E' $(\subseteq \mu^{h(\mu)})$ of some member $H^{r_k^s} \in \hat{\mathbf{C}}(\mathcal{H})$ where $r_k^s \geq r_i > r_j^s$. As a consequence,the constructive algorithm for determining $\hat{\mathbf{C}}(\mathcal{H})$ would not select E as a member of $\hat{\mathbf{E}}^{r_j^s}$, which is contradictory. Thus every member $\mu \in M$ has height $h(\mu) = r_j^s$. Hence $\mu^{h(\mu)} = E$ for every $\mu \in M$. With this knowledge and the method used in Construction 4.1 for determining $\hat{\mathbf{C}}(\mathcal{H})$, it follows that every member of M satisfies both conditions (2) and (3) of Theorem 4.19. Thus, the structure of $\hat{\mathbf{E}}^{r_j^s}$ given in (1) is established. More specifically, note that if condition (2) did not hold for some member $\mu \in M$, then $E = \mu^{h(\mu)}$ would contain $\nu^{h(\nu)}$ for some $\nu \in \mathcal{E}$ with $h(\nu) > h(\mu)$. In turn, this would imply $\mu^{h(\mu)}$ contains an edge $E' = \nu^{h(\nu)}$ which belongs to some $H^{r_i} \in \mathbf{C}(\mathcal{H})$ with $r_i > r_j^s$. Finally, either E' is or contains an edge E'' of some $\hat{H}^{r_k^s} \in \hat{\mathbf{C}}(\mathcal{H})$ where $r_k^s \geq r_i$. With $E'' \subseteq E$, the algorithm for determining $\hat{\mathbf{C}}(\mathcal{H})$ would not allow E to be a member of $\mathbf{E}^{r_j^s}$ since $r_j^s < r_k^s$. Of course, this contradicts the assumption that $E \in \hat{\mathbf{E}}^{r_j^s}$.

(2) Suppose μ is an edge of $\mathcal{H}$ which satisfies condition (2) of Theorem 4.19. From Corollary 4.20, it follows that $h(\mu) \in \mathbf{F}(\mathcal{H})$, say $h(\mu) = r_k$. Moreover, an argument in the proof of part (1) indicates that $\mu^{h(\mu)}$ contains no edge of $H^{r_i} \in \mathbf{C}(\mathcal{H})$ for $r_i > r_k$.

Thus, if μ is an edge of $\mathcal{H}$ which satisfies both conditions (2) and (3) of Theorem 4.19, then the algorithm for computing $\hat{\mathbf{C}}(\mathcal{H})$ clearly assigns $E = \mu^{h(\mu)}$ to $\hat{\mathbf{E}}^{r_k}$. Consequently, $r_k \in \hat{\mathbf{F}}(\mathcal{H})$. This result, combined with the characterization of $\hat{\mathbf{E}}^{r_j^s}$ given in (1), clearly establishes part (2).

(3) The argument given in the proof of part (2) indicates that the right side of (3) is contained in $\hat{\mathbf{F}}(\mathcal{H})$. On the other hand, the characterization of $\hat{\mathbf{E}}^{r_j^s}$ given in (1) implies that $\hat{\mathbf{F}}(\mathcal{H})$ is contained in the right side of (3). This establishes (3). ■

Lemma 4.22 *Let $\mathcal{H} = (X, \mathcal{E})$ be a fuzzy hypergraph and let $\hat{\mathbf{C}}(\mathcal{H}) = \{\hat{\mathbf{H}}^{r_j^s} = (\hat{X}^{r_j^s}, \hat{\mathbf{E}}^{r_j^s}) \mid r_j^s \in \hat{\mathbf{F}}(\mathcal{H})\}$. Then,*

(1) $\mathbf{F}(Tr(\mathcal{H})) = \hat{\mathbf{F}}(\mathcal{H})$.

(2) $\{\tau(x) \mid x \in supp(\tau),\ \tau \in Tr(\mathcal{H})\} = \hat{\mathbf{F}}(\mathcal{H})$.

(3) $\hat{X}^{r_j^s} = \{x \in X \mid \tau \in Tr(\mathcal{H}) \text{ and } \tau(x) = r_j^s\}$.

(4) $\cup\{supp(\tau) \mid \tau \in Tr(\mathcal{H})\} = \cup\{\hat{X}^{r_j^s} \mid r_j^s \in \hat{\mathbf{F}}(\mathcal{H})\}$.

(5) $\cup_{j=1}^{k}\hat{X}^{r_j^s} = \{x \in X \mid \tau \in Tr(\mathcal{H}) \text{ and } \tau(x) \geq r_k^s\}$.

Proof. (1), (2): It follows immediately from Theorem 4.19 that

$$\begin{aligned}\{\tau(x) \mid x \in \text{supp}(\tau), \tau \in Tr(\mathcal{H})\} = \{&h(\mu) \mid \mu \in \mathcal{E} \text{ and satisfies} \\ &\text{conditions (2) and (3) of Theorem 4.19}\}.\end{aligned} \tag{4.30}$$

On the other hand, by (2) of Lemma 4.10,

$$\mathbf{F}(Tr(\mathcal{H})) = \{\tau(x) \mid x \in \text{supp}(\tau),\ \tau \in Tr(\mathcal{H})\}, \tag{4.31}$$

and by (3) of Lemma 4.21,

$$\begin{aligned}\hat{\mathbf{F}}(\mathcal{H}) = \{&h(\mu) \mid \mu \in \mathcal{E} \text{ and satisfies conditions (2)} \\ &\text{and (3) of Theorem 4.19}\}.\end{aligned} \tag{4.32}$$

Clearly, (4.30)-(4.32) yield the desired result.

(3) Let $\hat{H}^{r_j^s} \in \hat{\mathbf{C}}(\mathcal{H})$ where $\hat{H}^{r_j^s} = (\hat{X}^{r_j^s}, \hat{\mathbf{E}}^{r_j^s})$. Then, according to part (1) of Lemma 4.21,

$$\begin{aligned}\hat{X}^{r_j^s} &= \cup\{E \mid E \in \hat{\mathbf{E}}^{r_j^s}\} = \cup\{\mu^{h(\mu)} \mid \mu \in \mathcal{E},\ h(\mu) = r_j^s, \text{ and } \mu \text{ satisfies} \\ &\quad \text{conditions (2) and (3) of Theorem 4.19}\}.\end{aligned} \tag{4.33}$$

However, according to Theorem 4.19,

$$\begin{aligned}\cup\{\mu^{h(\mu)} \mid \mu &\in \mathcal{E},\ h(\mu) = r_j^s, \text{ and } \mu \text{ satisfies conditions (2)} \\ &\quad \text{and (3) of 4.19}\} \\ &= \{x \in X \mid \tau \in Tr(\mathcal{H}) \text{ and } \tau(x) = r_j^s\}.\end{aligned} \tag{4.34}$$

Clearly, (4.33) and (4.34) yield (3).

(4), (5): The result here follows from properties (2) and (3) of Lemma 4.22. ■

Lemma 4.23 *Suppose $\mathcal{H}^s = (X^s, \mathcal{E}^s)$ is determined from the fuzzy hypergraph $\mathcal{H}$ according to the procedure stated in Construction 4.2 and let the core set of $\mathcal{H}^s$ be identified by $C(H^s) = \{(H^s)^{r_k^s} = ((X^s)^{r_k^s}, (E^s)^{r_k^s}) \mid r_k^s \in \hat{F}(H)\}$. Then,*

(1) $\mathbf{F}(\mathcal{H}^s) = \mathbf{F}(Tr(\mathcal{H}))$,

(2) $(\mathbf{E}^s)^{r_k^s} = \cup_{j=1}^{k} \hat{\mathbf{E}}^{r_j^s}$ *where* $\hat{\mathbf{E}}^{r_j^s}$ *is the edge set of* $\hat{H}^{r_j^s} = (\hat{X}^{r_j^s}, \hat{\mathbf{E}}^{r_j^s}) \in \hat{\mathbf{C}}(\mathcal{H})$,

(3) $(X^s)^{r_k^s} = \cup_{j=1}^{k} \hat{X}^{r_j^s} = \{x \in X \mid \tau \in Tr(\mathcal{H})$ *and* $\tau(x) \geq r_k^s\}$,

(4) $X^s = \mathbf{V}(\mathcal{H}^s) = \mathbf{V}(Tr(\mathcal{H}))$,

(5) $\mathcal{E}^s = \mathcal{E}(\mathcal{H}^s) = \{\sigma(E, r) \in \mathfrak{F}_{\wp}(X) \mid (E, r) \in \hat{\mathbf{E}}^{r_j^s} \times \{r_j^s\},\ r_j^s \in \hat{\mathbf{F}}(\mathcal{H})\}$,

where it is understood that $\sigma(E, r)$ represents the elementary fuzzy subset with support E and height r.

Proof. That $\mathbf{F}(\mathcal{H}^s) = \hat{\mathbf{F}}(\mathcal{H})$ follows from Constructions 4.1 and 4.2. This fact, together with Lemma 4.22(1), confirms property (1) of Lemma 4.23. Properties (3) and (4) of Lemma 4.23 follow easily from parts (5) and (4) of Lemma 4.22, respectively. Properties (2) and (5) of Lemma 4.23 are obvious from the definition of $\mathcal{H}^s$. ■

Lemma 4.24 *Let $\mathcal{H} = (X, \mathcal{E})$ be a fuzzy hypergraph. Then the following properties hold.*

(1) Every fuzzy transversal of $\mathcal{H}^s$ is a fuzzy transversal of $\mathcal{H}$.

(2) A fuzzy transversal τ of $\mathcal{H}$ is a fuzzy transversal of $\mathcal{H}^s$ if and only if supp$(\tau) \subseteq \mathbf{V}(\mathcal{H}^s)$.

Proof. Assume $0 < r_m^s < \cdots < r_1^s$.

(1) Let τ^s be a fuzzy transversal of $\mathcal{H}^s$. Since $\mathbf{V}(\mathcal{H}^s) \subseteq \mathbf{V}(\mathcal{H})$, supp$(\tau^s) \subseteq X$. Let $\mu \in \mathcal{E}$. We now show that

$$\mu^{h(\mu)} \cap (\tau^s)^{h(\mu)} \neq \emptyset. \tag{4.35}$$

Let $r_k = \wedge \{r_i \in \mathbf{F}(\mathcal{H}) \mid r_i \geq h(\mu)\}$. It follows, by definition of $\mathbf{F}(\mathcal{H})$, that $\mu^{h(\mu)}$ belongs to the edge set $\mathbf{E}^{r_k}$ of the core hypergraph H^{r_k} of $\mathcal{H}$.

Then, according to the construction of $\hat{\mathbf{C}}(\mathcal{H})$ and Lemma 4.21(1), there must exist some $r_j^s \in \hat{\mathbf{F}}(\mathcal{H})$ satisfying $r_j^s \geq r_k$ and a $\nu \in \mathcal{E}$ such that

(a) $h(\nu) = r_j^s \geq h(\mu)$,
(b) $\nu^{h(\nu)} \subseteq \mu^{h(\mu)}$,
(c) $\nu^{h(\nu)} \in \hat{\mathbf{E}}^{r_j^s}$, the edge set of $\hat{H}^{r_j^s} \in \hat{\mathbf{C}}(\mathcal{H})$.

Thus, the elementary fuzzy subset $\sigma(\nu^{h(\nu)}, r_j^s) \in \mathcal{H}^s$. Therefore, since τ^s is a fuzzy transversal of $\mathcal{H}^s$, it follows that $(\tau^s)^{h(\nu)} \cap \nu^{h(\nu)} \neq \emptyset$. This fact, coupled with properties (a) and (b) above, implies that property (4.35) holds.

(2) Let τ be a fuzzy transversal of $\mathcal{H}$. Then
$\mu^{h(\mu)} \cap \tau^{h(\mu)} \neq \emptyset$
for each edge $\mu \in \mathcal{E}(\mathcal{H})$. Therefore,

$$\nu^{h(\nu)} \cap \tau^{h(\nu)} \neq \emptyset \tag{4.36}$$

for every $\nu \in \mathcal{E}(\mathcal{H}^s)$, since

$$\{\nu^{h(\nu)} \mid \nu \in \mathcal{E}(\mathcal{H}^s)\} \subseteq \{\mu^{h(\mu)} \mid \mu \in \mathcal{E}(\mathcal{H})\}, \tag{4.37}$$

where (4.37) holds because by the construction of $\hat{\mathbf{C}}(\mathcal{H})$ from Construction 5.1, it follows that if $E \in \hat{\mathbf{E}}^{r_j^s}$, then E does not contain a t-cut ν^t of some edge ν of $\mathcal{H}$ where $t > r_j^s$.

Hence, if $E \in \hat{\mathbf{E}}^{r_j^s}$, then there is a $\mu \in \mathcal{E}(\mathcal{H})$ such that $\mu^{h(\mu)} = E$. For there is a $\mu \in \mathcal{E}(\mathcal{H})$ such that $E = \mu^{r_j^s}$. Since E does not contain a t-cut ν^t of some edge ν of $\mathcal{H}$ where $t > r_j^s$, it is evident that $h(\mu) = r_j^s$. Also, in this regard, recall Lemma 4.21(1). Therefore, by the construction process for determining $\mathcal{H}^s$ (see, in particular, part (5) of Lemma 4.23), it follows that
$\{\nu^{h(\nu)} \mid \nu \in \mathcal{E}(\mathcal{H}^s)\} = \{E \in \hat{\mathbf{E}}^{r_j^s} \mid r_j^s \in \hat{\mathbf{F}}(\mathcal{H})\} \subseteq \{\mu^{h(\mu)} \mid \mu \in \mathcal{E}(\mathcal{H})\}$.
By (4.36), it follows from Definition 4.13 that if τ is a fuzzy transversal of $\mathcal{H}$, then τ is a fuzzy transversal of $\mathcal{H}^s \Leftrightarrow \text{supp}(\tau) \subseteq \mathbf{V}(\mathcal{H}^s)$. ■

Theorem 4.25 *Let $\mathcal{H}$ be a fuzzy hypergraph. Then $Tr(\mathcal{H}^s) = Tr(\mathcal{H})$.*

Proof. Let $\tau^s \in Tr(\mathcal{H}^s)$. By Lemma 4.24(1), τ^s is also a fuzzy transversal of $\mathcal{H}$. Therefore, in view of Lemma 4.12, there exists a $\tau_1 \in Tr(\mathcal{H})$ such that $\tau_1 \subseteq \tau^s$. Since $\text{supp}(\tau_1) \subseteq \mathbf{V}(\mathcal{H}^s)$, Lemma 4.24(2) implies τ_1 is a fuzzy transversal of $\mathcal{H}^s$. Consequently $\tau_1 = \tau^s$ and so $Tr(\mathcal{H}^s) \subseteq Tr(\mathcal{H})$.

Conversely, suppose $\tau \in Tr(\mathcal{H})$. From Lemma 4.23(4), we know that $\text{supp}(\tau) \subseteq \mathbf{V}(\mathcal{H}^s)$. Therefore, according to Lemma 4.24(2), τ is a fuzzy transversal of $\mathcal{H}^s$. Hence, there is a $\tau_1^s \in Tr(\mathcal{H}^s)$ such that $\tau_1^s \subseteq \tau$. However, according to Lemma 4.24(1), τ_1^s is a transversal of $\mathcal{H}$ and so $\tau_1^s = \tau$. Thus $Tr(\mathcal{H}) \subseteq Tr(\mathcal{H}^s)$. ■

Example 4.7 *Consider the fuzzy hypergraph of Example 4.1. Recall that $r_1 = 0.9, r_2 = 0.4$, $\mathbf{E}^{r_1} = \{\{a,b\},\{b,c\}\}$, and $\mathbf{E}^{r_2} = \{\{a,b\},\{a,b,d\},\{b,c\}, \{b,c,d\},\{a,c,d\}\}$. No edge of H^{r_1} properly contains another edge of H^{r_1}.*

Hence $\hat{H}^{r_1} = H^{r_1}$. *For* H^{r_2}, $\{a,b,d\} \supseteq \{a,b\}$ *and* $\{b,c,d\} \supseteq \{b,c\}$. *Thus we remove* $\{a,b,d\}$ *and* $\{b,c,d\}$ *from* $\mathbf{E}^{r_2}$ *to obtain* $\{\{a,b\},\{b,c\},\{a,c,d\}\}$. *Now* $\{a,b\}$ *and* $\{b,c\}$ *improperly contain* $\{a,b\}$ *and* $\{b,c\}$ *of* $\hat{H}^{r_1}$, *respectively. Hence we remove* $\{a,b\}$ *and* $\{b,c\}$ *to obtain* $\hat{H}^{r_2} = \{\{a,c,d\}\}$. *It follows that* $r_1^s = r_1, r_2^s = r_2$. *Now* $\sigma(\{a,b\},r_1^s)(a) = \sigma(\{a,b\},r_1^s)(b) = 0.9, \sigma(\{a,b\},r_1^s)(c) = \sigma(\{a,b\},r_1^s)(d) = 0$, $\sigma(\{b,c\},r_1^s)(b) = \sigma(\{b,c\},r_1^s)(c) = 0.9, \sigma(\{b,c\},r_1^s)(a) = \sigma(\{b,c\},r_1^s)(d) = 0$, $\sigma(\{a,c,d\},r_2^s)(a) = \sigma(\{a,c,d\}, r_2^s)(c) = \sigma(\{a,c,d\},r_2^s)(d) = 0.4, \sigma(\{a,c,d\},r_2^s)(b) = 0$. *Then* $\mathcal{H}^s = (X^s,\mathcal{E}^s)$, *where* $X^s = \{a,b,c,d\}$ *and* $\mathcal{E}^s = \{\sigma(\{a,b\},r_1^s),\sigma(\{b,c\},r_1^s),\sigma(\{a,c,d\},r_2^s)\}$.

An inductive argument dependent upon the cardinality of $\mathbf{F}(\mathcal{H})$ was also developed to prove Theorem 4.25. The argument was supported by the notion of lower truncations $\mathcal{H}_{(t)}$ of $\mathcal{H}$ at level t (see Definitions 4.15 and 4.16). The rationale for this is clear: If $\mathbf{F}(\mathcal{H}) = \{r_1, \ldots, r_i, \ldots, r_n\}$, then

$\mathbf{F}(\mathcal{H}_{(r_i)}) = \{r_1, \ldots, r_i\}$,

and the t-level hypergraph $H^t_{(r_i)}$ of $\mathcal{H}_{(r_i)}$ satisfies

$$H^t_{(r_i)} = \begin{cases} H^t & \text{if } r_i \le t \le h(\mathcal{H}), \\ H^{r_i} & \text{if } 0 < t \le r_i. \end{cases}$$

The proof of our next theorem relies upon the fact that a subset $T \subseteq \mathbf{V}((H^s)^{r_k^s})$ is a transversal of $(H^s)^{r_k^s} \Leftrightarrow$ for each $j = 1, \ldots, k$, there is a subset $T_j \subseteq T$ such that T_j is a transversal of $H^{r_j^s} \in \mathbf{C}(\mathcal{H})$. This characterization is established by the following argument: Suppose $T \subseteq \mathbf{V}((H^s)^{r_k^s})$. Then,

T is a transversal of $(H^s)^{r_k^s} \in \mathbf{C}(\mathcal{H}^s)$

$\Leftrightarrow$ for each $j = 1, \ldots, k$, there is a subset $T_j' \subseteq T$ such that T_j is a transversal of $\hat{H}^{r_j^s} \in \hat{\mathbf{C}}(\mathcal{H})$

$\Leftrightarrow$ for each $j = 1, \ldots, k$, there is a subset $T_j \subseteq T$ such that T_j is a transversal of $H^{r_j^s} \in \mathbf{C}(\mathcal{H})$.

The first equivalence follows from the structure of the edge $\mathcal{E}^s$ of $\mathcal{H}^s$ as described in Lemma 4.23(5). Recall that $\mathcal{H}^s$ is elementary, thus ordered. The latter equivalence follows from the method of constructing $\hat{\mathbf{C}}(\mathcal{H})$ from $\mathbf{C}(\mathcal{H})$ as detailed in Construction 4.1.

Theorem 4.26 *(1) For every fuzzy hypergraph* $\mathcal{H}, Tr^*(\mathcal{H}) \subseteq Tr^*(\mathcal{H}^s)$.

(2) For some $\mathcal{H}, Tr^*(\mathcal{H}) \subset Tr^*(\mathcal{H}^s)$.

Proof. (1) Let $\mathcal{H}$ be a fuzzy hypergraph and let $\mathbf{F}(\mathcal{H}^s) = \{r_1^s, \ldots, r_m^s\}$ be such that $0 < r_m^s < \cdots r_1^s$. Suppose $\tau \in Tr^*(\mathcal{H})$. To show that $\tau \in Tr^*(\mathcal{H}^s)$ it is required to show that $\tau^t \in Tr((H^s)^t)$ for $0 < t \le h(\mathcal{H})$. However, since $\mathbf{F}(\mathcal{H}^s) = \mathbf{F}(Tr(\mathcal{H}))$ (see Lemma 4.23(1)) and $Tr(\mathcal{H})$ is sectionally elementary (see Definition 4.14 and Lemma 4.10(2)), it suffices to show that

$$\tau^{r_k^s} \in Tr((H^s)^{r_k^s}) \quad \text{for } k = 1, \ldots, m. \tag{4.38}$$

Since $\tau \in Tr^*(\mathcal{H})$, it follows that

(1) $\tau \in Tr(\mathcal{H})$,

(2) $\tau^{r_j^s} \in Tr(H^{r_j^s})$, $j = 1, \ldots, m$,

(3) $\tau^{r_k^s} \subseteq \cup_{j=1}^{k} \hat{X}^{r_j^s} = \mathbf{V}((H^s)^{r_k^s})$, $k = 1, \ldots, m$.

Part (3) follows from the fact that $\tau \in Tr(\mathcal{H})$ together with the properties stated in Lemma 4.23(3).

We claim that $\tau^{r_k^s}$ is a transversal of $(H^s)^{r_k^s}$, $k = 1, \ldots, m$, where $(H^s)^{r_k^s} \in \mathbf{C}(\mathcal{H}^s)$.

Clearly,

$$\tau^{r_1^s} \subseteq \tau^{r_2^s} \subseteq \cdots \subseteq \tau^{r_k^s}. \tag{4.39}$$

Each $\tau^{r_j^s}$, $j = 1, \ldots, k$, satisfies

$$\tau^{r_j} \in Tr(H^{r_j^s}), \tag{4.40}$$

according to property (2) above, and, by (3),

$$\tau^{r_k^s} \subseteq \mathbf{V}((H^s)^{r_k^s}). \tag{4.41}$$

Conditions (4.39)-(4.41) are precisely those required for $\tau^{r_k^s}$ to be a transversal of $(H^s)^{r_k^s}$, according to the statement preceding this theorem. Thus, the claim is established.

Property (2) implies that $\tau^{r_k^s} \setminus \{y\}$ is not a transversal of $H^{r_k^s}$ for each $y \in \tau^{r_k^s}$; but this implies that $\tau^{r_k^s} \setminus \{y\}$ is not a transversal of $(H^s)^{r_k^s}$ for each $y \in \tau^{r_k^s}$, according to the statement preceding this theorem. Hence, it is now clear that

$\tau^{r_k^s} \in Tr((H^s)^{r_k^s})$, for $k = 1, \ldots, m$,

and so we have the desired result.

(2) Since $\mathcal{H}^s$ is elementary, it is ordered. Thus $Tr^*(\mathcal{H}^s) \neq \emptyset$ by Theorem 4.8. However by Example 4.3, it is possible for $Tr^*(\mathcal{H}) = \emptyset$. ■

Corollary 4.27 *For every fuzzy hypergraph* $\mathcal{H}$, $Tr^*(\mathcal{H}) \subseteq Tr^*(\mathcal{H}^s) \subseteq Tr(\mathcal{H}^s) = Tr\ (\mathcal{H})$.

Proof. The proof follows from Theorem 4.25 and 4.26(1). ■

Corollary 4.28 *If* $\mathcal{H}$ *is a* T*-related fuzzy hypergraph, then* $Tr^*(\mathcal{H}) = Tr^*(\mathcal{H}^s) = Tr(\mathcal{H}^s) = Tr(\mathcal{H})$.

Proof. The proof follows from Theorem 4.13 and Corollary 4.27. ■

Corollary 4.29 *Let $\mathcal{H}$ be a fuzzy hypergraph. Then*

(1) $Tr^*(\mathcal{H}^s) = Tr(\mathcal{H}^s) \Leftrightarrow \mathcal{H}^s$ is T-related;

(2) $Tr^*(\mathcal{H}^s) = Tr(\mathcal{H}) \Leftrightarrow \mathcal{H}^s$ is T-related.

Proof. Since $\mathcal{H}^s$ is an elementary fuzzy hypergraph, $\mathcal{H}^s$ is an ordered fuzzy hypergraph. Therefore, (1) follows from Theorem 4.14. Part (2) follows easily from part (1) and Theorem 4.25. ∎

The next result provides us with our first unrestricted necessary condition that must exist when $Tr^*(\mathcal{H}) = Tr(\mathcal{H})$.

Corollary 4.30 *Let $\mathcal{H}$ be a fuzzy hypergraph where $Tr^*(\mathcal{H}) = Tr\ (\mathcal{H})$. Then $\mathcal{H}^s$ is T-related.*

Proof. Follows immediately from Corollaries 4.27 and 4.29 (1). ∎

Corollary 4.31 *If $\mathcal{H}$ is T-related, then $\mathcal{H}^s$ is T-related.*

Proof. If $\mathcal{H}$ is T-related, then $Tr^*(\mathcal{H}) = Tr(\mathcal{H})$, according to Theorem 4.13. Therefore, by Corollary 4.30, $\mathcal{H}^s$ is T-related. ∎

Neither Corollary 4.30 nor 4.31 enjoy a converse. In the case of Corollary 4.31, we use Example 4.4, which exhibits a fuzzy hypergraph $\mathcal{H}$ that is not T-related but satisfies $Tr^*(\mathcal{H}) = Tr(\mathcal{H})$. Applying Corollary 4.30 to this example reveals that $\mathcal{H}^s$ is T-related. However, since $\mathcal{H}$ is not T-related, Corollary 4.31 has no converse.

The following example shows why Corollary 4.30 has no converse.

Example 4.8 *Let $\mathcal{H} = (X, \mathcal{E})$ be a fuzzy hypergraph with an edge set $\mathcal{E}$ defined by the incidence matrix*

$$\begin{array}{c} \\ a \\ b \\ c \end{array} \begin{array}{c} \begin{array}{ccc} \mu_1 & \mu_2 & \mu_3 \end{array} \\ \begin{pmatrix} 0.9 & 0 & 0 \\ 0.4 & 0.4 & 0.4 \\ 0.2 & 0.9 & 0.2 \end{pmatrix} \end{array}.$$

$\mathbf{C}(\mathcal{H}) = \{H^{r_i} = (X_i, \mathbf{E}_i) \mid r_1 = 0.9,\ r_2 = 0.4,\ r_3 = 0.2\}$
with
$\mathbf{E}_1 = \{\{a\}, \{c\}\}, \qquad \mathbf{E}_2 = \{\{a, b\}, \{b, c\}, \{b\}\},$
and
$\mathbf{E}_3 = \{\{a, b, c\}, \{b, c\}\}.$
Therefore,

$$Tr(\mathcal{H}) = \begin{array}{c} \\ a \\ b \\ c \end{array} \begin{array}{c} \tau \\ \begin{pmatrix} 0.9 \\ 0.4 \\ 0.9 \end{pmatrix} \end{array} \text{ and } Tr^*(\mathcal{H}) = \emptyset.$$

However, the incidence matrix for $\mathcal{H}^s$ is

$$\begin{array}{c} \\ a \\ b \\ c \end{array} \begin{array}{c} \begin{array}{ccc} \nu_1 & \nu_2 & \nu_3 \end{array} \\ \left(\begin{array}{ccc} 0.9 & 0 & 0 \\ 0 & 0 & 0.4 \\ 0 & 0.9 & 0 \end{array} \right) \end{array}.$$

Since
$\mathbf{C}(\mathcal{H}^s) = \{(H^s)^{r_i^s} = ((X^s)^{r_i^s}, (\mathbf{E}^s)^{r_i^s}) \mid r_1^s = 0.9,\ r_2^s = 0.4\}$
with
$(\mathbf{E}^s)^{r_1^s} = \{\{a\}, \{c\}\}$ *and* $(\mathbf{E}^s)^{r_2^s} = \{\{a\}, \{b\}, \{c\}\}$,
we have that
$Tr^*(\mathcal{H}^s) = Tr(\mathcal{H}^s) = Tr(\mathcal{H})$.
Consequently, in view of Corollary 4.29(1), $\mathcal{H}^s$ *is T-related, while, as indicated earlier,* $Tr^*(\mathcal{H}) \neq Tr(\mathcal{H})$.

The remaining set of theorems consider the question: What sort of attributes are required of $\mathcal{H}$ to guarantee that $\mathcal{H}^s$ is a μ-tempered fuzzy hypergraph, $\mu \otimes H$, of some crisp hypergraph H? The feasibility in answering this question centers on Theorem 4.2.

In every case, the skeleton $\mathcal{H}^s$, associated with a fuzzy hypergraph $\mathcal{H}$, is elementary, ordered and support simple. (This is evident from Construction 4.2 of $\mathcal{H}^s$.) Thus, in view of Theorem 4.2, the answer to the above question hinges on recognizing properties which, if possessed by $\mathcal{H}$, are sufficient to guarantee that $\mathcal{H}^s$ is simply ordered. One such special property is provided below in Definition 4.22; this property is confirmed as acceptable in Theorem 4.34. But first it is useful to formalize the above discussion and provide detailed structural information about $\mathcal{H}^s$.

The next theorem utilizes considerable notation and results from Lemmas 4.22 and 4.23, some of which are restated here for convenience: $\mathcal{H}^s = (X^s, \mathcal{E}^s)$ and $\mathbf{F}(\mathcal{H}^s) = \{r_j^s \mid j = 1, \ldots, m\}$, also $\hat{\mathbf{C}}(\mathcal{H}) = \{\hat{H}^{r_j^s} = (\hat{X}^{r_j^s}, \hat{\mathbf{E}}^{r_j^s}) \mid r_j^s \in \mathbf{F}(\mathcal{H}^s)\}$ and $\mathbf{C}(\mathcal{H}^s) = \{(H^s)^{r_j^s} = ((X^s)^{r_j^s}, (\mathbf{E}^s)^{r_j^s}) \mid r_j^s \in \mathbf{F}(\mathcal{H}^s)\}$. Moreover, $(X^s)^{r_k^s} = \cup_{j=1}^k \hat{X}^{r_j^s}$, $k = 1, \ldots, m$. Thus, $(X^s)^{r_k^s} \setminus (X^s)^{r_{k-1}^s} = (\hat{X}^s)^{r_k^s} \setminus (X^s)^{r_{k-1}^s}$. Also, $X^s = \hat{X} = \cup_{j=1}^m \hat{X}^{r_j^s}$ and $\hat{\mathbf{E}} = \cup_{j=1}^m \hat{\mathbf{E}}^{r_j^s}$.

Theorem 4.32 *If* $\mathcal{H}^s$ *is simply ordered, then* $\mathcal{H}^s = \mu^s \otimes H^s$, *where* $H^s = (\hat{X}, \hat{E})$ *and* $\mu^s \colon \hat{X} \to (0, 1]$ *such that, for* $k = 1, \ldots, m$, $\mu^s(x) = r_k^s$ *if and only if* $x \in \hat{X}^{r_k^s} \setminus (X^s)^{r_{k-1}^s}$. *(Here it is understood that* $(X^s)^{r_0^s} = \emptyset$.*)*

Proof. Since $\mathcal{H}^s$ is simply ordered, $\hat{X}^{r_k^s} \setminus (X^s)^{r_{k-1}^s} \neq \emptyset$ for $j = 1, \ldots, m$, every $\mathcal{H}^s$ is elementary and support simple. Therefore, according to Theorem 4.2, if $\mathcal{H}^s$ is simply ordered then $\mathcal{H}^s = \mu^s \otimes H^s$. Clearly, $H^s = (\hat{X}, \hat{\mathbf{E}})$. To determine the structure of μ^s it suffices to notice that every edge $E \in \hat{H}^{r_k^s}$ satisfies $E \cap (\hat{X}^{r_k^s} \setminus (X^s)^{r_{k-1}^s}) \neq \emptyset$.

This follows at once from the assumption that $\mathcal{H}^s$ is simply ordered. The acceptability of the structure for μ^s, as stated in this theorem, now follows from Definition 4.12 and Lemma 4.23(5). ∎

We illustrate Theorem 4.32 by recalling the fuzzy hypergreaph $\mathcal{H} = (X, \mathcal{E})$ of Example 4.7. We see that $\hat{X}^{r_1^s} \setminus (X^s)^{r_0^s} = \hat{X}^{0.9} \setminus \emptyset = \hat{X}^{0.9} = \{a, c\}$ and $\hat{X}^{r_2^s} \setminus (X^s)^{r_1^s} = \hat{X}^{0.4} \setminus (X^s)^{0.9} = \{b\} \setminus \{a, c\} = \{b\}$. Thus $\mu^s(a) = \mu^s(c) = 0.9$ and $\mu^s(b) = 0.4$. Now $H^s = (\{a, b, c\}, \{\{a\}, \{b\}, \{c\}\})$ and $\mathcal{H}^s = \mu^s \otimes H^s$.

Definition 4.22 *Let $\mathcal{H}$ be a fuzzy hypergraph with $\mathbf{F}(\mathcal{H}) = \{r_1, \ldots, r_n\}$, indexed in the usual order $0 < r_m < \cdots < r_1$, and $\mathbf{C}(\mathcal{H}) = \{H^{r_i} = (X^{r_i}, \mathbf{E}^{r_i}) \mid i = 1, \ldots, n\}$. Then $\mathcal{H}$ is said to be* sequentially simple *if, for $i = 2, \ldots, n$, $E \in \mathbf{E}^{r_i} \setminus \mathbf{E}^{r_{i-1}}$ implies $E \setminus X^{r_{i-1}} \neq \emptyset$.*

Lemma 4.33 *If $\mathcal{H}$ is a sequentially simple fuzzy hypergraph, then the skeleton $\mathcal{H}^s$ of $\mathcal{H}$ is sequentially simple.*

Proof. Assume that $\mathbf{F}(\mathcal{H}) = \{r_1, \ldots, r_n\}$ where $0 < r_n < \cdots < r_1$ and $\mathbf{F}(\mathcal{H}^s) = \{r_1^s, \ldots, r_m^s\}$. Suppose
$E \in (\mathbf{E}^s)^{r_{j+1}^s} \setminus (\mathbf{E}^s)^{r_j^s}$.
Then, in view of Lemma 4.23(2), it is evident that

$$E \in \hat{\mathbf{E}}^{r_{j+1}^s}. \tag{4.42}$$

Therefore, since $\hat{\mathbf{E}}^{r_{j+1}^s} \subseteq \mathbf{E}^{r_{j+1}^s}$, it is clear that $E \in \mathbf{E}^{r_{j+1}^s}$. Now, as $\mathbf{F}(\mathcal{H}^s) = \hat{\mathbf{F}}(\mathcal{H})$ is a subsequence of $\mathbf{F}(\mathcal{H})$, there exist some $r_{k+1} \in \mathbf{F}(\mathcal{H})$ such that

$$r_{k+1} = r_{j+1}^s \quad \text{(of course } j \leq k\text{)}. \tag{4.43}$$

Thus,

$$E \in \mathbf{E}^{r_{k+1}}. \tag{4.44}$$

We claim that $E \in \mathbf{E}^{r_{k+1}} \setminus \mathbf{E}^{r_k}$. For suppose $E \notin \mathbf{E}^{r_{k+1}} \setminus \mathbf{E}^{r_k}$. Since E satisfies (4.44), this assumption would imply that

$$E \in \mathbf{E}^{r_k}, \tag{4.45}$$

where $r_{j+1}^s < r_k$. However, by the way $\hat{\mathbf{C}}(\mathcal{H})$ is constructed (see Construction 4.1), it would follow from (4.45) that $E \notin \hat{\mathbf{E}}^{r_{k+1}}$. This would imply, in view of (4.43), that $E \notin \mathbf{E}^{r_{j+1}^s}$ which contradicts (4.42).

With the claim established, it then follows from the hypothetical assumption concerning $\mathcal{H}$ that $E \nsubseteq X^{r_k} (= \mathbf{V}(H^{r_k}))$ which, in turn, implies that

$$E \nsubseteq (X^s)^{r_j^s} (= \mathbf{V}((H^s)^{r_j^s})) \tag{4.46}$$

since

$$(X^s)^{r_j^s} \subseteq X^{r_k}, \tag{4.47}$$

where (4.47) follows from the fact that $r_k \le r_j^s$ (since $r_{k+1} = r_{j+1}^s$) together with the obvious fact that, for $i = 1, \ldots, n-1, X_i = \mathbf{V}(H^i) \subseteq \mathbf{V}(H^{i+1}) = X_{i+1}$, and also from the method of constructing $\hat{\mathbf{C}}(\mathcal{H})$ along with the subsequent assembly of $\mathcal{H}^s$ according to Lemma 4.23(5). From (4.46) we have the desired result. ■

Theorem 4.34 *If $\mathcal{H}$ is a sequentially simple fuzzy hypergraph, then $\mathcal{H}^s = \mu^s \otimes H^s$ with μ^s and H^s defined in Theorem 4.32.*

Proof. By Lemma 4.33, $\mathcal{H}^s$ is sequentially simple since $\mathcal{H}$ is assumed to have this property. Therefore, since $\mathcal{H}^s$ is ordered, it is simply ordered, and the result follows from Theorem 4.32. ■

Corollary 4.35 *If $\mathcal{H} = \mu \otimes H$, then $\mathcal{H}^s = \mu^s \otimes H^s$ with μ^s and H^s defined in Theorem 4.32. Moreover, if H has no repeated edges, then $H^s = H$ if and only if H is simple.*

Proof. $\mathcal{H}$ is simply ordered by Theorem 4.2 and therefore $\mathcal{H}$ is sequentially simple. The first result now follows from Theorem 4.34. The proof of the remaining portion of the theorem follows from the process in determining $\mathcal{H}^s$ (see Construction 4.1 and 4.2) and from the construction of $\mu \otimes H$ as presented in Definition 4.12. In particular, if H is not simple and without repeated edges then there is a pair of edges E_1 and E_2 in H such that $E_1 \subset E_2$ and there is a unique pair of elementary edges $\sigma(E_1, r_1)$ and $\sigma(E_2, r_2)$ belonging to $\mu \otimes H$, where $r_1 \ge r_2 > 0$; this latter relation follows easily from the following conditions:

(a) $r_i = \wedge \{\mu(e) \mid e \in \mathbf{E}_i\}$, $i = 1, 2$,

(b) $E_1 \subset E_2$

(c) $\mu(e) > 0$ for all $e \in \mathbf{V}(H)$, see Definition 4.12. Therefore, in the determination of $\hat{\mathbf{C}}(\mu \otimes H)$, E_2 will be eliminated so that E_2 does not belong to H^s. Therefore, $H^s \subset H$, as the condition $H^s \subseteq H$ must hold. This proves that $H^s = H$ implies H is simple.

Conversely, suppose H is a simple crisp hypergraph with edge E; corresponding to E is the unique elementary edge $\sigma(E, r)$ of $\mu \otimes H$. Since E does not contain another edge E' of H, there cannot be another distinct elementary edge $\sigma(E', r')$ of $\mu \otimes H$ satisfying conditions:

(c) $E' \subseteq E$,

(d) $0 < r \le r'$.

Consequently, E is selected in the process of determining $\hat{\mathbf{C}}(\mu \otimes H)$. Hence as a result $E \in H^s$. This shows that $H \subseteq H^s$. Thus $H^s = H$, as it is always true that $H^s \subseteq H$. Thus the assertion that $H^s = H$ if and only if H is simple is established. ■

Proposition 4.36 *If $\mathcal{H}$ is a μ-tempered fuzzy hypergraph of a crisp hypergraph H, that is, $\mathcal{H} = \mu \otimes H$, then the skeleton $\mathcal{H}^s$ of $\mathcal{H}$ is a μ^s-tempered hypergraph of H^s, where H^s is a simple crisp hypergraph.* ■

We conclude this section with a working example of a fuzzy hypergraph. A fuzzy hypergraph conceptualization of a stock exchange might consider the publicly-traded companies in the exchange as the vertices and the sectors represented within the exchange as the edge set. The edge set might include, among others, the utility, energy, financial service, technology, industrial, leisure, and health science edges. Degree of membership within each sector could be determined by a variety of fuzzy strategies that would accept and integrate information supplied by one or several information sources. A portfolio (at time t) that includes all sectors at each level of importance would be considered a transversal of the fuzzy hypergraph as it exists at time t.

Portfolio management may utilize asset allocation stratagems which are designed to fuzzily identify sectors from the most desirable (ties permitted, of course) for investment at time t. Accordingly, under such management, a portfolio would normally include representations from only a proper subset of sectors at any given time; at such times t, the portfolio would, very likely, represent a transversal of a (proper) partial fuzzy hypergraph or, perhaps, of a suitably scaled partial fuzzy hypergraph. Scaling would occur, for example, if the degree of overall strength of each sector, as determined by asset allocation analysis, were utilized to reassign the degrees of membership within each edge as specified by some fuzzy logic procedure.

Using the above strategy, a well-designed minimal portfolio of stocks (within the exchange) would, most likely, represent a minimal fuzzy transversal of some determined (perhaps scaled) partial fuzzy hypergraph.

4.3 Coloring of Fuzzy Hypergraphs

A *k-coloring* C of a crisp hypergraph $H = (X, \mathbf{E})$ is a partition $\{S_1, \ldots, S_k\}$ of X into k subsets (colors) such that each edge $E \in \mathbf{E}$ with cardinality ≥ 2 intersects at least two of these subsets. That is, each edge which is not a loop (singleton edge) contains at least two of the k-colors. We now extend the concept of k-coloring to fuzzy hypergraphs. The first fuzzy extension below is trivial. It is valuable when $\mathcal{H}$ is elementary because it is totally indifferent to the effects of fuzziness. The objection is rectified in the second fuzzy extension which appears below.

Definition 4.23 *Let $\mathcal{H} = (X, \mathcal{E})$ be a fuzzy hypergraph. A* primitive k-coloring *(or simply a* p-coloring*), C, of $\mathcal{H}$ is a partition of X into k subsets (colors) such that the support of each fuzzy edge of $\mathcal{H}$ intersects at least two* *colors of* C*, except for edges which are "spikes," that is, fuzzy subsets with singleton support.*

The next definition features a *coloring* of the core set of $\mathcal{H}$ which is accordingly termed an *$\mathcal{L}$-coloring* of $\mathcal{H}$:

Definition 4.24 *Let* $\mathcal{H} = (X, \mathcal{E})$ *be a fuzzy hypergraph and let* $\mathbf{C}(\mathcal{H}) = \{H^{r_1}, H^{r_2}, \ldots, H^{r_n}\}$*. An* $\mathcal{L}$-coloring, C, *of* $\mathcal{H}$ *with* k *components is a partition of* X *into* k *subsets* $\{S_1, \ldots, S_k\}$ *such that* C *induces a coloring for each core hypergraph,* H^{r_i}*, of* $\mathcal{H}$*, that is, with* $H^{r_i} = (X_i, \mathbf{E}_i)$*, the restriction of* C *to* X_i*,* $\{S_1 \cap X_i, \ldots, S_k \cap X_i\}$*, is a coloring of* H^{r_i}*. (We allow a color set* S_i *to be empty.)*

It is clear that an $\mathcal{L}$-coloring of $\mathcal{H}$ is a primitive coloring of $\mathcal{H}$. However, as the next example shows, the converse is not generally true.

Example 4.9 *Suppose* $= (X, \mathcal{E})$ *is a fuzzy hypergraph and* $C(\mathcal{H}) = \{H^{r_1}, H^{r_2}, \ldots, H^{r_n}\}$*, where* $0 < r_n < \cdots < r_1$*. Assume the core hypergraph* $H^{r_n} = (X_n, \mathbf{E}_n)$ *is simple, and that there is an edge* S *of some core hypergraph* H^{r_i}*, with cardinality* $|S|$ *greater than one, which is not an edge of* H^{r_n}*. Then* S *is the* r_i*-cut of some fuzzy edge* ν*. Thus* $\hat{S} = \nu^{r_n}$ *is an edge of* H^{r_n} *and* $S \subset \hat{S}$*. Since* H^{r_n} *is simple, it follows that* $E \setminus \hat{S} \neq \emptyset$ *for every* $E \in \mathbf{E}_n \setminus \{\hat{S}\}$*. Hence* $E \setminus S \neq \emptyset$ *for each edge* $E \in \mathbf{E}_n$*. Assign the color blue to each* $s \in S$*. Let* C' *be a coloring (excluding the color blue) of the crisp hypergraph on* $X_n \setminus S$ *whose edges consist of all sets of the form* $E \setminus S$ *where* $E \in \mathbf{E}_n$*. Finally, let* C *be the coloring of* H^{r_n} *that coincides with* C' *on* $X_n \setminus S$ *and assigns the color blue to each* $s \in S$*. Then* C *is a primitive coloring of* $\mathcal{H}$*, but not a coloring of* H^{r_i} *when restricted to* X^{r_i}*. For example consider the fuzzy hypergraph* $\mathcal{H} = (X, \mathcal{E})$*, where* $X = \{a, b, c\}$ *and* $\mathcal{E} = \{\mu, \nu\}$ *such that* $\mu(a) = \mu(b) = \mu(c) = 0.4$ *and* $\nu(a) = \nu(b) = 0.9, \nu(c) = 0$*. We have that* H^{r_2} *is simple. There exists an edge of* S *of* H^{r_1} *such that* $|S| > 1$ *and* $S \notin \mathbf{E}_2$*, namely* $S = \{a, b\}$*. Now* $\hat{S} = \mu^{r_3} = \{a, b, c\}, S \subset \hat{S}, \mathbf{E}_2 = \{\{a, b, c\}\}$ *and* $E \setminus \hat{S} \neq \emptyset, \forall E \in \mathbf{E}_2 \setminus \{\hat{S}\}$*. We assign the color blue to* a *and* b*. Let* C' *be a coloring excluding blue on* $X_2 \setminus S = \{c\}$*. Let* C *be the coloring of* H^{r_2} *that coincides with* C' *on* $X_2 \setminus S$*, say red on* $\{c\}$ *and blue on* $\{a\}$ *and* $\{b\}$*. Then* C *is a primitive coloring of* $\mathcal{H}$ *because supp*$(\mu) = \{a, b, c\}$ *and* $\{a, b, c\} \cap C$ *intersects two colors of* C*, but is not a coloring of* H^{r_1} *when restricted to* X^{r_1}*. That is,* C *is not a* $\mathcal{L}$*-coloring because when* C *is restricted to* $X_1 = \{a, b\}$*, it is not a coloring of* H^{r_1}*. Note that* $\mathcal{H}$ *is not ordered.*

Under certain conditions primitive colorings and $\mathcal{L}$-colorings of $\mathcal{H}$ coincide. The next theorem gives an example of such a condition.

Theorem 4.37 *If $\mathcal{H} = (X, \mathcal{E})$ is an ordered fuzzy hypergraph and C is a primitive coloring of $\mathcal{H}$, then C is an $\mathcal{L}$-coloring of $\mathcal{H}$.* ∎

The next result is a partial converse of Theorem 4.37.

Theorem 4.38 *Let $\mathcal{H} = (X, \mathcal{E})$ be a fuzzy hypergraph and suppose $\mathbf{C}(\mathcal{H}) = \{H^{r_i} \mid i = 1, 2, \ldots, n\}$, where $0 < r_n < \cdots < r_1$. If H^{r_n} is a simple hypergraph and singleton edges do not appear in any core hypergraph of $\mathcal{H}$ and if each primitive coloring C of $\mathcal{H}$ is an $\mathcal{L}$-coloring of $\mathcal{H}$, then $\mathcal{H}$ is an ordered fuzzy hypergraph.*

Proof. Recall that $H^{r_i} = (X_i, \mathbf{E}_i)$ for $1 \leq i \leq n$. Assume H^{r_n} is simple and that $\mathcal{H}$ is not ordered. Then there exists a primitive coloring of $\mathcal{H}$ that is not an $\mathcal{L}$-coloring of $\mathcal{H}$ which is constructed as follows: Since $\mathcal{H}$ is assumed not ordered, it can then be assumed that there is some core hypergraph H^{r_i}, where $i \leq n-1$, such that some edge S of $\mathbf{E}_i$ is not an edge of $\mathbf{E}_{i+1}$. From the definition of core hypergraph, if follows that there is a fuzzy edge $\nu \in \mathcal{E}$ such that $\nu^{r_i} = S$. Let $\hat{S} = \nu^{r_{i+1}}$ and $T = \nu^{r_n}$. Then, $S \subset \hat{S} \subseteq T$. Since H^{r_n} is simple and $T \in \mathbf{E}_n$, it follows that $S \notin \mathbf{E}_n$; moreover, by hypothesis, $|S| \geq 2$. This is precisely the situation described in Example 4.9, therefore, there is a primitive coloring of $\mathcal{H}$ that is not an $\mathcal{L}$-coloring of $\mathcal{H}$. ∎

If prediction is necessary in a problem, $\mathcal{L}$-colorings of suitable fuzzy hypergraphs may provide an avenue toward analysis and resolution of the problem. Consider the following problem: Stability within high-tech fields is problematic as rapidly emerging technologies and applications compete with established methods and practices for dominance within such fields; as a result, obsolescence and incompatibilities between existing and emerging technologies and services are commonplace. Survivability, even for the largest players, demands awareness of emerging trends and, therefore, players must plan and adapt accordingly. From this viewpoint, the study of possible colorings of relevant fuzzy hypergraphs may assist in identifying fuzzy groups of existing and emerging technologies and services which are fuzzily compatible for the fuzzy foreseeable future.

For example, assume $\mathcal{H} = (V, \mathcal{E})$ is a fuzzy hypergraph, where V is either a crisp or fuzzy subset μ over a groundfield X of existing and emerging technologies and applications within one or more sectors of the economy and where the support set,

$\mathbf{E} = \{E \subseteq \text{supp}(V) \mid E = \text{supp}(\nu),\ \nu \in \mathcal{E}\}$,

of $\mathcal{E}$ contains all minimal incompatible subsets of supp(V) (i.e., with out loss of generality it is assumed that $\mathbf{E}$ is a simple crisp hypergraph over supp(V)). With the establishment of $\mathbf{E}$, one method to determine $\mathcal{E}$ would be to first determine a fuzzy subset $\omega : \mathbf{E} \to (0, 1]$, where degree of membership, $\omega(E)$, indicates the degree of incompatibility for each $E \in \mathbf{E}$; once established, ω could then be combined with the fuzzy vertex set μ, via a determined fuzzy logic process, to obtain the fuzzy edge set $\mathcal{E}$ of $\mathcal{H}$.

The "colors" of the $\mathcal{L}$-colorings would provide a collection of fuzzily compatible groups of technologies and services which could then be compared and critiqued by a player for acceptability or rejection with respect to the player's goals and objectives.

We note that to simplify the above presentation it was tacitly assumed that every member of $\mathbf{E}$ should have cardinality ≥ 2 and that the union of all members of $\mathbf{E}$ should be supp(V). However, one could easily envision the possibility that some members are compatible with all members of supp(V); in such cases the corresponding spikes, $\sigma(\{a\},\ \mu(a))$, would be desirable additions to the edge set $\mathcal{E}$ considering the fact that their supports could belong to any desired color in a valid coloring of $\mathcal{H}$ (with membership value $\mu(a)$, see concluding remarks in this section).

In the example to follow, we consider another way to interpret a coloring of a fuzzy hypergraph. We partition the set X of companies into sectors of compatible or similar companies, e.g., airlines. The partition is held fixed and we are interested in the intensity with which the partition is an $\mathcal{L}$-coloring of the fuzzy hypergraph $(X, \mathcal{E})$, where members of $\mathcal{E}$ denote technological companies which supply support to the members of X and whose membership may have any number of interpretations depending upon the problem.

Example 4.10 *Consider the hypergraph $\mathcal{H} = (X, \mathcal{E})$ where $X = \{a, b, c, d, e, f\}$ and $\mathcal{E} = \{\mu_1, \mu_2, \mu_3\}$, which is represented by the following incidence matrix:*

$$\begin{array}{c} \\ a \\ b \\ c \\ d \\ e \\ f \end{array} \begin{array}{c} \begin{array}{ccc} \mu_1 & \mu_2 & \mu_3 \end{array} \\ \begin{pmatrix} 0.9 & 0.7 & 0.7 \\ 0.9 & 0.9 & 0.4 \\ 0.7 & 0 & 0.9 \\ 0 & 0 & 0.9 \\ 0.4 & 0.4 & 0 \\ 0.4 & 0.7 & 0 \end{pmatrix} \end{array}.$$

Then the core hypergraphs are as follows:

$H^{0.9} = (\{a, b, c, d\}, \{\{a, b\}, \{b\}, \{c, d\}\})$

$H^{0.7} = (\{a, b, c, d, f\}, \{\{a, b, c\}, \{a, b, f\}, \{a, c, d\}\})$

$H^{0.4} = (\{a, b, c, d, e, f\}, \{\{a, b, c, d, e, f\}, \{a, b, e, f\}, \{a, b, c, d\}\})$

Suppose that the partition of X of interest is $C = \{\{a, b\}, \{c, d\}, \{e, f\}\}$. Then C is a coloring of $H^{0.7}$ and $H^{0.4}$, but not of $H^{0.9}$. Hence we conclude that C is an $\mathcal{L}$-coloring of $\mathcal{H}$ with intensity 0.7.

The *chromatic number* $\chi(H)$ of a crisp hypergraph is the minimal number, k, of colors needed to produce a k-coloring of H. A trivial extension of the crisp definition of a chromatic number (given below) can be useful for certain fuzzy hypergraph problems; however, it is insufficient for a variety of other questions. For this reason important non-trivial extensions of chromatic numbers appear later.

Definition 4.25 *The* p-chromatic number *of a fuzzy hypergraph* $\mathcal{H} = (X, \mathcal{E})$ *is the minimal number* $\chi_p(\mathcal{H})$, *of colors needed to produce a primitive coloring of* $\mathcal{H}$. *The* chromatic number *of* $\mathcal{H}$ *is the minimal number,* $\chi(\mathcal{H})$, *of colors needed to produce an* $\mathcal{L}$*-coloring of* $\mathcal{H}$.

The following result is easily established:

Theorem 4.39 *If* $\mathcal{H} = (X, \mathcal{E})$ *is an ordered fuzzy hypergraph and* $\mathbf{C}(\mathcal{H}) = \{H^{r_i} \mid i = 1, 2, \ldots, n\}$, *then* $\chi(H^{r_1}) \le \chi(H^{r_2}) \le \cdots \le \chi(H^{r_n}) = \chi(H)$, *where* $\chi(H^{r_i})$ *represents the minimal number of colors required to color the crisp hypergraph* H^{r_i}. ■

However, there is no general outline for the sequence of chromatic numbers attached to core sets. This can be seen by comparing Theorem 4.39 and the next example:

Example 4.11 *Suppose* $\mathcal{H} = (X, \mathcal{E})$ *is the fuzzy hypergraph where* $X = \{a, b, c, d, e, f\}$ *and* $\mathcal{E}$ *is determined by the incidence matrix*

$$\begin{array}{c} \\ a \\ b \\ c \\ d \\ e \\ f \end{array}
\begin{array}{c} \begin{array}{cccccccccc} \mu_1 & \mu_2 & \mu_3 & \mu_4 & \mu_5 & \mu_6 & \mu_7 & \mu_8 & \mu_9 & \mu_{10} \end{array} \\
\left(\begin{array}{cccccccccc}
0.9 & 0.9 & 0.9 & 0.9 & 0.9 & 0.9 & 0 & 0 & 0 & 0 \\
0.9 & 0.9 & 0.9 & 0 & 0 & 0 & 0.9 & 0.9 & 0.9 & 0 \\
0 & 0 & 0 & 0.9 & 0.9 & 0.9 & 0.9 & 0.9 & 0.9 & 0 \\
0.4 & 0 & 0 & 0.4 & 0 & 0 & 0.4 & 0 & 0 & 0.4 \\
0 & 0.4 & 0 & 0 & 0.4 & 0 & 0 & 0.4 & 0 & 0.2 \\
0 & 0 & 0.4 & 0 & 0 & 0.4 & 0 & 0 & 0.4 & 0.2
\end{array}\right) \end{array}$$

Clearly, $\mathbf{C}(\mathcal{H}) = \{H^{r_i} = (X_i, \mathbf{E}_i) \mid i = 1, 2, 3\}$, *where* $r_1 = 0.9$, $r_2 = 0.4$, $r_3 = 0.2$, $\mathbf{E}_1 = \{\{a, b\}, \{a, c\}, \{b, c\}\}$, $\mathbf{E}_2 = \{\{a, b, d\}, \{a, b, e\}, \{a, b, f\}, \{a, c, d\}, \{a, c, e\}, \{a, c, f\}, \{b, c, d\}, \{b, c, e\}, \{b, c, f\}, \{d\}\}$, *and* $\mathbf{E}_3 = (\mathbf{E}_2 \cup \{d, e, f\}) \setminus \{d\}$. *Consider* H^{r_1}. *Suppose* $\{S_1, S_2\}$ *is a coloring of* H^{r_1}. *Then* $\{a, b\} \cap S_i \neq \emptyset$, $\{a, c\} \cap S_i \neq \emptyset$ *and* $\{b, c\} \cap S_i \neq \emptyset$ *for* $i = 1, 2$. *Hence* $S_1 \cap S_2 \neq \emptyset$ *a contradiction. Thus* $\chi(H^{r_1}) = 3$. *We have that* $\{\{a, b, c\}, \{d, e, f\}\}$ *is a coloring for* H^{r_2} *and so* $\chi(H^{r_2}) = 2$. *For* H^{r_3}, *we see that for* $E \subseteq X$ *such that* $|E| = 3$, *either* E *or its complement is in* $\mathbf{E}_3$. *This fact can be used in the argument that* $\chi(H^{r_3}) = 3$.

The search for optimal colorings (for example, discovering a coloring with minimal chromatic number) can be complicated by the added dimension of fuzziness. Consider a hypothetical situation concerning managerial decisions of possible future events. The results of the process may produce a fuzzy hypergraph where an efficient top-down $\mathcal{L}$-coloring is desired - it is conceivable that management would like to obtain an optimal coloring of the present situation represented by the core hypergraph H^{r_1} and postpone coloring future core hypergraphs, until it is absolutely necessary to do so. However, a top-down method of coloring the core set generally does not work well if at all (except in special cases, some of which may appear among the $\mu \otimes H$ hypergraphs). On the other hand, if management is committed "long term" and the required "speculative" analysis is carefully crafted so that the resulting fuzzy hypergraph $\mathcal{H}$ is "realistic," then, with this extended fuzzy knowledge, there are effective coloring methods for use in long term analysis, but may necessarily be less than optimal in coloring H^{r_1}. One "long-term" coloring procedure centers on the idea of replacing $\mathcal{H}$ by a simpler structure similar to $\mathcal{H}^s$ as described previously. An advantage is that $\mathcal{H}^s$ is ordered. Therefore, by Theorem 4.37, any $\mathcal{L}$-coloring is equivalent to a primitive coloring of $\mathcal{H}^s$ and thus a crisp hypergraph coloring problem. Usually, in coloring problems, the conversion from $\mathcal{H}$ to $\mathcal{H}^s$ cannot be done directly since, in the $(\cdot)^s$ procedure, edges in $\mathcal{H}$ which are not spikes may be replaced by representatives which are spikes in $\mathcal{H}^s$. Generally, spikes are not significant in coloring problems since their supports are singletons; therefore, colorings of $\mathcal{H}^s$ may not be relevant to colorings of $\mathcal{H}$. To avoid this situation, first remove all spikes and all terminal spikes from edges of $\mathcal{H}$ before applying the $(\cdot)^s$ procedure. The conversion procedure is expressed explicitly as follows:

Definition 4.26 *A* spike reduction *of $\mu \in \mathfrak{F}\wp(X)$, denoted by μ^-, is defined by $\mu^-(x) = \vee\{r \mid |\mu^r| \geq 2, r \leq \mu(x)\}$, where $\vee\emptyset = 0$.*

In particular, if μ is a spike, then $\mu^- = \chi_\emptyset$.

Definition 4.27 *Let $\mathcal{H} = (X, \mathcal{E})$ be a fuzzy hypergraph and let $\mathcal{H}^- = (X^-, \mathcal{E}^-)$, where $\mathcal{E}^- = \{\mu^- \mid \mu \in \mathcal{E}\}$ and $X^- = \bigcup_{\mu^- \in \mathcal{E}^-} supp(\mu^-)$.*

In Example 4.11, $(\mu_{10})^{0.4} = \{d\}$. Hence $\mu_{10}^-(a) = \mu_{10}^-(b) = \mu_{10}^-(c) = 0$ and $\mu_{10}^-(d) = \mu_{10}^-(e) = \mu_{10}^-(f) = 0.2$. We see that $\mu_{10} \neq \mu_{10}^-$. Since $\mu_{10}^- \neq \emptyset$, μ_{10} is not a spike.

If each edge of $\mathcal{H}$ is a spike, then $\mathcal{E}^- = \emptyset$ (i.e., $\mathcal{H}^-$ is not a fuzzy hypergraph). This special case has no real coloring problem associated with it, so we exclude it from further consideration and always assume $\mathcal{H}^-$ exists.

Definition 4.28 *Let $\mathcal{H}^\Delta$ denote the skeleton of $\mathcal{H}^-$; that is, $\mathcal{H}^\Delta = (\mathcal{H}^-)^s$.*

Theorem 4.40 *For each fuzzy hypergraph $\mathcal{H}$ for which $\mathcal{H}^-$ exists, every primitive p-coloring of $\mathcal{H}^\Delta$ is an $\mathcal{L}$-coloring of $\mathcal{H}^\Delta$ and conversely.*

Proof. Since $\mathcal{H}^{\Delta}$ is an ordered fuzzy hypergraph, the result follows from Theorem 4.37. ■

Definition 4.29 *Let $\mathcal{H}_1 = (X_1, \mathcal{E}_1)$ and $\mathcal{H}_2 = (X_2, \mathcal{E}_2)$ be a pair of fuzzy hypergraphs such that $X_1 \subseteq X_2$. Suppose $C_1 = \{S_1, S_2, \ldots, S_k\}$, where $\cup_{i=1}^{k} S_i = X_1$, and $S_i \neq \emptyset$, for $i = 1, \ldots, k$, is an $\mathcal{L}$-coloring (or p-coloring) of $\mathcal{H}_1$. Then C_2 is a* stable $\mathcal{L}$-coloring (*or* p-coloring) extension *of C_1 to $\mathcal{H}_2$ if $C_2 = \{S_1', S_2', \ldots, S_k'\}$ is an $\mathcal{L}$-coloring (or p-coloring) of $\mathcal{H}_2$ which satisfies*

(1) $\cup_{i=1}^{k} S_i' = X_2$,

(2) $S_i \subseteq S_i'$ *for* $i = 1, \ldots, k$.

The proof of the following theorem follows from the construction of $\mathcal{H}^{\Delta}$.

Theorem 4.41 *Suppose that $\mathcal{H} = (X, \mathcal{E})$ is a fuzzy hypergraph and that $\mathcal{H}^{-}$ exists. Then every $\mathcal{L}$-coloring of $\mathcal{H}$ is a color stable extension of some $\mathcal{L}$-coloring of $\mathcal{H}^{\Delta}$. Conversely, any $\mathcal{L}$-coloring of $\mathcal{H}^{\Delta}$ which is extended in any manner that does not introduce another color is a color stable extended $\mathcal{L}$-coloring of $\mathcal{H}$.*

Example 4.12 *Consider the fuzzy hypergraph $\mathcal{H}$ of Example 4.1. Then $\mathcal{H}^{-} = \mathcal{H}$ and so $\mathcal{H}^{\Delta} = \mathcal{H}^{s}$. Now $\mathcal{H}^{s}$ is given in Example 4.7. We see that every $\mathcal{L}$-coloring of $\mathcal{H}$ is a color stable extension of some $\mathcal{L}$-coloring of $\mathcal{H}^{\Delta}$. In fact every $\mathcal{L}$-coloring of $\mathcal{H}^{\Delta}$ is an $\mathcal{L}$-coloring of $\mathcal{H}$ since $\mathbf{E}_1 = \{\{a,b\},\{a,c\}\} = \mathbf{E}_1^s$, $\mathbf{E}_2^s = \{\{a,b\},\{b,c\},\{a,c,d\}\}$, and $\mathbf{E}_2 = \mathbf{E}_2^s \cup \{\{a,b,d\}, \{b,c,d\}\}$.*

Theorem 4.42 *Let $\mathcal{H} = (X, \mathcal{E})$ be a fuzzy hypergraph for which $\mathcal{H}^{-}$ exists. Then each $\mathcal{L}$-coloring of $\mathcal{H}$ is a stable extension of some p-coloring of $\mathcal{H}^{\Delta}$. Conversely, each p-coloring of $\mathcal{H}^{\Delta}$, where all colors are assumed to be non-empty, can be extended to a partitioning of the vertex set of $\mathcal{H}$ in any manner which does not introduce a new color and the result will be a stably extended $\mathcal{L}$-coloring of $\mathcal{H}$.*

Proof. The proof follows from the construction of $\mathcal{H}^{\Delta}$ and Theorem 4.40. ■

Theorem 4.42 makes it clear that the development of $\mathcal{L}$-colorings of $\mathcal{H}$ can be reduced to the problem of finding primitive p-colorings of $\mathcal{H}^{\Delta}$, which can be labor saving. However, it should be noted that colorings of $\mathcal{H}^{s}$ may not be relevant to colorings of $\mathcal{H}$ for, in the procedure to determine $\mathcal{H}^{s}$, edges in $\mathcal{H}$ which are not spikes may be replaced by representatives which are spikes in $\mathcal{H}^{s}$. We illustrate this in the following example.

Example 4.13 *Let $\mathcal{H} = (X, \mathcal{E})$ be a fuzzy hypergraph of Example 4.10. Then it follows that $\mathcal{H}^s = (X^s, \mathcal{E}^s)$ where $X^s = \{b, c, d\}$ and $\mathcal{E}^s = \{\sigma(\{b\}, 0.9), \sigma(\{c, d\}, 0.9)\}$. Hence $\{\{b, d\}, \{c\}\}$ and $\{\{b, c\}, \{d\}\}$ are $\mathcal{L}$-colorings of $\mathcal{H}^s$. Clearly the chromatic number $\chi(\mathcal{H}^s) = 2$. Note the creation of the spike $\sigma(\{b\}, 0.9)$ in $\mathcal{H}^s$. We now consider the spike reduction in $\mathcal{H}$. We have $\mathcal{H}^- = (X^-, \mathcal{E}^-)$, where $X^- = \{a, b, c, d, e.f\}$ and $\mathcal{E}^- = \{\mu_1^-, \mu_1^-, \mu_1^-\}$ which is represented by the following incidence matrix:*

$$\begin{array}{c} \\ a \\ b \\ c \\ d \\ e \\ f \end{array} \begin{array}{c} \begin{array}{ccc} \mu_1^- & \mu_2^- & \mu_3^- \end{array} \\ \begin{pmatrix} 0.9 & 0.7 & 0.4 \\ 0.9 & 0.7 & 0.4 \\ 0 & 0 & 0.9 \\ 0 & 0 & 0.9 \\ 0.4 & 0.4 & 0 \\ 0.4 & 0.4 & 0 \end{pmatrix} \end{array}.$$

Thus $(\mathcal{H}^-)^{0.9} = (\{a, b, c, d\}, \{\{a, b\}, \{c, d\}\}) = (\mathcal{H}^-)^{0.7}$ and $(\mathcal{H}^-)^{0.4} = (\{a, b, c, d, e, f\}, \{\{a, b, e, f\}, \{a, b, c, d\}\})$. Then $\mathcal{H}^\Delta = (\mathcal{H}^-)^s = (X^\Delta, \mathcal{E}^\Delta)$, where $X^\Delta = \{a, b, c, d\}$ and $\mathcal{E}^\Delta = \{\sigma(\{a, b\}, 0.9), \sigma(\{a, b\}, 0.9)\}$. Hence $\{\{a, c\}, \{b, d\}\}$ and $\{\{a, d\}, \{b, c\}\}$ are $\mathcal{L}$-colorings of $\mathcal{H}^\Delta$. Clearly $\chi(\mathcal{H}^\Delta) = 2$.

C. Berge gives an interesting example of a waste management problem in [2, p.115]. We extend this problem and place it within the context of fuzzy hypergraphs. The following example also illustrates the usefulness of $\mu \otimes H$ within the theory of fuzzy hypergraphs.

Example 4.14 *Suppose that waste management is required to design a long-term cost efficient system of disposal sites for chemical waste by-products produced by a chemical plant. It is required that hazardous combinations of stored waste products should never be stored together. Suppose also that the design of waste sites for present needs should be highly adaptable to future augmentation as new waste by-products are produced. We show how a fuzzy hypergraph can be used to illustrate essential features of the problem.*

Assume that the directors of the plant have made best educated guesses regarding future production possibilities. Let X represent the set of chemical waste by-products now being produced together with those which may be produced in the future and let μ be a fuzzy subset of X that gives the degree of possibility of producing each waste by-product in X. Those currently being produced have degree 1 of possibility.

In addition, let $H = (X, \mathbf{E})$ be the (crisp) hypergraph where the collection of hazardous combinations of waste by-products forms the edge set $\mathbf{E}$. For simplicity, it is assumed that no edge of H is a singleton. $\mathcal{H} = \mu \otimes H$ functions as model for representing the fuzzy context of this example.

Every k-coloring $\{S_1, \ldots, S_k\}$ of H^{r_1}, where $\mathbf{C}(\mathcal{H}) = \{H^{r_1}, H^{r_2}, \ldots, H^{r_n}\}$ is the core set of $\mathcal{H}$ and $r_1 = 1$, is a possible solution for the current waste disposal problem where each dump site collects only those by-products that belong to a particular set S_i. The longer viewpoint where future disposal requirements are also considered leads naturally to consideration of $\mathcal{L}$-colorings associated with $\mathcal{H} = \mu \otimes H$.

Consider the μ–tempered fuzzy hypergraph $\mathcal{H}$ given in Example 4.2. Consider also a, b, c, d as waste products where the members of $E^{0.4}$, $\{a, b\}, \{a, c\}$, $\{b, c\}$, and $\{b, d\}$ yield hazardous waste combinations. It is easy to determine that $\chi(\mathcal{H}) = 3$ and $\{\{a, d\}, \{b\}, \{c\}\}$ and $\{\{a\}, \{b\}, \{c, d\}\}$ are the only 3-colorings of $E^{0.4}$. In this fuzzy hypergraph environment, a number of basic questions related to $\mathcal{L}$-colorings can be considered:

(i) What is the value of $\chi(\mathcal{H})$?

(ii) How can a minimal $\mathcal{L}$-coloring of $\mathcal{H}$ be constructed?

(iii) Do certain structural attributes exist which, if satisfied , guarantee that k-conservative colorings of H^{r_1} (to be defined later) can be extended color stably to $\mathcal{L}$-colorings of $\mathcal{H}$?

(iv) Can fuzzy chromatic evaluations of $\mathcal{L}$-colorings be designed to distinguish among acceptable colorings, those which are superior in a specific criterion (or criteria) such as a long-term cost effective strategy for developing waste disposal sites?

As previously shown, $\mathcal{H}^{\Delta}$ can be used to reformulate questions (i) and (ii) into related questions involving crisp hypergraphs. Questions (iii) and (iv) will be examined in succeeding sections.

Before examining Questions (iii) and (iv), we consider the following question: How sensitive is μ in the formulation of $\mathcal{H} = \mu \otimes H$? There may be other fuzzy subsets ν which satisfy $\nu \otimes H = \mu \otimes H$ (or $\mathbf{C}(\nu \otimes H) = \mathbf{C}(\mu \otimes H)$). Such inquiries lead to the following definitions concerning the relationship between the members of $\mathfrak{F}\wp(X)$.

Definition 4.30 *Two fuzzy hypergraphs $\mathcal{H}^{(1)}$ and $\mathcal{H}^{(2)}$ are said to be* **C**-related, *if they are* sequentially identical; *that is, if the fundamental sequences, $F(H^{(i)}) = \{r_j^{(i)} \mid j = 1, \ldots, n\}$, with $r_1^{(i)} > r_2^{(i)} > \cdots > r_n^{(i)}$ and $i = 1, 2$, have identical length n and for $j = 1, 2, \ldots, n$, $(H^{(1)})^{r_j^{(1)}} = (H^{(2)})^{r_j^{(2)}}$, where $(H^{(i)})^{r_j^{(i)}} \in \mathbf{C}(\mathcal{H}^{(i)}), i = 1, 2$.*

Definition 4.31 *Two fuzzy subsets μ_1 and μ_2 on X are said to be* H-related *with respect to a crisp hypergraph H on X if $\mu_1 \otimes H$ and $\mu_2 \otimes H$ are* **C**-related. *In addition μ_1 and μ_2 are said to be* identically H-related *if $\mu_1 \otimes H = \mu_2 \otimes H$.*

Clearly, H-relatedness and identically H-relatedness are equivalence relations in $\mathfrak{F}\wp(X)$.

We note that the H-related equivalence class containing the fuzzy subset μ described in the above example would reveal alternative fuzzy hypotheses generally compatible with waste management decisions based upon $\mu \otimes H$.

β-degree Coloring Procedures

Finding $\chi(H)$ for crisp hypergraphs is a highly nontrivial problem. Using β degrees and a step-by-step coloring of the vertices, Berge establishes a valuable upper bound for $\chi(H)$. We use Berge's technique to partially answer Question (iii), which was raised on the previous page.

We begin our study by reviewing some essential definitions and a basic theorem that appears in [2].

Recall that, in a crisp hypergraph $H = (X, \mathbf{E})$, the *star in* H *of a vertex* x is the set $H(x) = \{E \in \mathbf{E} \mid x \in E\}$.

Definition 4.32 *Suppose* $H = (X, \mathbf{E})$ *is a crisp hypergraph. Let* $x \in X$. *A* β_H-star *of* x, *denoted by* $H^\beta(x)$, *is a subset of the star,* $H(x)$, *of* x *such that*

(1) if $E \in H^\beta(x)$, *then* $|E| \geq 2$;

(2) if $E, E' \in H^\beta(x)$, *then* $E \cap E' = \{x\}$.

The β_H-degree, $d_H^\beta(x)$, *of* x *is defined by* $d_H^\beta(x) = \vee \{|H^\beta(x)| \mid H^\beta(x)$ *is a* β_H*-star of* $x\}$, *where* $|H^\beta(x)|$ *denotes the number of edges in* $H^\beta(x)$.

Note that if a vertex x has no β_H-stars, then $d_H^\beta(x) = 0$.

The symbols $\Delta^\beta(H)$ and $\delta^\beta(H)$, defined by

$\Delta^\beta(H) = \vee \{d_H^\beta(x) \mid x \in X\}$

and

$\delta^\beta(H) = \wedge \{d_H^\beta(x) \mid x \in X\}$,

represent the *maximum* and *minimum* β_H*-degree* of H.

Definition 4.33 *Let* $H = (X, \mathbf{E})$ *be a crisp hypergraph and let* $A \subseteq X$. *The* partial hypergraph, H/A, *of* H circumscribed by A *is defined by the edge set* $\mathbf{E}(H/A) = \{E \in \mathbf{E}(H) \mid E \subseteq A\}$. H/A *is said to be* filled *if* $A = \cup\{E \mid E \in \mathbf{E}(H/A)\}$.

Clearly H/A in Definition 4.33 has vertex set $\bigcup_{E \in \mathbf{E}(H)} E$.

Example 4.15 *Let* $X = \{a, b, c, d, e, f\}$ *and* $\mathbf{E} = \{E_1, E_2, E_3, E_4\}$, *where* $E_1 = \{a, b\}, E_2 = \{a, c\}, E_3 = \{a, d, e\}$ *and* $E_4 = \{a, d, f\}$. *Then the (crisp)* *hypergraph* $(H, \mathbf{E})$ *is the star* $H(a)$. *The* β_H*-stars of* a *with* 2 *or more elements are* $\{E_1, E_2\}, \{E_1, E_3\}, \{E_1, E_4\}, \{E_2, E_3\}, \{E_2, E_4\}, \{E_1, E_2, E_3\}$, *and* $\{E_1, E_2, E_4\}$. *Hence* $d^{\beta}_H(a) = 3$. *It is clear that* $\Delta^{\beta}(H) = 3$ *and* $\delta^{\beta}(H) = 1$. *Let* $A = \{a, b, c\}$. *Then* $H/A = (\{a, b, c\}, \{E_1, E_2\})$ *and we see that* H/A *is filled.*

To establish some fuzzy coloring results related to the next theorem, we provide a slightly modified version of Berge's proof of the following result.

Theorem 4.43 *(Berge). If* $H = (X, \mathbf{E})$ *is a crisp hypergraph, then* $\chi(H) \leq \vee \{\delta^{\beta}(H/A) \mid A \subseteq X\} + 1$.

Proof. We first β_H-order X; assume $|X| = n$.

Step 1: Select a point x_1 in X, such that

$$d^{\beta}_H(x_1) = \delta^{\beta}(H).$$

For $k \leq n$, assume $x_1, \ldots, x_{k-1}$ have been chosen. Let $A_k = X \setminus \{x_1, \ldots, x_{k-1}\}$.

Step k: (i) If H/A_k exists (i.e., $\{E \in \mathbf{E} \mid E \subseteq A_k\} \neq \emptyset\}$), choose a point x_k in the vertex set, $\mathbf{V}(H/A_k)$, of the partial hypergraph H/A_k such that

$$d_{H/A_k}(x_k) = \delta^{\beta}(H/A_k).$$

(ii) If H/A_k does not exist, label the $(n-k+1)$ members of A_k: x_k, $x_{k+1}, \ldots, x_n$ in any order. This completes the β_H-ordering of X.

Suppose the algorithm terminates in case (ii) of step k. Then $A_k = X \setminus \{x_1, \ldots, x_{k-1}\}$ contains no edge of H. In this case, let $\mathbf{R}^{\beta}_H(X)$ denote an arbitrary ordering $\{x_k, x_{k+1}, \ldots, x_n\}$ of A_k. Let $\mathbf{Q}^{\beta}_H(X)$ represent the linearly ordered set $\{x_1, \ldots, x_{k-1}\}$ determined by the previous $(k-1)$-steps of the β_H-ordering algorithm. We let

$$\mathbf{L}^{\beta}_H(X) = \mathbf{Q}^{\beta}_H(X) + \mathbf{R}^{\beta}_H(X)$$

denote the β_H-ordering of the vertices of X. Observe that if $\mathbf{R}^{\beta}_H(X) = \emptyset$, case (ii) did not occur at any step in the execution of the algorithm.

Now we color the vertices in X step by step in reverse β_H-ordering, starting with x_n. We assume without loss of generality that $\mathbf{R}^{\beta}_H(X) \neq \emptyset$ and $\mathbf{Q}^{\beta}_H(X) = \{x_1, \ldots, x_{k-1}\}$. First, one color is used to color $\mathbf{R}^{\beta}_H(X)$. Next assume the set of vertices $\{x_{i+1}, x_{i+2}, x_{i+3}, x_{k-1}\}$ in $\mathbf{Q}^{\beta}_H(X)$ have been colored so that each edge in H that has been fully colored, and is not a loop, has at least two colors. To color x_i we consider the edges of $(H/A_i)(x_i)$ (the edges of $H(x_i)$ contained in $A_i = X \setminus \{x_1, \ldots, x_{i-1}\}$). These edges have been totally colored except for vertex x_i. We need to color x_i so that each edge in $(H/A_i)(x_i)$, which is not a loop, has at least two colors. We now determine if we need a new color for x_i, or can we use a color already used.

If every non-loop member of $(H/A_i)(x_i)$ contains at least two differently colored vertices distinct from x_i, then x_i can be colored with any existing color (color already used). Suppose then that some edges of $(H/A_i)(x_i)$ are monochromatic (if x_i is excluded). Since any two such edges with different colors have only x_i in common, we claim that the number of different colors appearing among the monochromatic edges (excluding x_i) of $(H/A_i)(x_i)$ is less than or equal to $\delta^\beta(H/A_i)$. To see this, consider the following argument. Let $M(x_i)$ be a an arbitrary subset of the subset of edges of star $H/A_i(x_i)$ which have been colored monochromatically (excepting x_i which is still uncolored) wherein $M(x_i)$ satisfies the property that each edge of $M(x_i)$ exhibits a different (monochromatic) color. Clearly, $M(x_i)$ is a β-star of x_i in hypergraph H/A_i. Thus,

$$|M(x_i)| \leq d^\beta_{H/A_i}(x_i) = \delta^\beta(H/A_i),$$

where $d^\beta_{H/A_i}(x_i)$ is the β_{H/A_i}-degree of x_i in the hypergraph H/A_i and $\delta^\beta(H/A_i)$ is the minimal β_{H/A_i}-degree of H/A_i; the above equality follows from the algorithm used to β_H-order X. Consequently, the above claim is established. Hence, we conclude that if at least $\delta^\beta(H/A_i) + 1$ colors have already been used to color $= X \setminus \{x_1, \ldots, x_i\}$, then no new color is required to color x_i appropriately. Consequently, a simple inductive argument is now obvious. ■

We illustrate the proof of the Theorem 4.44 in the following example.

Example 4.16 *Let $H = (X, \mathbf{E})$ be the hypergraph in Example 4.15.*

Step 1: $\delta^\beta(H) = 1$. $d^\beta_H(f) = 1$. Let $x_i = f$.

Step 2 (i): $A_2 = X \setminus \{f\} = \{a,b,c,d,e\}$. H/A_2 exists. $\mathbf{V}(H/A_2) = A_2$. Now $\delta^\beta(H/A_2) = 1 = d_{H/A_2}(e)$. Let $x_2 = e$.

Step 3 (i): $A_3 = X \setminus \{e,f\} = \{a,b,c,d\}$. H/A_3 exists. $\mathbf{V}(H/A_3) = \{a,b,c\}$. Now $\delta^\beta(H/A_3) = 1 = d_{H/A_3}(c)$. Let $x_3 = c$.

Step 4 (i): $A_4 = X \setminus \{c,e,f\} = \{a,b,d\}$. H/A_4 exists. $\mathbf{V}(H/A_4) = \{a,b\}$. Now $\delta^\beta(H/A_4) = 1 = d_{H/A_4}(b)$. Let $x_4 = b$.

Step 5 (ii): $A_5 = X \setminus \{b,c,e,f\} = \{a,d\}$. H/A_5 does not exist. Let $x_5 = d$ and $x_6 = a$. Now A_5 contains no edge of H. $\mathbf{R}^\beta_H(H) = \{x_5, x_4\} = \{d,a\}$ and $\mathbf{Q}^\beta_H(H) = \{x_1,x_2,x_3,x_4\} = \{f,e,c,b\}$.

Color a, d red. Now color b blue. E_1 is fully colored with two colors. Next consider $c = x_3$. $(H/A_3)(x_3) = \{\{a,c\}\}$. $\delta^\beta(H/A_3) + 1 = 2$ colors have already been used to color $X \setminus \{f,e,c\}$. Hence no new color is required to cover c. Continuing in this manner we also see that no new color is needed to color e and f. We have $\chi(H) = 2 = 1 + 1 = \vee\{\delta^\beta(H/A) \mid A \subseteq X\} + 1$.

The following example from [11] is another example where the upper bound of $\chi(H)$ in Theorem 4.43 is obtained by $\chi(H)$. For instance, consider the binary graph G with 5 vertices and 10 binary edges as depicted in the

FIGURE 4.1 The upper bound of $\chi(H)$ is obtained by $\chi(H)$.

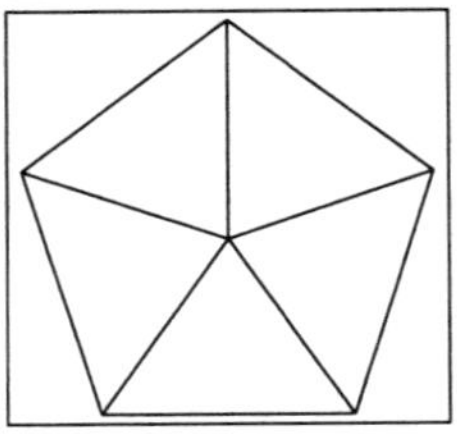

diagram given in Figure 4.1. Clearly the chromatic number $\chi(G) = 4$ while $\vee\{\delta^\beta(G/A) \mid A \subseteq \mathbf{V}(G)\} = 3$.

Remark 2. The phrase *β-degree coloring method*, shall denote the coloring method just described. With respect to this coloring method an important point to remember is that if $\delta^\beta(H/A_i)$ is less than the number of colors used to color A_{i+1}, then no new color is needed to color x_i. More specifically, if $i < j$ and $\delta^\beta(H/A_i) \leq \delta^\beta(H/A_j)$ and at least $\delta^\beta(H/A_j) + 1$ colors have been used to color A_{i+1}, then some color already used to color A_{i+1} can be used to color x_i.

Definition 4.34 *Any k-coloring of $H = (X, \mathbf{E})$ which uses no more than $\vee\{\delta^\beta(H/A) \mid A \subseteq X\} + 1$ colors is called a* conservative coloring *(or a* conservative k-coloring*) of H.*

According to Theorem 4.43, a k-coloring of H by the β-degree coloring method is a conservative coloring of H.

A conservative coloring of H that uses precisely $\vee\{\delta^\beta(H/A) \mid A \subseteq X\} + 1$ colors is referred to as a *maximum conservative coloring* (or simply a *conservative $\bar{k}$-coloring*) of H.

The main fuzzy results of this section depend upon the crisp material already presented and upon the following definitions and lemmas. In the definitions to follow, we assume H is a crisp hypergraph on X and B is a non-empty subset of X.

Definition 4.35 *The* star of B in H, *denoted by $H(B)$, is the partial hypergraph of H defined by $H(B) = \cup\{H(x) \mid x \in B\}$, where $H(x)$ denotes the star of x in H. Let*

$\delta^\beta(H\blacksquare B) = \wedge\{d_H^\beta(x) \mid x \in B\}$,

$\Delta^\beta(H\blacksquare B) = \vee\{d_H^\beta(x) \mid x \in B\}$.

Lemma 4.44 $d_{H(B)}^\beta(x) = d_H^\beta(x)$, $\forall x \in B$ *and* $\delta^\beta(H(B)) \leq \delta^\beta(H\blacksquare B) \leq \Delta^\beta(H\blacksquare B) \leq \Delta^\beta(H(B))$.

Proof. From the definition of $H(B)$, we have for every $x \in B$ that $H(B)(x) = H(x)$. Hence,

$d^{\beta}_{H(B)}(x) = d^{\beta}_{H}(x)$, $\forall x \in B$.

Moreover, since $B \subseteq \mathbf{V}(H(B))$, it follows that

$\delta^{\beta}(H(B)) \leq \delta^{\beta}(H\blacksquare B) \leq \Delta^{\beta}(H\blacksquare B) \leq \Delta^{\beta}(H(B))$. ■

Definition 4.36 *Let $|B| = m$. A β_H-ordering of B, denoted by $\mathbf{L}^{\beta}_{H}(B)$, is a linear ordering of B determined by the following algorithm:*

Step 1: Select x_1 in B such that

$d^{\beta}_{H}(x_1) = \delta^{\beta}(H\blacksquare B)$.

For $k \leq m$, assume $x_1, \ldots, x_{k-1}$ have been selected from B. Let $A_k = X \setminus \{x_1, \ldots, x_{k-1}\}$ and $B_k = B \setminus \{x_1, \ldots, x_{k-1}\}$.

Step k: k(i) If H/A_k exists, and $B'_k = \mathbf{V}(H/A_k) \cap B_k \neq \emptyset$, then choose x_k in B'_k such that

$d^{\beta}_{H/A_k}(x_k) = \delta^{\beta}(H/A_k\blacksquare B'_k) = \wedge\{d^{\beta}_{H/A_k}(x) \mid x \in B'_k\}$.

k(ii) If H/A_k does not exist or $B'_k = \emptyset$, label the $(m-k+1)$ members of B_k: $x_k, x_{k+1}, \ldots, x_m$ in any order. This completes the β_H-ordering of B.

Often it is useful to decompose $\mathbf{L}^{\beta}_{H}(B)$ into an ordered sum

$\mathbf{L}^{\beta}_{H}(B) = \mathbf{Q}^{\beta}_{H}(B) + \mathbf{R}^{\beta}_{H}(B)$,

where $\mathbf{Q}^{\beta}_{H}(B)$ represents a linear ordering of those members of β which are selected recursively first by step 1 and then by step k(i) of the algorithm in Definition 4.36, and where $\mathbf{R}^{\beta}_{H}(B)$ represents an arbitrary, but fixed, ordering of the remaining members, if any, of B which satisfy the hypothesis of step k(ii) of the algorithm in Definition 4.36.

Remark 3. Suppose that $\{x_1, \ldots, x_{k-1}\} \subseteq B$. If $B_k \neq \emptyset$, then the edges of $H(B)$ that are contained in $A_k = \mathbf{V}(H) \setminus \{x_1, \ldots, x_{k-1}\}$ are precisely the edges of $H(B)$ that are contained in

$A^B_k = \mathbf{V}(H(B)) \setminus \{x_1, \ldots, x_{k-1}\}$.

Thus it follows that the edges of $H(B)/A^B_k$ are precisely the edges of $H(B)$ that belong to H/A_k.

Hence

$B''_k = \mathbf{V}(H(B)/A^B_k) \cap B_k = \mathbf{V}(H/A_k) \cap B_k = B'_k$

and, for every $x \in B'_k$,

$d^{\beta}_{H/A_k}(x) = d^{\beta}_{H(B)/A^B_k}(x)$.

Therefore,

$\delta^{\beta}(H/A_k\blacksquare B'_k) = \delta^{\beta}(H(B)/A^B_k\blacksquare B''_k)$.

Thus every $\beta_{H(B)}$-ordering of B is a β_H-ordering of B and conversely. In addition, since $d^{\beta}_{H(B)}(x) = d^{\beta}_{H}(x)$ for all $x \in B$, it follows that

$\delta^{\beta}(H(B)\blacksquare B) = \delta^{\beta}(H\blacksquare B)$

and

$\Delta^\beta(H(B)\blacksquare B) = \Delta^\beta(H\blacksquare B)$.

Certain results which appear later coordinate more than one application of the β_H-coloring method as described in the proof of Theorem 4.43. To arrange such matters effectively, a relaxed β_H-degree coloring scheme is introduced through the following algorithm.

Algorithm 4.3. Let
$M_H(X) = (x_1, x_2, \dots, x_n)$
be a linear ordering of $X(= \mathbf{V}(H))$. The vertices of H are colored sequentially in reverse $H^h(X)$-order. We begin coloring x_n arbitrarily. We then continue according to the following recursive rule which is directed by $M_H(x)$:

Rule M. Suppose the vertices $C_{i+1} = \{x_n, x_{n-1}, \dots, x_{i+1}\}$ have already been colored so that no nonsingleton edge of H contained in C_{i+1} is monochrome in color. Suppose also that we are ready to color x_i. Let $C_i = C_{i+1} \cup \{x_i\}$ and consider the star, $(H/C_i)(x_i)$, of x_i, the edges of which belong to hypergraph H/C_i. If this star does not exist or if each nonsingleton edge of $(H/C_i)(x_i)$ already exhibits at least two members with different colors, then color the vertex point x_i with any existing color (used in coloring C_{i+1}). On the other hand if there are nonsingleton edges of $(H/C_i)(x_i)$ which are monochrome in color (except for x_i, which has not yet been colored) and the set of colors used to color them is less than the set of colors which appear in the coloring of C_{i+1}, then color x_i with one of the existing colors in C_{i+1} so that all nonsingleton edges in $(H/C_i)(x_i)$ have at least two colors. However, if the number of colors among the monochromatic colored edges, excepting x_i, in $(H/C_i)(x_i)$ equals the number of colors already used in coloring C_{i+1}, then x_i must be colored with a new color (i.e., a color that does not appear in the coloring of C_{i+1}).

Note that if Algorithm 4.3 is directed by a β_H-ordering of $\mathbf{V}(H)$, then the result is a coloring that can be reproduced by the β_H-degree coloring method as described in the proof of Theorem 4.43.

Definition 4.37 *A linear ordering, $M_H(X)$, of $X = \mathbf{V}(H)$ is called a* conservative ordering *of X if when applying Rule M in Algorithm 4.3, this ordering yields a conservative k-coloring of H.*

Definition 4.38 *Let $H(X, \mathbf{E})$ be a crisp hypergraph on X and let $\{Y_1, Y_2\}$ be a nontrivial partition of X. Suppose a k-coloring, K_1, of H/Y_1 is extended to a full k-coloring, K, of H, where the total number of new colors introduced to k-color H does not exceed $0 \vee (\hat{\kappa}_2 - |K_1|)$, where $\hat{\kappa}_2 = 1 + \vee\{\delta^\beta(H/[X \setminus A]\blacksquare Y_2 \setminus A) \mid A \subset Y_2\}$ and $|K_1|$ represents the number of colors in K_1. In this case K is called a* weakly conservative k-coloring extension of K_1 to H with respect to Y_2 *(or simply, a* weakly conservative k-coloring *of H* with respect to Y_2*). On the other hand, if a k-coloring,*

K_1, of $H \setminus Y_1$ is extended to a full k-coloring, K, of H where the total number of new colors introduced to color H does not exceed $0 \vee (\hat{\kappa}_2 - |K_1|)$, where $\kappa_2 = 1+ \vee \{\delta^\beta(H/[X \setminus A]) \mid A \subset Y_2\}$, then K is called a strongly conservative k-coloring extension of K_1 to H with respect to Y_2 *(or simply, a* strongly conservative k-coloring of H with respect to Y_2*).*

Lemma 4.45 *Let Y_1 be a proper subset of X. Assume $H = (X, \mathbf{E})$ is a crisp hypergraph on X and K_1 is a conservative k-coloring of H/Y_1. Then a strongly conservative k-coloring extension of K_1 to H with respect to $Y_2 = X \setminus Y_1$ is a conservative k-coloring of H.*

Proof. From the hypothesis and Definition 4.34,

$$|K_1| \leq 1+ \vee\{\delta^\beta([H/Y_1]/A) \mid A \subseteq \mathbf{V}(H/Y_1)\}$$
$$= 1+ \vee\{\delta^\beta(H/A) \mid A \subseteq \mathbf{V}(H/Y_1)\}$$
$$= 1+ \vee\{\delta^\beta(H/A) \mid A \subseteq Y_1\} = \kappa_1.$$

Let K be a strongly conservative k-coloring extension of K_1 to H with respect to $Y_2 = X \setminus Y_1$. Then, by Definition 4.38,

$$|K| \leq |K_1| + 0 \vee (\hat{\kappa}_2 - |K_1|)$$
$$= |K_1| \vee \kappa_2 \leq \kappa_1 \vee \kappa_2$$
$$\leq 1+ \vee\{\delta^\beta(H/A) \mid A \subseteq X\},$$

since

$$\kappa_1 = 1+ \vee\{\delta^\beta(H/A) \mid A \subseteq Y_1\},$$

and

$$\kappa_2 = 1+ \vee\{\delta^\beta(H/[X \setminus A]) \mid A \subset Y_2 = X \setminus Y_1\}.$$

Therefore, by Definition 4.34, K is a conservative k-coloring of H. ■

Definition 4.39 *Let $H_1 = (X_1, \mathbf{E})$ be a crisp hypergraph on X_1, where $X_1 \subseteq X$. Then, a* post-extended β_{H_1}-ordering *of X, written $\vec{\mathbf{L}}^\beta_{H_1}(X)$, is a linear ordering of X expressed by the ordered sum*

$$\vec{\mathbf{L}}^\beta_{H_1}(X) = \mathbf{L}^\beta_{H_1}(X_1) + \mathbf{S}^\beta_{H_1}(X \setminus X_1),$$

where

$$\mathbf{L}^\beta_{H_1}(X_1) = \mathbf{Q}^\beta_{H_1}(X_1) + \mathbf{R}^\beta_{H_1}(X_1)$$

is a β_{H_1}-ordering of X_1 (see Definition 4.36) and $\mathbf{S}^\beta_{H_1}(X \setminus X_1)$ is an arbitrary, but fixed ordering of $X \setminus X_1$. Alternatively, $\vec{\mathbf{L}}^\beta_{H_1}(X)$ is the ordered sum

$$\vec{\mathbf{L}}^\beta_{H_1}(X) = \vec{\mathbf{Q}}^\beta_{H_1}(X) + \vec{\mathbf{R}}^\beta_{H_1}(X),$$

where

$$\vec{\mathbf{Q}}^\beta_{H_1}(X) = \mathbf{Q}^\beta_{H_1}(X),$$

and

$$\vec{\mathbf{R}}^\beta_{H_1}(X) = \mathbf{R}^\beta_{H_1}(X_1) + \mathbf{S}^\beta_{H_1}(X \setminus X_1).$$

A pre-extended β_{H_1}-ordering *of X, written $\overleftarrow{\mathbf{L}}^\beta_{H_1}(X)$, is expressed by the ordered sum*

$$\overleftarrow{\mathbf{L}}^\beta_{H_1}(X) = \mathbf{S}^\beta_{H_1}(X \setminus X_1) + \mathbf{L}^\beta_{H_1}(X_1)$$

or, alternatively, by the ordered sum

$$\overleftarrow{\mathbf{L}}^{\beta}_{H_1}(X) = \mathbf{S}^{\beta}_{H_1}(X \setminus X_1) + \mathbf{Q}^{\beta}_{H_1}(X_1) + \mathbf{R}^{\beta}_{H_1}(X_1).$$

In the above definition, if $X_1 = X$, then $\overleftarrow{\mathbf{L}}^{\beta}_{H_1}(X) = \mathbf{L}^{\beta}_{H_1}(X) = \vec{\mathbf{L}}^{\beta}_{H_1}(X)$.

Lemma 4.46 *Let $H = (X, \mathbf{E})$ be a crisp hypergraph on X and assume $\{Y_1, Y_2\}$ is a non-trivial partition of X with $|Y_1| = n_1$, $|Y_2| = n_2$ and $|X| = n = n_1 + n_2$. Let $M_H(X)$ be a linear ordering of X that corresponds to the partition $\{Y_1, Y_2\}$ such that $M_H(X)$ is the ordered sum*

$$M_H(X) = \mathbf{L}^{\beta}_{H}(Y_2) + \overleftarrow{\mathbf{L}}^{\beta}_{H/Y_1}(Y_1),$$

where

$$\mathbf{L}^{\beta}_{H}(Y_2) = \{x_1, \ldots, x_{n_2}\}$$

is a β_H-ordering of Y_2 (see Definition 4.36) and

$$\overleftarrow{\mathbf{L}}^{\beta}_{H/Y_1}(Y_1) = \{x_{n_2+1}, \ldots, x_n\} = S^{\beta}_{H/Y_1}(Y_1 \setminus V(H/Y_1)) + \mathbf{L}^{\beta}_{H/Y_1}(V(H/Y_1))$$

is a pre-extended β_{H/Y_1}-ordering of Y_1 (see Definition 4.39) wherein

$$S^{\beta}_{H/Y_1}(Y_1 \setminus V(H/Y_1)) = \{x_{n_2+1}, \ldots, x_{n_2+s}\},$$

if $Y_1 \setminus \mathbf{V}(H/Y_1) \neq \emptyset$,

is an arbitrary, but fixed, ordering of $Y_1 \setminus \mathbf{V}(H/Y_1)$ and $\mathbf{L}^{\beta}_{H/Y_1}(\mathbf{V}(H/Y_1))$ is a β_{H/Y_1}-ordering of the vertex set $\mathbf{V}(H/Y_1)$ expressed in the form:

$$\mathbf{L}^{\beta}_{H/Y_1}(V(H/Y_1)) = \{x_{n_2+s+1}, \ldots, x_n\}$$

provided $\mathbf{V}(H/Y_1) \neq \emptyset$; if H/Y_1 is filled (see Definition 4.33), then $s = 0$ and $Y_1 \setminus \mathbf{V}(H/Y_1) = \emptyset$.

Then, $M_H(X)$ satisfies the following properties:

Property (i). A k-coloring of H determined by Algorithm 4.3 under the direction of $M_H(X)$ will use at most $\kappa_1 \vee \hat{\kappa}_2$ colors, where

$$\kappa_1 = 1 + \vee\{\delta^{\beta}(H/A) \mid A \subseteq Y_1\}$$

and

$$\hat{\kappa}_2 = 1 + \vee\{\delta^{\beta}(H/[X \setminus A] \blacksquare Y_2 \setminus A) \mid A \subset Y_2\}.$$

Property (ii). Every k-coloring, K_1, of H/Y_1 that is extended to a k-coloring of H by Algorithm 4.3 under the direction of the ordered sum

$$\mathbf{L}^{\beta}_{H}(Y_2) + \mathbf{S}^{\beta}_{H/Y_1}(Y_1 \setminus \mathbf{V}(H/Y_1)) = \{x_1, \ldots, x_{n_2}, \ldots, x_{n_2+s}\} \tag{4.48}$$

is a weakly conservative k-coloring of H with respect to Y_2 (see Definition 4.38). In particular, suppose K_1 is a k-coloring of H/Y_1 which satisfies $|K_1| \geq \hat{\kappa}_2$, where $|K_1|$ represents the number of colors in K_1, then no further colors are required to extend K_1 to a k-coloring of H if the extension is produced by Algorithm 4.3 under the direction of sequence $(x_1, \ldots, x_{n_1}, \ldots, x_{n_2+s})$ described in (4.48) above. Additionally, suppose a crisp hypergraph $H = (X, \mathbf{E})$ satisfies, with respect to a nontrivial partition $\{Y_1, Y_2\}$ of X, either

Condition (i) $\hat{\kappa}_2 - 1 \leq \delta^{\beta}(H/Y_1)$ (where $\hat{\kappa}_2$ is determined above) or

Condition (ii) $\Delta^\beta(H\blacksquare Y_2) \leq \delta^\beta(H/Y_1)$ *and, in addition,* H/Y_1 *satisfies*
Condition (iii) H/Y_1 *is filled (see Definition 4.33).*
Then, two conclusions follow:
Conclusion (i). Any k-coloring, K_1*, of* H/Y_1 *that is extended to a full k-coloring of* H *through the application of Algorithm 4.3 directed by sequence* $(x_1, \ldots, x_{n_2}, \ldots, x_{n_2+s})$*, as described above in (4.48), is a strongly conservative k-coloring of* H *with respect to* Y_2 *(see Definition 4.38). In particular, if* K_1 *is a k-coloring of* H/Y_1 *which satisfies* $|K_1| \geq \kappa_2$*, where*
$\kappa_2 = 1+ \vee\{\delta^\beta(H/[X \setminus A]) \mid A \subset Y_2\}$,
then no further colors are required to extend K_1 *to a k-coloring of* H *if the extension is produced by Algorithm 4.3 under the direction of sequence* $(x_1, \ldots, x_{n_2}, \ldots, x_{n+s})$ *described earlier in (4.48).*
Conclusion (ii). The linear ordering, $M_H(x)$*, of* X *given by the ordered sum*
$M_H(X) = \mathbf{L}^\beta_H(Y_2) + \overleftarrow{\mathbf{L}}^\beta_{H/Y_1}(Y_1) = \{x_1, \ldots, x_{n_2}, x_{n_2+1}, \ldots, x_n\}$
is a conservative linear ordering of X *(see Definition 4.37).*

Proof. The proof consists in examining the production of a k coloring of H by use of Algorithm 4.3 via the linear ordering $M_H(X)$. We first divide $M_H(X)$ into an ordered sum of subsequences each of which, in turn, will be analyzed from the standpoint of determining an upper bound of how many new colors may possibly be required to color the subsequence.

Recall that $M_H(X)$ is the ordered sum
$M_H(X) = \mathbf{L}^\beta_H(Y_2) + \overleftarrow{\mathbf{L}}^\beta_{H/Y_1}(Y_1)$,
where, respectively, according to Definitions 4.36 and 4.39,

$$\mathbf{L}^\beta_H(Y_2) = \mathbf{Q}^\beta_H(Y_2) + \mathbf{R}^\beta_H(Y_2) \tag{4.49}$$

and, with the understanding that $\hat{Y}_1$ represents $\mathbf{V}(H/Y_1)$,
$\overleftarrow{\mathbf{L}}^\beta_{H/Y_1}(Y_1) = \mathbf{S}^\beta_{H/Y_1}(Y_1 \setminus \hat{Y}_1) + \mathbf{Q}^\beta_{H/Y_1}(\hat{Y}_1) + \mathbf{R}^\beta_{H/Y_1}(\hat{Y}_1)$.
Therefore, $M_H(X)$ can be expressed as the ordered sum:
$M_H(X) = \mathbf{Q}^\beta_H(Y_2)+\mathbf{R}^\beta_H(Y_2)+\mathbf{S}^\beta_{H/Y_1}(Y_1\setminus\hat{Y}_1)+\mathbf{Q}^\beta_{H/Y_1}(\hat{Y}_1)+\mathbf{R}^\beta_{H/Y_1}(\hat{Y}_1)$.

For the remainder of this proof we assume that neither $\mathbf{R}^\beta_H(Y_2)$, nor $\hat{Y}_1$, nor $(Y_1\setminus\hat{Y}_1)$, nor $\mathbf{R}^\beta_{H/Y_1}(\hat{Y}_1)$ is empty. Of course it then follows that neither $\mathbf{Q}^\beta_H(Y_2)$ nor $\mathbf{Q}^\beta_{H/Y_1}(\hat{Y}_1)$ is empty.

Recall from Algorithm 4.3 that $M_H(X)$ directs rule M so that the vertices of X are colored sequentially in the order which is opposite (or reverse to) the ordering $M_H(X)$. Thus, for example, the last member in sequence $M_H(X)$ is the first vertex to be colored, the second to last member in sequence $M_H(X)$ is the second vertex to be colored. This operation continues backwards, step by step, through the linear ordering $M_H(X)$ until the last vertex to be colored (by rule M) is the first member of sequence $M_H(X)$. Therefore $M_H(X)$ will direct a coloring of X by first coloring

Y_1 according to rule M as directed by the linear ordering $\overleftarrow{\mathbf{L}}^{\beta}_{H/Y_1}(Y_1)$, of Y_1. More specifically, Y_1 is colored in three consecutive stages: first the members of $\mathbf{R}^{\beta}_{H/Y_1}(\hat{Y}_1)$ are colored, next the members of $\mathbf{Q}^{\beta}_{H/Y_1}(\hat{Y}_1)$ are colored, and then the members of $Y \setminus \hat{Y}_1$ are colored with respect to the arbitrarily selected ordering $\mathbf{S}^{\beta}_{H/Y_1}(Y_1 \setminus \hat{Y}_1)$ to complete the coloring of Y_1.

The members of $\mathbf{R}^{\beta}_{H/Y_1}(\hat{Y}_1)$, where $\hat{Y}_1 = \mathbf{V}(H/Y_1)$, satisfy the hypothesis stated in step k(ii) of the algorithm which appears at the beginning of the proof of Theorem 4.43 when the algorithm is applied to the hypergraph H/Y_1. Therefore, no edge of H is contained in the set of members of $\mathbf{R}^{\beta}_{H/Y_1}(\hat{Y}_1)$. This follows because no edge of H/Y_1 is contained in the set of members of $\mathbf{R}^{\beta}_{H/Y_1}(\hat{Y}_1)$ and that every edge of H contained in Y_1 belongs to H/Y_1 (see Definition 4.33). Hence, according to rule M in Algorithm 5.3, all members of $\mathbf{R}^{\beta}_{H/Y_1}(\hat{Y}_1)$ will be assigned the same color.

Recall that $\mathbf{Q}^{\beta}_{H/Y_1}(\hat{Y}_1)$ is the non-residual portion of a β_{H/Y_1}-ordering of the vertex set $\hat{Y}_1$ $(= \mathbf{V}(H/Y_1))$, which is an ordering as specified in the proof of Theorem 4.43. Therefore, an upper estimate on the number of different colors introduced into the sequential coloring of $\mathbf{Q}^{\beta}_{H/Y_1}(\hat{Y}_1)$ is immediate from Theorem 4.43. For rule M directed by $\mathbf{L}^{\beta}_{H/Y_1}(\hat{Y}_1)$ follows precisely the same pattern of selecting colors as discussed in the proof of Theorem 4.43. Hence, no more than

$1 + \vee\{\delta^{\beta}([H/Y_1]/A) \mid A \subseteq \hat{Y}_1(= \mathbf{V}(H/Y_1))\} = 1 + \vee\{\delta^{\beta}(H/A) \mid A \subseteq Y_1\}$

colors are used to obtain a k-coloring of H/Y_1 by application of rule M (in Algorithm 4.3) according to the linear ordering, $\mathbf{L}^{\beta}_{H/Y_1}(\hat{Y}_1)$, of $\hat{Y}_1$.

We next, observe that no edge of H/Y_1 contains members of $Y_1 \setminus \hat{Y}_1(= Y_1 \setminus \mathbf{V}(H/Y_1))$. Therefore, no edge of H that contains members of $Y_1 \setminus \hat{Y}_1$ is contained in Y_1. Thus rule M as directed by $M_H(X)$ will color all members of $Y_1 \setminus \hat{Y}_1$ with colors that have been used to color Y_1. Hence, in Algorithm 4.3, rule M directed by $M_H(X)$ will color Y_1 with no more than

$$\kappa_1 = 1 + \vee\{\delta^{\beta}(H/A) \mid A \subseteq Y_1\} \tag{4.50}$$

colors.

(Before we begin to color Y_2, we note that with the coloring of Y_1 a k-coloring of H/Y_1 has been completed. Thus every edge in H not yet completely colored must have members belonging to Y_2.)

To complete the k-coloring of H through the use of Algorithm 4.3 directed by $M_H(X)$ it remains to color Y_2 by rule M directed by sequence $\mathbf{L}^{\beta}_{H}(Y_2)$. Recall from (4.49) that the members of Y_2 are divided into two ordered subsets. The first subset of Y_2 to be colored is the set of members belonging to $\mathbf{R}^{\beta}_{H}(Y_2)$. Then the task of coloring H is completed by coloring the members of $\mathbf{Q}^{\beta}_{H}(Y_2)$.

The members of $\mathbf{R}_H^\beta(Y_2)$ form the remainder of Y_2 determined by the β_H-ordering, $\mathbf{L}_H^\beta(Y_2)$, of Y_2. Thus the members of $\mathbf{R}_H^\beta(Y_2)$ satisfy the hypothesis stated in step k(ii) of the algorithm appearing in Definition 4.36. Consequently no edge of H containing members of $\mathbf{R}_H^\beta(Y_2)$ can be contained in $Y_1 \cup \{x_{k_2+1}, \ldots, x_{n_2}\}$, where it is understood that

$$\mathbf{Q}_H^\beta(Y_2) = \{x_1, \ldots, x_{k_2}\} \tag{4.51}$$

and

$$\mathbf{R}_H^\beta(Y_2) = \{x_{k_2+1}, \ldots, x_{n_2}\}. \tag{4.52}$$

Thus, every edge of H not belonging to H/Y_1 must contain members of $\mathbf{Q}_H^\beta(Y_2)$. Hence Algorithm 4.3 directed by $M_H(X)$ will introduce no new colors in coloring the members of $\mathbf{R}_H^\beta(Y_2)$. Therefore, no more than κ_1 (see (4.50) above) colors are used to color $Y_1 \cup \{x_{k_2+1}, \ldots, x_{n_2}\}$ though the utilization of Algorithm 4.3 directed by $M_H(X)$. Finally, the members of $\mathbf{Q}_H^\beta(Y_2)$ are ready to be colored.

Now $\mathbf{Q}_H^\beta(Y_2) = \{x_1, \ldots, x_{k_2}\}$ given in (4.51) is the non residual portion of a β_H-ordering of Y_2 developed according to the algorithm stated in Definition 4.36. Therefore, since x_1 satisfies the property specified in step 1, the β_H-degree of x_1 satisfies

$$d_H^\beta(x_1) = \wedge\{d_H^\beta(x) \mid x \in Y_2\} = \delta^\beta(H\blacksquare Y_2). \tag{4.53}$$

For $1 < k \le k_2$ with $A_k = X\backslash\{x_1, \ldots, x_{k-1}\}$ and $B_k = Y_2\backslash\{x_1, \ldots, x_{k-1}\}$, the β_{H/A_k}-degree of x_k satisfies,

$d_{H/A_k}^\beta(x_k) = \wedge\{d_{H/A_k}^\beta(x) \mid x \in B_k \cap \mathbf{V}(H/A_k)\}$.

by step $k(i)$. We now show that

$$d_{H/A_k}^\beta(x_k) \le \hat{\kappa}_2 - 1 \quad \text{(see (4.55) below)}. \tag{4.54}$$

Let $D_k = B_k \setminus \mathbf{V}(H/A_k)$. Since $B_k \subseteq A_k$, it follows that $D_k \subseteq A_k$. Thus, $D_k \subseteq A_k \setminus \mathbf{V}(H/A_k)$.Hence, according to Definition 4.33,

$H/A_k = H/[A_k \setminus D_k] = H/\hat{A}_k$,

where $\hat{A}_k = A_k \setminus D_k = X \setminus [\{x_1, \ldots, x_{k-1}\} \cup D_k]$. In addition, $Y_2 \setminus [\{x_1, \ldots, x_{k-1}\} \cup D_k] = B_k \cap \mathbf{V}(H/A_k)$. Hence it follows that

$$\begin{aligned} d_{H/A_k}^\beta(x_k) &= \wedge\{d_{H/\hat{A}_k}^\beta(x) \mid x \in B_k \cap \mathbf{V}(H/A_k)\} \\ &= \wedge\{d_{H/(X\backslash[\{x_1, \ldots, x_{k-1}\}\cup D_k])}(x) \mid x \in Y_2\backslash[\{x_1, \ldots, x_{k-1}\}\cup D_k]\} \\ &= \delta^\beta(H/(X\backslash[\{x_1, \ldots, x_{k-1}\}\cup D_k])\blacksquare Y_2\backslash[\{x_1, \ldots, x_{k-1}\}\cup D_k]) \\ &\le \vee\{\delta^\beta(H/[X \setminus A]\blacksquare Y_2 \setminus A) \mid A \subset Y_2\}. \end{aligned}$$

This gives the desired result.

By (4.53) and (4.54) and the requirements stipulated in rule M of Algorithm 4.3, it follows that when rule M is directed by $\mathbf{Q}_H^\beta(Y_2)$ the essential β-degree argument used to prove the upper estimate in Theorem 4.43 can be reapplied here in this given situation. Hence at most $\hat{\kappa}_2$ colors are required to color $\{x_1, \ldots, x_{k_2}\}$, where $\hat{\kappa}_2$ satisfies

$$\hat{\kappa}_2 = 1 + \vee\{\delta^\beta(H/[X \setminus A]\blacksquare Y_2 \setminus A) \mid A \subset Y_2\}. \tag{4.55}$$

With the completion of the above analysis it is clear that Algorithm 4.3 directed by $M_H(X)$ produces a k-coloring of H that uses at most

$$\kappa_1 \vee \hat{\kappa}_2 \tag{4.56}$$

colors where κ_1 and $\hat{\kappa}_2$ are defined, respectively, in (4.50) and (4.55)

Furthermore, it has been established that if K_1 is a k-coloring of H/Y_1 then at most

$0 \vee (\hat{\kappa}_2 - |K_1|)$

new colors are required to extend K_1 to a k-coloring of H produced by Algorithm 4.3 under the direction of ordered sum

$\mathbf{L}_H^\beta(Y_2) + \mathbf{S}(Y_1 \setminus \mathbf{V}(H/Y_1)) = \{x_1, \ldots, x_{n_2}, \ldots, x_{n_2+s}\}$.

Note that the above equation also appears in the statement of Lemma 4.46. In other words, it has been established, in view of Definition 4.38, that Algorithm 4.3 directed by this ordered sum in the above equation produces a weakly conservative k-coloring extension of K_1 to H with respect to Y_2 for any k-coloring, K_1, of H/Y_1. We now see that $M_H(X)$ satisfies both properties (i) and (ii) as stated in Lemma 4.46.

To complete the proof of Lemma 4.46 it is necessary to establish the following claim:

If the partition $\{Y_1, Y_2\}$ of X satisfies either
(i) $\hat{\kappa}_2 - 1 \leq \delta^\beta(H/Y_1)$, where $\hat{\kappa}_2$ is as defined in (4.55)
or
(ii) $\Delta^\beta(H\blacksquare Y_2) \leq \delta^\beta(H/Y_1)$
and
(iii) H/Y_1 is filled (see Definition 4.33) then $\hat{\kappa}_2 = \kappa_2$, where

$$\kappa_2 = 1 + \vee\{\delta^\beta(H/[X \setminus A]) \mid A \subset Y_2\}. \tag{4.57}$$

The proof of the claim is as follows: Since H/Y_1 is filled, $\mathbf{V}(H/Y_1) = Y_1$. Thus for any subset A of Y_2 such that $|A| < |Y_2|$, $X \setminus A$ is partitioned by a pair of non-empty sets $\{Y_2 \setminus A, \mathbf{V}(H/Y_1)\}$. Therefore,

$$\begin{aligned}
&\delta^\beta(H/[X \setminus A]) \\
&\quad = \wedge\{d^\beta_{H/[X\setminus A]}(x) \mid x \in X \setminus A\} \\
&\quad = \wedge\ [\{d^\beta_{H/[X\setminus A]}(x) \mid x \in Y_2 \setminus A\} \cup \{d_{H/[X\setminus A]}(x) \mid x \in \mathbf{V}(H/Y_1)\}] \\
&\quad \geq \wedge\ [\{d^\beta_{H/[X\setminus A]}(x) \mid x \in Y_2 \setminus A\} \cup \{d^\beta_{H/Y_1}(x) \mid x \in \mathbf{V}(H/Y_1)\}] \\
&\qquad (\text{since } d^\beta_{H/Y_1}(x) \leq d^\beta_{H/[X\setminus A]}(x) \ \text{ for all } x \in \mathbf{V}(H/Y_1)\})
\end{aligned}$$

$$= \wedge \{d^{\beta}_{H/[X\setminus A]}(x) \mid x \in Y_2 \setminus A\}$$

since, due to either (i) or (ii),

$$\begin{aligned}
&\wedge\{d^{\beta}_{H/[X\setminus A]}(x) \mid x \in Y_2 \setminus A\} \\
&\quad = \delta^{\beta}(H/[X \setminus A]\blacksquare Y_2 \setminus A) \\
&\quad \leq \delta^{\beta}(H/Y_1) = \wedge \{d^{\beta}_{H/Y_1}(x) \mid x \in \mathbf{V}(H/Y_1)\}.
\end{aligned}$$

Hence for every $A \subset Y_2$

$$\delta^{\beta}(H/[X \setminus A]) \geq \delta^{\beta}(H/[X \setminus A]\blacksquare[Y_2 \setminus A]).$$

Now, for every $A \subset Y_2$,

$$\delta^{\beta}(H/[X \setminus A]) \leq \delta^{\beta}(H/[X \setminus A]\blacksquare[Y_2 \setminus A]).$$

Therefore $\hat{\kappa}_2 = \kappa_2$.

From this claim, we can show that the following assertion holds. Now that our claim is established, the following assertion holds. If H/Y_1 is filled and either condition (i) or (ii) in the above claim is satisfied, then $M_H(X)$ is a conservative ordering of X (see Definition 4.37). This follows since we have already shown that if Algorithm 4.3 is directed by $M_H(X)$ to obtain a k-coloring of H, then no more than $\kappa_1 \vee \hat{\kappa}_2$ colors are used (see(4.56)). With $\hat{\kappa}_2 = \kappa_2$ it follows that at most $\kappa_1 \vee \hat{\kappa}_2$ colors are needed to color X. However, as seen from (4.50) and (4.57),

$$\vee\{\kappa_1, \kappa_2\} \leq 1 + \vee\{\delta^{\beta}(H/A) \mid A \subseteq X\}.$$

Therefore, $M_H(X)$ is a conservative ordering of X and the proof of conclusion (ii) is completed.

Finally conclusion (i) follows by the earlier establishment of properties (i) and (ii) together with the proof of the above claim. ■

We now examine the problem of extending partial $\mathcal{L}$-colorings to full $\mathcal{L}$-colorings for a certain fuzzy hypergraphs including the μ-tempered fuzzy hypergraphs, $\mu \otimes H$, of a crisp hypergraph H.

Lemma 4.47 *Let $\mathcal{H} = (X, \mathcal{E})$ be a sequentially simple fuzzy hypergraph with core set $\mathbf{C}(\mathcal{H}) = \{H^{r_i} = (X_i, \mathbf{E}_i) \mid i = 1, \ldots, n\}$, and the members of $\mathbf{F}(\mathcal{H})$ ordered $r_n < \cdots < r_1$. Suppose $E \in \mathbf{E}_{j+k} \setminus \mathbf{E}_j$ for some $j < n$ and $k \in \{1, \ldots < n-j\}$. Then $E \nsubseteq X_j$.*

We only sketch how a proof of Lemma 4.47 would go. Suppose $E \in \mathbf{E}_{j+3} \setminus \mathbf{E}_j$. Then either $E \in \mathbf{E}_{j+2}$ or $E \notin \mathbf{E}_{j+2}$. In the latter case, $E \in \mathbf{E}_{j+3} \setminus \mathbf{E}_{j+2}$, which implies that $E \nsubseteq X_{j+2}$ and thus, $E \subseteq X_j$ since $X_j \subseteq X_{j+2}$. Hence assume $E \in \mathbf{E}_{j+2}$. Then either $E \in \mathbf{E}_{j+1}$ or $E \notin \mathbf{E}_{j+1}$. In the latter case, $E \in \mathbf{E}_{j+2} \setminus \mathbf{E}_{j+1}$, which implies that $E \nsubseteq X_{j+1}$, and therefore, $E \nsubseteq X_j$. Thus assume $E \in \mathbf{E}_{j+1}$. Then, since $E \notin \mathbf{E}_j$, it follows that $E \in \mathbf{E}_{j+1} \setminus \mathbf{E}_j$. Hence $E \not\subseteq X_j$. Therefore, we see that $E \not\subseteq X_j$.

Definition 4.40 *Let $\mathcal{H} = (X, \mathcal{E})$ be a fuzzy hypergraph on X with core set $\mathbf{C}(\mathcal{H}) = \{H^{r_i} = (X_i, \mathbf{E}_i) \mid i = 1, \ldots, n\}$, where, $\mathbf{E}(H^{r_i}) = \mathbf{E}_i$ is the crisp edge set of the core hypergraph H^{r_i}. Let $\mathbf{E}(\mathcal{H})$ denote the* crisp edge set *of $\mathcal{H}$ defined by $\mathbf{E}(\mathcal{H}) = \cup\{\mathbf{E}_i \mid \mathbf{E}_i = \mathbf{E}(H^{r_i}),\ H^{r_i} \in \mathbf{C}(\mathcal{H})\}$. $\mathbf{E}(\mathcal{H})$ is a crisp hypergraph on X or, more precisely, the edge set of the crisp hypergraph, $H(\mathcal{H}) = (X, \mathbf{E}(\mathcal{H}))$, on X called the* core's (crisp) aggregate hypergraph *of $\mathcal{H}$ or, alternatively called the* (crisp) aggregate hypergraph of $\mathcal{H}$.

Lemma 4.48 *For every fuzzy hypergraph $\mathcal{H}$, a k-coloring of $H(\mathcal{H})$ is a $\mathcal{L}$-coloring of $\mathcal{H}$ and conversely.* ■

Definition 4.41 *If $\mathcal{H}$ is a fuzzy hypergraph, then every $\mathcal{L}$-coloring of $\mathcal{H}$ which is a conservative k-coloring of $H(\mathcal{H})$, is said to be a* conservative $\mathcal{L}$-coloring *of $\mathcal{H}$.*

Definition 4.42 *Let $H^{r_j} = (X_j, \mathbf{E}_j)$ be a (crisp) core hypergraph of a fuzzy hypergraph $\mathcal{H} = (X, \mathcal{E})$ where $X_j^c = X \setminus X_j \neq \emptyset$. Furthermore, suppose K is an $\mathcal{L}$-coloring of the upper truncated fuzzy hypergraph $\mathcal{H}^{(r_j)}$ which is obtained by extending a k-coloring, K_j, of H^{r_j}. If K is a weakly (or strongly) conservative k-coloring extension of K_j to the (crisp) core aggregate hypergraph $H(\mathcal{H}^{(r_j)})$ of $\mathcal{H}^{(r_j)}$ with respect to X_j^c (see Definition 4.38), then K is called a* weakly (or strongly) conservative $\mathcal{L}$-coloring extension *of K_j to $\mathcal{H}^{(r_j)}$ with respect to X_j^c.*

In the above situation, we also say: "K_1 is *extended weakly (or, strongly) conservative with respect to X_j^c, to a $\mathcal{L}$-coloring K of $\mathcal{H}^{(r_j)}$*".

Theorem 4.49 *Let $\mathcal{H} = (X, \mathcal{E})$ be a sequentially simple fuzzy hypergraph on X with core set $\mathbf{C}(\mathcal{H}) = \{H^{r_j} = (X_j, \mathbf{E}_j) \mid j = 1, \ldots, n\}$. Let $H^{r_j} = (X_j, \mathbf{E}_j)$ be a core hypergraph of $\mathcal{H}$ for which the complement $X_j^c = X \setminus \mathbf{V}(H^{r_j}) \neq \emptyset$. Then any k-coloring K_j of H^{r_j} can be extended, weakly conservative with respect to X_j^c, to an $\mathcal{L}$-coloring K of $\mathcal{H}^{(r_j)}$, which is called the upper r_j-level truncation of $\mathcal{H}$ (see Definition 4.16). The extension K can be produced by allowing K_j to be extended, weakly conservative with respect to X_j^c, to a k-coloring of the core aggregate hypergraph $H(\mathcal{H}^{(r_j)})$ by a procedure described in Lemma 4.46. Therefore, whenever $|K_1|$ equals or exceeds $\hat{\kappa}_2^{(j)}$, where $\hat{\kappa}_2^{(j)} = 1 + \vee\{\delta^\beta(H(\mathcal{H}^{(r_j)})/[X \setminus A]\blacksquare X_j^c \setminus A) \mid A \subset X_j^c\}$, the k-coloring, K_1, of H^{r_j} can be extended to an $\mathcal{L}$-coloring of $\mathcal{H}^{(r_j)}$ without introducing any new colors by a procedure set forth in Lemma 4.46. If, in addition, complement $X_j^c = X \setminus \mathbf{V}(H^{r_j})$ satisfies*

Condition j: $\Delta^{\beta}(H(\mathcal{H}^{(r_j)})\blacksquare X \setminus \mathbf{V}(H^{r_j})) \leq \delta^{\beta}(H^{r_j})$
then any k-coloring, K_j, *of* H^{r_j} *can be extended, strongly conservative with respect to* X_j^c, *to an* $\mathcal{L}$*-coloring of* $\mathcal{H}^{(r_j)}$ *where no new colors are introduced if* $|K_1| \geq \kappa_2^{(j)}$, *where* $\kappa_2^{(j)} = 1 + \vee\{\delta^{\beta}(H(\mathcal{H}^{(r_j)})/[X \setminus A]) \mid A \subset X_j^c\}$. *(This can be accomplished by extending* K_1 *to a k-coloring of the core aggregate hypergraph* $H(\mathcal{H}^{(r_j)})$ *by a procedure described in Lemma 4.46.) In particular, if* K_1 *is a conservative k-coloring of* H^{r_j}, *where* $X_j^c = X \setminus \mathbf{V}(H^{r_j})$ *is non empty and satisfies condition j, then* K_1 *can be extended to a conservative* $\mathcal{L}$*-coloring of* $\mathcal{H}^{(r_j)}$ *by a procedure described in Lemma 4.46.*

Proof. Theorem 4.49 is a direct result of Lemma 4.46 through an interpretation provided by Lemma 4.48. More specifically, the (crisp) core aggregate hypergraph $H(\mathcal{H}^{(r_j)})$ is identified with H appearing in Lemma 4.46. The partition $\{Y_1, Y_2\}$, stated in Lemma 4.46, is understood here to satisfy $Y_1 = X_j$, $Y_2 = X_j^c$. Since $\mathcal{H}$ is sequentially simple, $Y_1 = \mathbf{V}(H^{r_j})$. Thus $H^{r_j} = H(\mathcal{H}^{(r_j)})/Y_1$ by Lemma 4.47, and H^{r_j} is filled.

With the above associations and observations the results of Theorem 4.49 follow readily from the statement of Lemma 4.46. ■

Corollary 4.50 *Suppose* $\mathcal{H} = (X, \mathcal{E})$ *is a simply ordered fuzzy hypergraph and suppose* $H^{r_j} \in \mathbf{C}(\mathcal{H})$. *Assume* $X_j^c = X \setminus \mathbf{V}(H^{r_j}) \neq \emptyset$ *and*
Condition j: $\Delta^{\beta}(H(\mathcal{H}^{(r_j)})\blacksquare X_j^c) \leq \delta^{\beta}(H^{r_j})$,
is satisfied. Then any k-coloring of H^{r_j} *with at least* $\kappa_2^{(j)}$ *colors, where* $\kappa_2^{(j)}$ *is defined in Theorem 4.49, can be extended to an* $\mathcal{L}$*-coloring of* $\mathcal{H}$ *without the addition of any new colors by a procedure described in Lemma 4.46. Moreover, any conservative k-coloring of* H^{r_j} *can be extended to a conservative* $\mathcal{L}$*-coloring of* $\mathcal{H}$ *by a procedure described in Lemma 4.46.*

Proof. Since $\mathcal{H}$ is ordered, the lower truncation, $\mathcal{H}_{(r_j)}$, of $\mathcal{H}$ at level r_j is ordered (see Definition 4.16). Thus, any k-coloring of H^{r_j} is an $\mathcal{L}$-coloring of $\mathcal{H}_{(r_j)}$ (see Theorem 4.37).

Hence, since
$\mathbf{C}(\mathcal{H}_{(r_j)}) \cup \mathbf{C}(\mathcal{H}^{(r_j)}) = \mathbf{C}(\mathcal{H})$,
the desired results follow from Theorem 4.49, which is applicable here since condition j is assumed and $\mathcal{H}$ is sequentially simple. ■

Since $\mathcal{H} = \mu \otimes H$ is simply ordered, Corollary 4.50 is applicable, provided condition j is valid for some core hypergraph H^{r_j}. Moreover, since $\mathcal{H}$ is sequentially simple, the edge set of star $H(\mathcal{H})(X_j^c)$, in the core aggregate hypergraph $H(\mathcal{H})$ of $\mathcal{H}$, is precisely the full set of "new" edges that belong to those $H^{r_{j+k}}$'s that "lie below" H^{r_j}; that is the edge set of star, $H(\mathcal{H})(X_j^c)$, of X_j^c in $H(\mathcal{H})$ satisfies
$\mathbf{E}(H(\mathcal{H})(X_j^c)) = \{E \mid E \in \mathbf{E}_{j+k} \setminus \mathbf{E}_j,\ k = 1, \ldots, n-j\}$,

where $\mathbf{E}_i$ is the edge set of $H^{r_i} \in \mathbf{C}(\mathcal{H})$. Thus, all new edges introduced "below" H^{r_j} and only those edges are involved in the determination of $\Delta^{\beta}(H(\mathcal{H})(X_j^c)\blacksquare X_j^c) = \Delta^{\beta}(H(\mathcal{H})\blacksquare X_j^c)$.

Theorem 4.51 *If $\mathcal{H}$ is a sequentially simple fuzzy hypergraph, then $\mathcal{H}^-$, $\mathcal{H}^s$ and $\mathcal{H}^{\Delta}$ are sequentially simple fuzzy hypergraphs as well.*

Proof. Clearly $\mathcal{H}^-$ is sequentially simple and, by Lemma 4.33, the skeleton of $\mathcal{H}$, $\mathcal{H}^s$, is sequentially simple as well. Therefore $\mathcal{H}^{\Delta} = (\mathcal{H}^-)^s$ must be sequentially simple provided $\mathcal{H}^-$ exists. ∎

Since $\mathcal{H}^{\Delta}$ is sequentially simple whenever $\mathcal{H}$ is sequentially simple (provided $\mathcal{H}^-$ exists), the results of Theorem 4.49 apply to $\mathcal{H}^{\Delta}$ whenever $\mathcal{H}$ is sequentially simple. In some cases, moreover, $\mathcal{H}^{\Delta}$ is considerably less complicated than $\mathcal{H}$. If this situation is true and $\mathcal{H}$ is sequentially simple, it may be advantageous to work with $\mathcal{H}^{\Delta}$, rather than with $\mathcal{H}$, to compute $\mathcal{L}$-coloring extensions of given k-colorings of specific core hypergraphs.

Chromatic Values of Fuzzy Colorings

We now respond to question (iv) in Example 4.14. We do this by introducing a *chromatic value* to each fuzzy coloring of $\mathcal{H}$. We also present an equivalent definition of an $\mathcal{L}$-coloring, which takes the form of a fuzzy hypergraph. Finally, we end the section with a discussion, through an example, on how chromatic values provide fuzzy information.

Definition 4.43 *Let $\mathcal{H} = (X, \mathcal{E})$ be a fuzzy hypergraph on X and suppose $\Gamma = \{\gamma_i \in \mathfrak{F}\wp(X) \mid i = 1, \ldots, m\}$ is a finite subset of $\mathfrak{F}\wp(X)$. Then Γ is called a* fuzzy coloring *of $\mathcal{H}$ if the following properties are satisfied.*

(1) $\forall x \in X, \vee\{\gamma_i(x) \mid i = 1, \ldots, m\} = \vee\{\mu(x) \mid \mu \in \mathcal{E}\}$,

(2) $\gamma_i \cap \gamma_j = \chi_{\emptyset}$ *if* $i \neq j$,

(3) Γ^c *is a coloring of* H^c *for* $0 < c \leq h(\mathcal{H})$.

We note that Γ is sequentially elementary with respect to $\mathbf{F}(\mathcal{H})$.

There is a one-to-one correspondence between the $\mathcal{L}$-colorings of $\mathcal{H}$ and the fuzzy colorings of $\mathcal{H}$, under the condition that all members (color sets) of a coloring are non-empty. For suppose Γ is a fuzzy coloring of $\mathcal{H} = (X, \mathcal{E})$. Then property (3) in Definition 4.43 implies the r_n-cut, Γ^{r_n}, of Γ, where r_n is the smallest value in the fundamental sequence, $\mathbf{F}(\mathcal{H})$, of $\mathcal{H}$, is a k-coloring of the core aggregate hypergraph, $H(\mathcal{H})$, of $\mathcal{H}$ which in turn implies Γ^{r_n} is an $\mathcal{L}$-coloring of $\mathcal{H}$ (see Lemma 4.48).

Conversely, suppose $C = \{S_1, \ldots, S_k\}$ is an $\mathcal{L}$-coloring of $\mathcal{H}$. Then C is a crisp coloring of the core aggregate hypergraph, $H(\mathcal{H})$, of $\mathcal{H}$, $\cup_{i=1}^k S_k = X$

and $S_i \cap S_j = \emptyset$ if $i \neq j$. Now with each S_i associate a fuzzy subset $\gamma_i \in \mathfrak{F}\wp(X)$, with support S_i, defined by

$$\gamma_i(x) = \begin{cases} \vee\{\mu(x) \mid \mu \in \mathcal{E}\} & \text{if } x \in S_i, \\ 0 & \text{otherwise.} \end{cases}$$

It is easily shown that $\Gamma = \{\gamma_1, \ldots, \gamma_k\}$ is a fuzzy coloring of $\mathcal{H}$.

Clearly each of the above associations is a one-to-one mapping and each is the inverse of the other. Thus each association yields a one-to-one correspondence between the set of fuzzy colorings of $\mathcal{H}$ and the set of $\mathcal{L}$-colorings of $\mathcal{H}$.

Among several chromatic values that can be assigned to Γ, two primary examples are given. The first example assigns a chromatic value to Γ, designated $\Lambda_f(\Gamma)$, which averages the chromatic cardinalities $|\Gamma^c|$ through an apportionment dependent upon those fuzzy levels, c, where new color increments first appear (as one descends the scale of fuzziness). The second example assigns a chromatic value to Γ, denoted $\bar{\Lambda}_f(\Gamma)$, which is a weighted average that is dependent upon the "fuzzy duration" between successive new color increments in Γ.

Before we introduce the definition of $\Lambda_f(\Gamma)$ below in Definition 4.46, we first introduce the following two definitions:

Definition 4.44 *Let $\gamma \in \mathfrak{F}\wp(X)$. Define the fuzzy subset $\gamma_{(\sigma)}$ of X by $\forall x \in X$,*

$$\gamma_{(\sigma)}(x) = \begin{cases} h(\gamma) & \textit{if } \gamma(x) = h(\gamma), \\ 0 & \textit{otherwise.} \end{cases}$$

$\gamma_{(\sigma)}$ is called the *elementary center*

Definition 4.45 *Let $\Gamma = \{\gamma_i \in \mathfrak{F}\wp(X) \mid i = 1, \ldots, k\}$. Then $\Gamma_{(\sigma)}$, called the* elementary center *of Γ, is defined by*

$$\Gamma_{(\sigma)} = \{\gamma_{1(\sigma)}, \ldots, \gamma_{k(\sigma)}\},$$

where $\gamma_{i(\sigma)}$ is the elementary center of γ_i.

Definition 4.46 *Let $\Gamma_{(\sigma)}$ be the elementary center of a fuzzy coloring Γ of $\mathcal{H}$ with fundamental sequence $\mathbf{F}(\Gamma_{(\sigma)}) = \{t_1^\Gamma, t_2^\Gamma, \ldots, t_q^\Gamma\}$, where $t_1^\Gamma > t_2^\Gamma > \cdots > t_q^\Gamma$ and let f be a monotonic increasing function on the interval $[0,1]$ such that $f(0) = 0$ and $f(1) = 1$. Such an f is called a* scaling function.

We assign a chromatic value $\Lambda_f(\Gamma)$ to Γ, called the *Λ_f-chromatic valuation* of Γ (or, alternatively, the *f-chromatic value of* Γ) as follows:

$$\Lambda_f(\Gamma) = f(t_1^\Gamma)\frac{\left(\sum_{j=1}^{q} f(t_j^\Gamma)\, |\Gamma_{(\sigma),t_j^\Gamma}|\right)}{\left(\sum_{j=1}^{q} f(t_j^\Gamma)\right)},$$

where $|\Gamma_{(\sigma),t_j^\Gamma}|$ is cardinality of the t_j^Γ-cut of $\Gamma_{(\sigma)}$ (i.e., $|\Gamma_{(\sigma),t_j^\Gamma}|$ represents the number of edges in $\Gamma_{(\sigma),t_j^\Gamma}$).

We also assign another chromatic value $\bar{\Lambda}_f(\Gamma)$ to Γ, called the $\bar{\Lambda}_f$*-chromatic valuation of* Γ (or, alternatively, the $\bar{f}$*-chromatic value of* Γ) as follows:

$$\bar{\Lambda}_f(\Gamma) = \sum_{j=1}^{q} (f(t_j^\Gamma) - f(t_{j+1}^\Gamma))\, |\Gamma_{(\sigma),t_j^\Gamma}|,$$

where t_{q+1}^Γ is understood to be zero.

If r and $s \in [0,1]$ and f is a scaling function, then $|f(s) - f(r)|$ will be called an "f-fuzzy duration." This simple tool can be useful in situations where a given property is invariant over all fuzzy degrees of membership between r and s. We also use the convention: If f is the identity map, then $\Lambda(\Gamma) = \Lambda_f(\Gamma)$ and $\bar{\Lambda}(\Gamma) = \bar{\Lambda}_f(\Gamma)$.

It is of interest to consider when values of f are referenced against some fixed non-zero value of f. A typical point of reference for a scaling function f relative to a given fuzzy hypergraph $\mathcal{H}$ would be $f(r_n)$, where r_n is the smallest value of $\mathbf{F}(\mathcal{H})$. For a coloring Γ, a reasonable reference point might be $f(t_q^\Gamma)$, which appears in the formulation of $\Lambda_f(\Gamma)$. Such scalings, when carefully adjusted to particular circumstances, can enhance the usefulness of chromatic valuations of fuzzy colorings; for example in the determination of relative costs attached to competing long range proposals (i.e., fuzzy colorings, Γ), which may arise in situations typified by Example 4.14.

Indeed, since scaling functions like $f_1(x) = x^3$ de-emphasize lower fuzzy degrees while scaling functions like $f_2(x) = x^{1/3}$ emphasize lower fuzzy degrees, it would be reasonable to assume, in situations like prototype Example 3.1, that managements which stress short term goals would probably select f_1 over f_2, while managements that stress long term goals would be more inclined to favor f_2 over f_1. In this way, chromatic valuations can be customized to fit stated requirements.

Consider the situation where the fuzzy hypergraphs $\widehat{\mathcal{H}} = (X, \widehat{\mathcal{E}})$ and $\mathcal{H}(X, \mathcal{E})$ satisfy the property that there exists a satisfying $0 < a < 1$ such that

$$\widehat{\mathcal{E}} = \{\nu = a\mu \mid \mu \in \mathcal{E}\},$$

where $\nu = a\mu$ means: $\nu(x) = a\mu(x)$, for all $x \in X$. In this case we write $a\mathcal{H}$ for $\widehat{\mathcal{H}}$. With $\mathbf{F}(\mathcal{H}) = \{r_1, \ldots, r_n\}$ it follows that the fundamental sequence $\mathbf{F}(\widehat{\mathcal{H}}) = \{\hat{r}_1, \ldots, \hat{r}_n\}$ of $\widehat{\mathcal{H}}$ satisfies

$$\hat{r}_i = ar_i, \quad i = 1, \ldots, n,$$

$$\begin{aligned}\mathbf{C}(\widehat{\mathcal{H}}) &= \{\widehat{H}^{\hat{r}_i} \mid i = 1, \ldots, n\}\\ &= \{H^{r_i} \mid i = 1, \ldots, n\} = \mathbf{C}(\mathcal{H}).\end{aligned}$$

Clearly, the set of fuzzy colorings Γ on $\mathcal{H}$ correspond uniquely with the fuzzy colorings $\widehat{\Gamma}$ of $\widehat{\mathcal{H}}$ under the correspondence $\widehat{\Gamma} = a\Gamma$. Hence if the initial factor $f(t_1^\Gamma)$ were eliminated from the formulation of $\Lambda_f(\Gamma)$, then, in the special case where f is the identity map on $[0,1]$,

$\Lambda(\Gamma)$ would equal $\Lambda(a\Gamma) = \Lambda(\widehat{\Gamma})$.

In fact

$\Lambda(a\Gamma) = a\Lambda(\Gamma)$.

and $\bar{\Lambda}(a\Gamma) = a\bar{\Lambda}(\Gamma)$.

For any fuzzy coloring Γ of $\mathcal{H}$, the f-factor $f(t_1^\Gamma)$ in $\Lambda_f(\Gamma)$ scales the valuation $\Lambda_f(\Gamma)$ according to the maximum degree of fuzziness in $\mathcal{H}$ as interpreted by the scaling function f. This sensitizes the valuation of $\Lambda_f(\Gamma)$ to the range of fuzziness in $\mathcal{H}$.

The concept of f-chromatic values of Γ leads naturally to the following definitions.

Definition 4.47 *Let $\mathcal{H}$ be a fuzzy hypergraph and let f denote a scaling function. Then*

$\chi_f(\mathcal{H}) = \wedge\{\Lambda_f(\Gamma) \mid \Gamma$ *is a fuzzy coloring of* $\mathcal{H}\}$

and

$\bar{\chi}_f(\mathcal{H}) = \wedge\{\bar{\Lambda}_f(\Gamma) \mid \Gamma$ *is a fuzzy coloring of* $\mathcal{H}\}$

are called, respectively, the Λ_f-chromatic number *and the* $\bar{\Lambda}_f$-chromatic number *of $\mathcal{H}$. In the special case where f is the identity map on $[0,1]$, $\chi_f(\mathcal{H})$, or $\bar{\chi}_f(\mathcal{H})$ will sometimes be called* linear chromatic numbers *of $\mathcal{H}$.*

Theorem 4.52 *For every fuzzy hypergraph $\mathcal{H}$ and every scaling function $f : [0,1] \to [0,1]$,*

$\chi_f(\mathcal{H}) \le \chi(\mathcal{H}), \bar{\chi}_f(\mathcal{H}) \le \chi(\mathcal{H})$ *and*

$\chi(\mathcal{H}) = \wedge\{|\Gamma| \mid \Gamma$ *is a fuzzy coloring of* $\mathcal{H}\} = \wedge\{|K| \mid K$ *is an* $\mathcal{L}$*-coloring of* $\mathcal{H}\}$,

where, $|\Gamma|$ is the number of edges in Γ and $|K|$ is the number of colors in K. ■

The following example illustrates how the linear chromatic value $\Lambda(\Gamma)$ depends upon the fundamental sequence $\mathbf{F}(\Gamma_{(\sigma)})$ of $\Gamma_{(\sigma)}$. In particular for f the identity map on $[0,1]$ we contrast the fuzzy attributes of $\Lambda(\Gamma)$ and $\chi_f(\mathcal{H})$, with its dependency upon $\mathbf{F}(\Gamma_{(\sigma)})$, against the nonfuzzy attributes of $\chi(\mathcal{H})$.

Example 4.17 *Consider the elementary fuzzy hypergraph $\mathcal{H} = (X, \mathcal{E})$, where $\mathbf{F}(\mathcal{H}) = \{r_1, r_2\}$, with $r_1 > r_2$, and an edge set $\mathcal{E} = \{\sigma(E_i, r_1) \mid i = 1, \ldots, 6\} \cup \{\sigma(E_i, r_2) \mid i = 7, \ldots, 15\}$, where each support E_i, $i = 1, \ldots, 15$ has cardinality two. Recall that $\sigma(E, r)$ is the elementary fuzzy subset on X defined by*

$$\sigma(E,r)(x) = \begin{cases} r & \textit{if } x \in E, \\ 0 & \textit{otherwise.} \end{cases}$$

FIGURE 4.2 Vertex and edge sets of $\mathcal{H}$.

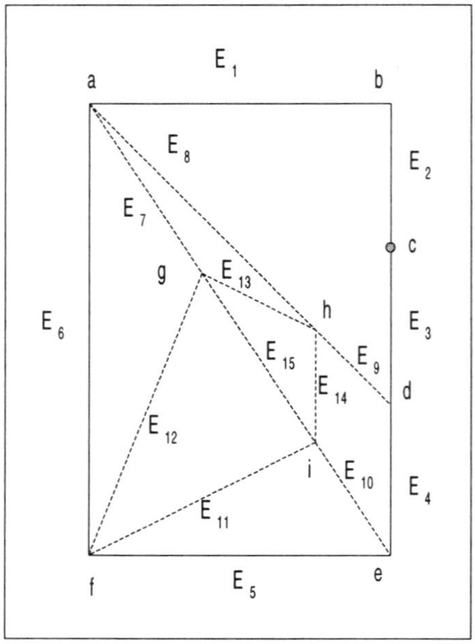

Figure 4.2 illustrates the vertex and edge set of $\mathcal{H}$. *Edges of height* r_1 *are indicated by solid-lined arcs; edges of height* r_2 *are indicated by dash-lined arcs.*

Clearly, $X = \mathbf{V}(\mathcal{H}) = \{a, b, \ldots, i\}$ *and* $\mathbf{C}(\mathcal{H}) = (H^{r_1}, H^{r_2})$, *where*
$H^{r_1} = (\{a, \ldots, f\}, \{E_1, \ldots, E_6\})$
and
$H^{r_2} = (X, \{E_1, \ldots, E_{15}\})$.

Since $\mathcal{H}$ *is elementary, it is ordered; Thus by Theorem 4.37, every primitive coloring of* $\mathcal{H}$ *is an* $\mathcal{L}$*-coloring of* $\mathcal{H}$. *Therefore,* $\chi(\mathcal{H}) = 3$ *since* H^{r_2} *has the following primitive coloring:* $C_1 = \{B, G, W\}$, *where* $B = \{a, d, i\}$, $G = \{b, f, h\}$ *and* $W = \{c, e, g\}$. *Notice, however, when* C_1 *is restricted to* H^{r_1}, *it remains a 3-color set,* $C_1^{'} = \{B', C', W'\}$, *where* $B' = \{a, d\}$, $G' = \{b, f\}$ *and* $W' = \{c, e\}$.

If this problem were the representation of a waste management problem, the fact that C_1' *is not a minimal coloring of* H^{r_1} *might suggest that alternative solutions (other than* C_1*) might be more efficient from certain specific viewpoints, such as cost analysis. Indeed, it is reasonable to ask whether or not some minimal coloring,* C_2' *of* H^{r_1}, *which is expandable to an* $\mathcal{L}$*-coloring* C_2 *of* $\mathcal{H}$ *is more (cost effective than solution* C_1 *(even though* C_2 *is not a minimal* $\mathcal{L}$*-coloring of* $\mathcal{H}$*)).*

Suppose f *is the identity map. Under the assumption that an optimal cost effective solution,* Γ, *satisfies* $\Lambda_f(\Gamma) = \chi_f(\mathcal{H})$, *it is interesting to compare* $\Lambda_f(\Gamma_1)$ *with* $\Lambda_f(\Gamma_2)$, *where* Γ_1 *and* Γ_2 *are the fuzzy colorings of* $\mathcal{H}$, *where*

C_2 is given below. This is what we intend to do. Let C_2 be the following primitive coloring of H^{r_2}: $C_2 = \{B, G, R, W, Y\}$, where $B = \{a, c, e\}$, $G = \{g\}$, $R = \{h\}$, $W = \{b, d, f\}$ and $Y = \{i\}$. The restriction, C_2', of C_2 to H^{r_1} is a 2-color set, $C_2' = \{B, W\}$.

For $i = 1, 2$, we have $\Lambda_f(\Gamma_i) = r_1(k_i)/(r_1 + r_2)$, where $k_1 = 3r_1 + 3r_2$ and $k_2 = 2r_1 + 5r_2$. To compare $\Lambda_f(\Gamma_i)$, $i = 1, 2$, it suffices to compare k_i, $i = 1, 2$. By changing scales so that $r_2 \leftrightarrow 1$ and $r_1 \leftrightarrow 1+\varepsilon$, $\varepsilon > 0$ (i.e., $r_1/r_2 = 1+\varepsilon$), it suffices to compare $\hat{k}_1 = 3+3(1+\varepsilon)$ and $\hat{k}_2 = 5+2(1+\varepsilon)$. It may be noted that as $r_2 \to 0$, ε could approach ∞ if the relative sizes of r_1 and r_2 are altered sufficiently.

Clearly, $\hat{k}_2 \leq \hat{k}_1 \Leftrightarrow 1 \leq \varepsilon$. Thus $\Lambda_f(\Gamma_2) \leq \Lambda_f(\Gamma_1) \Leftrightarrow r_1 \geq 2r_2$ and equality holds $\Leftrightarrow r_1 = 2r_2$. In particular, if $r_1 = 1$, then $\Lambda_f(\Gamma_2) < \Lambda_f(\Gamma_1) \Leftrightarrow r_2 < \frac{1}{2}$.

In fact if $r_1 = 1$, then

$$\chi_f(\mathcal{H}) = \begin{cases} \Lambda_f(\Gamma_1) = 3 & \text{if } r_2 \geq \frac{1}{2}, \\ \Lambda_f(\Gamma_2) & \text{if } r_2 < \frac{1}{2}. \end{cases}$$

Moreover, as $r_2 \to 0$, $\chi(\mathcal{H}) = \Lambda_f(\Gamma_1)$ remains 3 while $\Lambda_f(\Gamma_2) \to 2$.

To clarify the above evaluation of the linear chromatic number $\chi_f(\mathcal{H})$, it suffices to see that if H^{r_1} is minimally colored with 2 colors then there is really only one $\mathcal{L}$-coloring extension to $\mathcal{H}$ which is given by C_2.

From this example, it is clear that the coloring problem associated with $\chi_f(\mathcal{H})$ is a non-traditional coloring problem. Moreover, it is an example of a fuzzy coloring problem with applications.

Suppose that we restrict the scaling functions to a 1-parameter family such as $S = \{f_p(x) = x^p \mid p > 0\}$. Then consider the question: How sensitive is the value of $\chi_f(\mathcal{H})$ or $\Lambda_f(\Gamma)$ to perturbations in S about a specific member f_{p_0} of S? For example, given $\varepsilon > 0$, what is the maximum $\delta > 0$ for which $\Lambda_{f_p}(\Gamma)$ is a member of the ε-neighborhood, $N_\varepsilon(\Lambda_{f_{p_0}}(\Gamma))$, of $\Lambda_{f_{p_0}}(\Gamma)$ provided p is a member of the δ-neighborhood, $N_\delta(p_0)$, of p_0? It would be helpful to evaluate the first derivative of $\Lambda_{f_p}(\Gamma)$ with respect to p and develop sensitivity analysis from this information.

4.4 Intersecting Fuzzy Hypergraphs

Let H be a crisp hypergraph. H is said to be *intersecting* if the edges of H are pairwise non-disjoint. Every vertex x of H has a star, $H(x)$, which is the set of edges in H containing x. The edges in $H(x)$ are, of course, pairwise joined. This is the property that distinguishes *intersecting families* of edges (which may be full or partial hypergraphs as the situation warrants). For multigraphs (hypergraphs, with repeated edges, where all edges have cardinality equal or less than two), the only intersecting families of edges are stars and triangles (perhaps with multiple edges). However, the following example shows that the general case is more extensive.

Example 4.18 *Let H be represented by the following incidence matrix.*

$$\begin{array}{c} \\ a \\ b \\ c \\ d \\ e \\ f \end{array} \begin{array}{c} \begin{array}{cccc} \mu_1 & \mu_2 & \mu_3 & \mu_4 \end{array} \\ \begin{pmatrix} 1 & 0 & 1 & 0 \\ 1 & 1 & 0 & 0 \\ 0 & 1 & 1 & 0 \\ 1 & 0 & 0 & 1 \\ 0 & 1 & 0 & 1 \\ 0 & 0 & 1 & 1 \end{pmatrix} \end{array}$$

Clearly, H is an intersecting hypergraph which is neither a star nor triangle.

There are several ways to define intersecting fuzzy hypergraphs.

Definition 4.48 *A fuzzy hypergraph $\mathcal{H} = (X, \mathcal{E})$ is said to be* intersecting *if, for each pair of fuzzy edges $\{\mu_1, \mu_2\} \subseteq \mathcal{E}$, $\mu_1 \cap \mu_2 \neq \chi_\emptyset$, where $\chi_\emptyset$ is the fuzzy subset which is identically equal to zero on X.*

Definition 4.49 *Let $\mathcal{H} = (X, \mathcal{E})$ be a fuzzy hypergraph and suppose $\mathbf{C}(\mathcal{H}) = \{H^{r_1}, H^{r_2}, \ldots, H^{r_n}\}$. If H^{r_i} is an intersecting hypergraph for each $i =$ 1, 2, ..., n, then $\mathcal{H}$ is* $\mathcal{L}$-intersecting.

Theorem 4.53 *$\mathcal{H} = (X, \mathcal{E})$ be a fuzzy hypergraph and suppose $\mathbf{C}(\mathcal{H}) = \{H^{r_1}, H^{r_2}, \ldots, H^{r_n}\}$. Then $\mathcal{H}$ is intersecting if and only if $H^{r_n} = (X, \mathbf{E}^{r_n})$ is intersecting.*

Proof. $\mathcal{H}$ is intersecting $\Leftrightarrow$ $\text{supp}(\mathcal{H}) = \{\text{supp}(\mu) \mid \mu \in \mathcal{E}\}$ is intersecting. The desired result follows since $\mathbf{E}^{r_n} = \text{supp}(\mathcal{H})$ for every fuzzy hypergraph $\mathcal{H}$. ■

Theorem 4.54 *Let $\mathcal{H} = (X, \mathcal{E})$ be an ordered fuzzy hypergraph and let $\mathbf{C}(\mathcal{H}) = \{H^{r_1}, H^{r_2}, \ldots, H^{r_n}\}$. Then $\mathcal{H}$ is intersecting if and only if $\mathcal{H}$ is $\mathcal{L}$-intersecting.*

Proof. If $\mathcal{H}$ is $\mathcal{L}$-intersecting, then $\mathcal{H}$ is intersecting. This follows easily from Definition 4.49 and Theorem 4.53. In fact, $\mathcal{H}$ is $\mathcal{L}$-intersecting $\Rightarrow$ H^{r_n} is intersecting $\Leftrightarrow$ $\mathcal{H}$ is intersecting. Conversely, suppose that $\mathcal{H}$ is intersecting. It follows that H^{r_n} is also intersecting. If E and F are edges in H^{r_i}, then since $H^{r_i} \subseteq H^{r_n}$, E and F are also edges of the intersecting hypergraph H^{r_n}, and hence intersect. ■

Example 4.19 *Consider the fuzzy hypergraph $\mathcal{H} = (X, \mathcal{E})$ which has the following incidence matrix:*

$$\begin{array}{c} \\ a \\ b \end{array} \begin{pmatrix} \mu_1 & \mu_2 \\ 0.7 & 0.9 \\ 0.5 & 0.3 \end{pmatrix}$$

Then $\mathbf{E}^{0.3} = \{\{a,b\}\}$, $\mathbf{E}^{0.5} = \{\{a,b\},\{a\}\}, \mathbf{E}^{0.7} = \{\{a\}\}, \mathbf{E}^{0.9} = \{\{a\}\}$. *Thus* $\mathbf{E}^{0.9} = \mathbf{E}^{0.7} \sqsubset \mathbf{E}^{0.5} \sqsubset \mathbf{E}^{0.3}$. *Note also* $\mathbf{E}^{0.3} \sqsubset \mathbf{E}^{0.5}$. *We have* $r_1 = 0.9$, $r_2 = 0.5$, *and* $r_3 = 0.3$ *and* $H^{r_1} = (\{a\},\{\{a\}\}), H^{r_2} = (\{a,b\},\{\{a,b\},\{a\}\})$, $H^{r_3} = (\{a,b\},\{\{a,b\}\})$. *Now* $\mathbf{E}^{r_1} = \{\{a\}\} \subset \mathbf{E}^{r_2} = \{\{a,b\},\{a\}\} \not\subseteq \mathbf{E}^{r_3} = \{\{a,b\}\}$. *Thus* $\mathcal{H}$ *is not ordered.* $\mathcal{H}$ *is however* $\mathcal{L}$*-intersecting. (*H^{r_1} *and* H^{r_3} *are vacuously intersecting hypergraphs.)*

Example 4.20 *Consider the fuzzy hypergraph* $\mathcal{H} = (X,\mathcal{E})$ *which has the following incidence matrix:*

$$\begin{array}{c} \\ a \\ b \\ c \end{array} \begin{pmatrix} \mu_1 & \mu_2 \\ 0.9 & 0 \\ 0.4 & 0.4 \\ 0 & 0.9 \end{pmatrix}$$

Then $\mathbf{E}^{0.4} = \{\{a,b\},\{b,c\}\}$, $\mathbf{E}^{0.9} = \{\{a\},\{c\}\}$. *Thus* $\mathbf{E}^{0.9} \sqsubset \mathbf{E}^{0.4}$. *We have* $r_1 = 0.9$, $r_2 = 0.4$ *and* $H^{r_1} = (\{a,c\},\{\{a\},\{c\}\}), H^{r_2} = (\{a,b,c\},\{\{a,b\},\{b,c\}\})$. *Now* $\mu_1 \cap \mu_2 \neq \emptyset$ *and so* $\mathcal{H}$ *is intersecting. However* $\mathcal{H}$ *is not* $\mathcal{L}$*-intersecting since* H^{r_1} *is not intersecting. Note that* $\mathcal{H}$ *is not ordered since* $\mathbf{E}^{r_1} \not\subseteq \mathbf{E}^{r_2}$

Theorem 4.55 *Suppose* $\mathcal{H} = (X,\mathcal{E})$ *is an ordered, intersecting fuzzy hypergraph. Then each fuzzy edge* ν *of* $\mathcal{H}$ *contains a member of* $Tr^*(\mathcal{H}^{(h(\nu))})$, *where* $\mathcal{H}^{(h(\nu))}$ *is the upper truncation of* $\mathcal{H}$ *at level* $h(\nu)$. *In particular,* ν *is a fuzzy transversal of* $\mathcal{H}^{(h(\nu))}$.

Proof. Let $\mathbf{C}(\mathcal{H}) = \{H^{r_1}, H^{r_2}, \ldots, H^{r_n}\}$, where $0 < r_n < \cdots < r_1$, and suppose $\nu \in \mathcal{E}$. We assume without loss of generality that $r_1 = h(\nu)$. Since $\mathcal{H}$ is ordered and $\nu^{r_1} \neq \emptyset$, it follows that $\nu^{r_1} \in H^{r_n}$. Since H is intersecting, H^{r_n} is intersecting. Therefore, ν^{r_1} is a transversal of H^{r_n}. Let T_n be a minimal transversal of H^{r_n} contained in ν^{r_1}. Since $\mathcal{H}$ is ordered, it follows from Lemma 4.7 that there is a nested sequence of sets $T_n \supseteq \cdots \supseteq T_i \supseteq \cdots \supseteq T_1$, such that T_i is a minimal transversal of H^{r_i} for every $r_i \in \mathbf{F}(\mathcal{H})$. Let σ_i be the elementary fuzzy subset with support T_i and height r_i, for $i = 1, \ldots, n$. Then clearly, $\bigcup_{i=1}^{n} \sigma_i \in Tr^*(\mathcal{H})$ and $\tau \subseteq \nu$. ■

Example 4.21 *Let* $\mathcal{H} = (X,\mathcal{E})$ *be the fuzzy hypergraph, where* $X = \{a,b,c\}$ *and* $\mathcal{E} = \{\mu_1, \mu_2, \mu_3\}$ *which is represented by the following incidence matrix.*

$$\begin{array}{c} \\ a \\ b \\ c \end{array} \begin{pmatrix} \mu_1 & \mu_2 & \mu_3 \\ 0.9 & 0.7 & 0.4 \\ 0 & 0.7 & 0.4 \\ 0 & 0 & 0.4 \end{pmatrix}$$

Then $r_1 = 0.9, r_2 = 0.7, r_3 = 0.4$ *is the fundamental sequence of* $\mathcal{H}$ *and* $\mathbf{E}_1 = \{\{a\}\}, \mathbf{E}_2 = \{\{a\}, \{a,b\}\}, \mathbf{E}_3 = \{\{a\}, \{a,b\}, \{a,b,c\}\}$. *Hence* $\mathcal{H}$ *is an ordered, intersecting fuzzy hypergraph. Now* $h(\mu_2) = 0.7$, $\mu_1^{(0.7)}(a) = 0.7$, $\mu_1^{(0.7)}(b) = \mu_1^{(0.7)}(c) = 0, \mu_2^{(0.7)} = \mu_2, \mu_3^{(0.7)} = \mu_3$. *Clearly,* $\mu_1^{(0.7)}$ *is a minimal transversal of* $\mathcal{H}^{(0.7)}$, $\mu_2 \supseteq \mu_1^{(0.7)}$ *and* μ_2 *is a fuzzy transversal of* $\mathcal{H}^{(0.7)}$.

Several results in crisp hypergraph theory involve interaction between intersecting hypergraphs H, minimal transversal sets $Tr(H)$ and chromatic numbers $\chi(H)$. They invariably center on the simple fact that $\chi(H) = 2 \Leftrightarrow$ there is a transversal of H which covers only the singleton edges of H. Many of these results have fuzzy counterparts (see [2, pp. 46-48] for example). We present such a result below. It depends upon the following crisp result found in [2, p. 46].

Lemma 4.56 *If* $H = (X, \mathbf{E})$ *is a simple, intersecting (crisp) hypergraph such that* $\chi(H) > 2$, *then* $\mathbf{E}$ *is equal to the set of all minimal transversals of* H.

Corollary 4.57 *Suppose the conditions of Lemma 4.56 hold with* $\chi(H) > 2$. *Then* H *has no loops.* ■

Theorem 4.58 *Let* $\mathcal{H}$ *be an ordered, intersecting, fuzzy hypergraph with* $\mathbf{C}(\mathcal{H}) = \{H^{r_1}, H^{r_2}, \ldots, H^{r_n}\}$, *where* $0 < r_n < \cdots < r_1$. *Suppose that* $\chi(H^{r_1}) > 2$ *and* H^{r_n} *is simple. Then for each* $r_i \in \mathbf{F}(\mathcal{H})$,

$$Tr^*(\mathcal{H}^{(r_i)}) = \{\sigma(E, r_i) | E \in H^{r_i}\},$$

where $\sigma(E, r_i)$ *is an elementary fuzzy subset with support* E *and height* r_i.

Proof. By hypothesis it follows that H^{r_i} is simple, intersecting and $\chi(H^{r_i}) > 2$ for each $H^{r_i} \in \mathbf{C}(\mathcal{H})$. Thus, by Lemma 4.56, the edge set of $H^{r_i} = Tr(H^{r_i})$ for every $r_i \in \mathbf{F}(\mathcal{H})$. Hence the desired result follows. ■

Definition 4.50 *A fuzzy hypergraph is said to be* strongly intersecting *if for any two edges* μ_1 *and* μ_2, *both* μ_1 *and* μ_2 *contain a common spike of height* $h = h(\mu_1) \wedge h(\mu_2)$.

Theorem 4.59 *Let* $\mathcal{H}$ *be fuzzy hypergraph. Then* $\mathcal{H}$ *is strongly intersecting if and only if* $\mathcal{H}$ *is* $\mathcal{L}$*-intersecting.*

Proof. Suppose that $\mathcal{H}$ is strongly intersecting. Let E and E' be edges of $H^{r_j} \in \mathbf{C}(\mathcal{H})$. Then there exist two edges μ and μ' of $\mathcal{H}$ such that $\mu^{r_j} = E$ and $(\mu')^{r_j} = E'$. Since $\mathcal{H}$ is strongly intersecting, both μ and μ' contain a common spike σ_x where $h(\sigma_x) \geq r_j$. Thus, $\text{supp}(\sigma_x) = \{x\} \subseteq E \cap E'$. Hence, H^{r_j} is intersecting and $\mathcal{H}$ is $\mathcal{L}$-intersecting.

Conversely, suppose that $\mathcal{H}$ is $\mathcal{L}$-intersecting. Let ν and ν' be edges of $\mathcal{H}$. Let $c = h(\nu) \wedge h(\nu')$ and let $E = \nu^c$, $E' = (\nu')^c$. Then, both E and E' belong to $H^c = H^{r_j}$, where $r_{j+1} < c \leq r_j$, where we assume $r_{n+1} = 0$. Since H^{r_j} is intersecting, there exists $x \in E \cap E'$. Consequently, there is a spike σ_x with support $\{x\}$ and height c which is contained in both ν and ν'. Hence, $\mathcal{H}$ is strongly intersecting. ■

Our next fuzzy result depends upon the following crisp result.

Theorem 4.60 *(Berge [2]). If H is a crisp intersecting hypergraph, then* $\chi(H) \leq 3$.

Proof. We may assume without loss of generality that $n(H) > 2$ (i.e., the number of vertices is > 2), and assume H has no repeated edges (for if it does, delete the "extra" edges). If H has a loop, say $\{x\}$, then $H = H(x)$ is a star and consequently, $\chi(H) = 2$. Therefore, it suffices to consider only hypergraphs without loops or repeated edges and with order $n(H) > 2$. Assume H has these properties. Delete from H all edges which properly contain another edge of H and call the resulting partial hypergraph H^δ. Clearly, $\chi(H) = \chi(H^\delta)$. Moreover, H^δ is simple, intersecting and without loops.

We now show that $\chi(H^\delta) \leq 3$. Assume,without loss of generality, that H^δ has at least two edges. Pick an edge $E \in H^\delta = (X^\delta, \mathbf{E}^\delta)$. Clearly, $X^\delta \setminus E \neq \emptyset$ since H^δ is simple (with no repeated edges) and $|\mathbf{E}^\delta| \geq 2$. Let $y \in E$ and note that $E \setminus \{y\} \neq \emptyset$ since $|E| \geq 2$. (Recall that we deleted any possible loop earlier.) Since all edges of H^δ other than E intersect both E and $X^\delta \setminus E$, it follows that $\{X^\delta \setminus E,\ E \setminus \{y\},\ \{y\}\}$ is a coloring of H^δ. ■

Certain "partial intersections" of $\mathcal{H}$ are useful. The following definitions are especially important.

Definition 4.51 *A fuzzy graph $\mathcal{H}$ is said to be essentially intersecting if $\mathcal{H}^-$ is intersecting. $\mathcal{H}$ is said to be essentially strongly intersecting if $\mathcal{H}^-$ is strongly intersecting.*

The $(\cdot)^s$ process, described in Construction 4.2, when applied to $\mathcal{H}^-$ provides a basic construction required in several investigations to follow.

Definition 4.52 *Let $\mathcal{H}^\Delta = (\mathcal{H}^-)^s$. Then $\mathcal{H}$ is called Δ-intersecting if $\mathcal{H}^\Delta$ is intersecting.*

For notational consistency, we shall always assume $\mathbf{F}(\mathcal{H}^{\Delta}) = \{t_1^s, \ldots, t_k^s\}$, where $t_1 = t_1^s > \cdots > t_k^s$, and assume $\mathbf{F}(\mathcal{H}^-) = \{t_1, \ldots, t_m\}$ with $t_1 > \cdots > t_m$.

Theorem 4.61 *If $\mathcal{H}^s$ is intersecting, then $\mathcal{H}$ is strongly intersecting.*

Proof. Let $C(\mathcal{H}) = \{H^{r_i} = (X_i, \mathbf{E}_i) \mid i = 1, \ldots, n\}$ be the set of core hypergraphs of $\mathcal{H}$ and consider the core's *aggregate hypergraph* $H(\mathcal{H}) = (X, \mathbf{E}(\mathcal{H}))$, where $\mathbf{E}(\mathcal{H}) = \cup\{\mathbf{E}_i \mid i = 1, \ldots, n\}$. In addition, let $(H^s)^{r_m^s} = (X_m^s, \mathbf{E}_m^s)$ represent the core hypergraph of $\mathcal{H}^s$ associated with the smallest member r_m^s of $\mathbf{F}(\mathcal{H})$. From the construction of $\mathcal{H}^s$, it follows that every edge belonging to $\mathbf{E}(\mathcal{H})$ contains an edge of $\mathbf{E}_m^s$. Hence

H^s is intersecting $\Rightarrow \mathcal{H}$ is strongly intersecting.

For, indeed, if $\mathcal{H}^s$ is intersecting, then, according to Theorem 4.53, $(H^s)^{r_m^s}$ is intersecting, and therefore, the family of (crisp) edges $E(\mathcal{H})$ is intersecting as well. ■

The converse of Theorem 4.61 is not true in general.

That $\mathcal{H}^s$ need not be intersecting when $\mathcal{H}$ is strongly intersecting can be seen from the following possibility: Let $\{\nu_1, \nu_2\}$ be a pair of edges in $\mathcal{H}^s$. Then there exists a corresponding pair of edges $\{\mu_1, \mu_2\}$ in $\mathcal{H}$ such that

(1) $h(\nu_1) = h(\mu_1)$ and $(\mu_1)^{h(\mu_1)} = (\nu_1)^{h(\nu_1)}$,

(2) $h(\nu_2) = h(\mu_2)$ and $(\mu_2)^{h(\mu_2)} = (\nu_2)^{h(\nu_2)}$.

Under the assumption that $h(\nu_1) < h(\nu_2)$ and the assumption that $\mathcal{H}$ is strongly intersecting, it follows that

$(\mu_1)^{h(\mu_1)} \cap (\mu_2)^{h(\mu_1)} \neq \emptyset$.

However, it is possible that

$(\mu_2)^{h(\mu_2)} \subset (\mu_2)^{h(\mu_1)}$

and, therefore, it is also possible that

$(\mu_1)^{h(\mu_1)} \cap (\mu_2)^{h(\mu_2)} = \emptyset$.

In view of properties (1) and (2), if this latter possibility were to occur, then $\nu_1 \cap \nu_2 = \emptyset$ would be valid thereby demonstrating that $\mathcal{H}^s$ is not an intersecting fuzzy hypergraph. ■

Corollary 4.62 *If $\mathcal{H}$ is Δ-intersecting, then $\mathcal{H}$ is essentially strongly intersecting.* ■

The converse of Corollary 4.62 is not true in general.

Theorem 4.63 *If $\mathcal{H}$ is ordered and essentially intersecting, then $\chi(\mathcal{H}) \leq 3$.*

Proof. We may assume $\mathcal{H}^-$ exists, for otherwise $\chi(\mathcal{H}) = 1$. Let $(H^-)^{r_m} \in \mathbf{C}(\mathcal{H}^-)$, where r_m is the smallest value of $\mathbf{F}(\mathcal{H}^-)$. Since $\mathcal{H}^-$ is intersecting, it follows from Theorem 4.53 that $(H^-)^{rm}$ is a crisp intersecting hypergraph. Therefore by Theorem 4.60, $\chi((H^-)^{rm}) \leq 3$. Moreover, since $\mathcal{H}$ is ordered, $\mathcal{H}^-$ is ordered as well. Hence, since a coloring of $(H^-)^{rm}$ must be a primitive coloring of $\mathcal{H}^-$ (see Definition 4.23), it follows from Theorem 4.37 that a coloring of $(H^-)^{rm}$ is a $\mathcal{L}$-coloring of $\mathcal{H}^-$. Therefore, since $\chi((H^-)^{rm}) \leq 3$, it follows from Definition 4.25 that $\chi(\mathcal{H}^-) \leq 3$. Finally, since $\chi(\mathcal{H}) = \chi(\mathcal{H}^-)$, we have the desired result. ■

Corollary 4.64 *If $\mathcal{H}$ is elementary and essentially intersecting, then $\chi(\mathcal{H}) \leq 3$.*

Proof. Since $\mathcal{H}$ is ordered, the result follows from Theorem 4.63. ■

Corollary 4.65 *If $\mathcal{H}$ is of the form $\mu \otimes H$ and essentially intersecting, then $\chi(\mathcal{H}) \leq 3$.*

Proof. The result follows from Corollary 4.64 since $\mathcal{H}$ is elementary (see Theorem 4.2). ■

Corollary 4.66 *If $\mathcal{H}$ is a Δ-intersecting fuzzy hypergraph, then $\chi(\mathcal{H}) \leq 3$.*

Proof. Now $\mathcal{H}^\Delta$ is intersecting. Since $\mathcal{H}^\Delta$ is also elementary, it follows from Corollary 4.64 that $\chi(\mathcal{H}^\Delta) \leq 3$. Since $\chi(\mathcal{H}^\Delta) = \chi(\mathcal{H}^-) = \chi(\mathcal{H})$, the result is established. ■

Unless $\mathcal{H} = \mathcal{H}^-$, some $\mathcal{L}$-colorings of the skeleton, $\mathcal{H}^s$, of $\mathcal{H}$ may not be extendible to $\mathcal{L}$-colorings of $\mathcal{H}$, or if extendible, then possibly not without the use of new colors. Therefore, unless $\mathcal{H} = \mathcal{H}^-$ it may happen that $\chi(\mathcal{H}^s) < \chi(\mathcal{H})$.

Characterization of Strongly Intersecting Hypergraphs

Definition 4.53 *Suppose $\mathcal{H} = \{\nu_i \in \mathfrak{F}\wp(X) \mid i = 1, \ldots, m\}$ is a finite collection of fuzzy subsets of X and let $c \in (0,1]$. Then $\mathcal{H}|_c = \{\nu \in \mathfrak{F}\wp(X) \mid h(\nu) = c\}$ denotes the* set of edges in $\mathcal{K}$ of height c. *In particular, $\mathcal{H}|_c$ denotes the partial hypergraph of $\mathcal{H} = (X, \mathcal{E})$ with edge set $\mathcal{E}|_c$, provided $\mathcal{E}|_c \neq \emptyset$.*

Definition 4.54 *Let $\mathcal{H}_i = (X_i, \mathcal{E}_i)$, $i = 1, 2$, be fuzzy hypergraphs. Then $\mathcal{H}_1 \preceq \mathcal{H}_2$ if every edge of $\mathcal{H}_1$ contains an edge of $\mathcal{H}_2$. (If H^i, $i = 1, 2$, are crisp hypergraphs, then $H^1 \preceq H^2$ if every edge of H^1 contains an edge of H^2.)*

Lemma 4.67 *(Berge [1]). A crisp hypergraph H is intersecting if and only if $H \preceq Tr(H)$.*

Proof. If H is intersecting, every edge is a transversal and must contain a minimal transversal of H. Thus, $H \preceq Tr(H)$. Conversely, since every transversal of H intersects all edges of H, $H \preceq Tr(H)$ implies H is intersecting. ∎

Theorem 4.68 *$\mathcal{H}$ is strongly intersecting if and only if $H^{r_i} \preceq Tr(H^{r_i})$ for every $H^{r_i} \in \mathbf{C}(\mathcal{H})$.*

Proof. By Theorem 4.59, Definition 4.49, and Lemma 4.67, it follows that $\mathcal{H}$ is strongly intersecting $\Leftrightarrow$ $\mathcal{H}$ is $\mathcal{L}$-intersecting $\Leftrightarrow$ H^{r_i} is intersecting for all $H^{r_i} \in \mathbf{C}(\mathcal{H}) \Leftrightarrow H^{r_i} \preceq Tr(H^{r_i})$ for all $H^{r_i} \in \mathbf{C}(\mathcal{H})$. ∎

Theorem 4.69 *$\mathcal{H}$ is a strongly intersecting fuzzy hypergraph if and only if for every $r_i \in \mathbf{F}(\mathcal{H}), (\mathcal{H}^{(r_i)})|_{r_i} \preceq Tr(\mathcal{H}^{(r_i)})$.*

Proof. Suppose for every $r_i \in \mathbf{F}(\mathcal{H})$, $(\mathcal{H}^{(r_i)})|_{r_i} \preceq Tr(\mathcal{H}^{(r_i)})$. For each $H^{r_i} \in \mathbf{C}(\mathcal{H})$, the edge set $\mathbf{E}(H^{r_i}) = \{\mu^{r_i} \mid \mu \in (\mathcal{H}^{(r_i)})|_{r_i}\} \preceq \{\tau^{r_i} \mid \tau \in Tr(\mathcal{H}^{(r_i)})\} = Tr(\mathbf{E}(H^{r_i}))$. Hence, $H^{r_i} \preceq Tr(H^{r_i})$, $\forall H^{r_i} \in \mathbf{C}(\mathcal{H})$ and by Theorem 4.68, $\mathcal{H}$ is strongly intersecting.

Conversely, suppose $\mathcal{H}$ is strongly intersecting. Let $\mu \in \mathcal{H}|_{r_1}$, where r_1 is the largest member of $\mathbf{F}(\mathcal{H})$. Let $H^{r_j} \in \mathbf{C}(\mathcal{H})$. We now show that μ^{r_j} is a transversal of H^{r_j}. For suppose $E \in H^{r_j}$. Then there is an edge ν of $\mathcal{H}$ such that $\nu^{r_j} = E$. Since $\mathcal{H}$ is strongly intersecting, there is a spike σ_x with height

$$h(\sigma_x) = h(\mu) \wedge h(\nu) = h(\nu) \geq r_j,$$

and support $\{x\}$, which is contained in both μ and ν. Hence, $x \in E \cap \mu^{r_j}$. Thus μ is a transversal of $\mathcal{H}$ and therefore contains a member of $Tr(\mathcal{H})$. Therefore $(\mathcal{H}^{(r_i)})\,|_{r_i} \preceq Tr(\mathcal{H}^{(r_1)})$.

It follows from Theorem 4.59 that $\mathcal{H}$ is $\mathcal{L}$-intersecting. Consequently, by Theorem 4.59 again, it follows that every $\mathcal{H}^{(r_i)}$ must be strongly intersecting. Hence

$$(\mathcal{H}^{(r_i)})\,|_{r_i} \preceq Tr(\mathcal{H}^{(r_i)})$$

for each $r_i \in \mathbf{F}(\mathcal{H})$. ∎

Corollary 4.70 *Let $\mathcal{H}$ be a fuzzy hypergraph with $\mathbf{C}(\mathcal{H}) = \{H^{r_i} \mid r_i \in \mathbf{F}(\mathcal{H})\}$. Then $H^{r_i} \preceq Tr(H^{r_i})$, for every $H^{r_i} \in \mathbf{C}(\mathcal{H})$ if and only if $(\mathcal{H}^{(r_i)})\,|_{r_i} \preceq Tr(\mathcal{H}^{(r_i)})$ for every $r_i \in \mathbf{F}(\mathcal{H})$.*

Proof. The proof follows immediately from Theorems 4.68 and 4.69. ∎

Theorem 4.71 *$\mathcal{H}$ is strongly intersecting if and only if $\mathcal{H}_{(r_i)}$ is intersecting $\forall r_i \in \mathbf{F}(\mathcal{H})$.*

Proof. By applying Theorem 4.53 to $\mathcal{H}_{(r_i)}$ and by Theorem 4.59, the following chain of equivalencies hold: $\mathcal{H}_{(r_i)}$ is intersecting for every $r_i \in \mathbf{F}(\mathcal{H}) \Leftrightarrow \mathbf{E}(H^{r_i})$ is intersecting for each $H^{r_i} \in \mathbf{C}(\mathcal{H}) \Leftrightarrow \mathcal{H}$ is $\mathcal{L}$-intersecting $\Leftrightarrow \mathcal{H}$ is strongly intersecting. ∎

Simply Ordered Intersecting Hypergraphs

In Theorems 4.82 and 4.83, we obtain some fuzzy properties similar to the crisp results in Theorem 4.72.

Theorem 4.72 *(Berge [2]). Let $H = (X, \mathbf{E})$ be a simple (crisp) hypergraph of order ≥ 2 (i.e., $|X| \geq 2$). Then, $Tr(H) = H$ if and only if H is intersecting and $\chi(H) > 2$.*

Note that a simple intersecting hypergraph with a loop has order 1 (i.e., the vertex set is a singleton). Thus, the above theorem excludes hypergraphs with loops. Its proof follows since H is intersecting $\Leftrightarrow$ every edge of H contains a minimal transversal (see Lemma 4.67) and since also if H has at most one loop, then $\chi(H) > 2 \Leftrightarrow$ every transversal contains at least one edge which is not a loop.

Definition 4.55 *A fuzzy hypergraph is said to be* non-trivial *if it has at least one edge μ such that $|supp(\mu)| \geq 2$.*

Definition 4.56 *A fuzzy hypergraph, $\mathcal{H}$, is said to be* sequentially simple *if $\mathbf{C}(\mathcal{H}) = \{H^{r_i} = (X^{r_i}, \mathbf{E}^{r_i}) \mid r_i \in \mathbf{F}(\mathcal{H})\}$ satisfies the property that if $E \in \mathbf{E}^{r_{i+1}} \setminus \mathbf{E}^{r_i}$, then $E \not\subseteq X^{r_i}$, where $r_n < \ldots < r_1$. $\mathcal{H}$ is said to be* essentially sequentially simple *if $\mathcal{H}^-$ is sequentially simple.*

Note, if $\mathcal{H}$ is simply ordered, then $\mathcal{H}$ is sequentially simple. However the converse is not true.

Theorem 4.73 *Suppose $\mathcal{H}$ is a nontrivial essentially sequentially simple fuzzy hypergraph which is Δ-intersecting. Then $\chi((H^-)^{t_1}) > 2$ if and only if $Tr(\mathcal{H}^-) = \mathcal{H}^\Delta|_{t_1}$ where t_1 is the largest member of $\mathbf{F}(\mathcal{H}^-)$.*

Proof. Suppose that $\chi((H^-)^{t_1}) > 2$. Since $\mathcal{H}$ is nontrivial, the order of $(H^\Delta)^{t_1}$ is greater than or equal to two; and, since $\mathcal{H}^-$ is sequentially simple, $\mathcal{H}^\Delta$ is locally simple (i.e., all members of $\mathbf{C}(\mathcal{H}^\Delta)$ are simple crisp hypergraphs.) Moreover, since $\mathcal{H}^\Delta$ is intersecting and ordered, it follows from Theorem 4.54 that every member $(H^\Delta)^{t_j^s} \in \mathbf{C}(\mathcal{H}^\Delta)$ is simple and intersecting. Since $\mathcal{H}^\Delta$ is ordered, $\chi((H^-)^{t_1}) = \chi((H^\Delta)^{t_1})$. Since also $\chi((H^-)^{t_1}) > 2$, it follows that $\chi((H^\Delta)^{t_j^s}) > 2$ for all members of

$\mathbf{C}(\mathcal{H}^\Delta)$. (Note, whenever $\mathcal{H}$ is ordered, then $\chi(H^{r_i}) \leq \chi(H^{r_{i+1}})$; this relationship, however, is not necessarily true if $\mathcal{H}$ is not ordered.) Therefore, by Theorem 4.72,

$$(H^\Delta)^{t_j^s} = Tr((H^\Delta)^{t_j^s})$$

for all $(H^\Delta)^{t_j^s} \in \mathbf{C}(\mathcal{H}^\Delta)$. Since H^Δ is ordered, every edge of $(H^\Delta)^{t_1}$ is an edge of each core hypergraph of $\mathcal{H}^\Delta$ and so $Tr(\mathcal{H}^\Delta) = \mathcal{H}^\Delta|_{t_1}$. (Recall, with respect to this last argument, that t_1, the largest member of $\mathbf{F}(\mathcal{H}^-)$, equals t_1^s, the largest member of $\mathbf{F}(\mathcal{H}^\Delta)$.) By Theorem 4.25 and Definition 4.52, $Tr(\mathcal{H}^-) = Tr(\mathcal{H}^\Delta)$ and since $\chi((H^-)^{t_1}) > 2$, $Tr(\mathcal{H}^-) = \mathcal{H}^\Delta|_{t_1}$.

Suppose that $Tr(\mathcal{H}^-) = \mathcal{H}^\Delta|_{t_1}$. Since $Tr(\mathcal{H}^-) = Tr(\mathcal{H}^\Delta)$ we now assume $Tr(\mathcal{H}^\Delta) = \mathcal{H}^\Delta|_{t_1}$. Based upon this assumption, it follows that

$$(H^\Delta)^{t_1} = (\mathcal{H}^\Delta|_{t_1})^{t_1} = (Tr(\mathcal{H}^\Delta)^{t_1} = Tr((H^\Delta)^{t_1}),$$

where the last equality in the above chain is established in Proposition 4.11. Moreover, since $(H^\Delta)^{t_1}$ is simple and $\mathcal{H}$ is nontrivial, the order of $(H^\Delta)^{t_1}$ is greater than or equal to two. Thus it follows from Theorem 4.72 that $\chi((H^\Delta)^{t_1}) > 2$. Hence $\chi((H^-)^{t_1}) > 2$ since $\chi((H^-)^{t_1}) = \chi((H^\Delta)^{t_1})$. ■

Corollary 4.74 *Suppose $\mathcal{H}$ is a nontrivial fuzzy hypergraph of the form $\mu \otimes H$ and is Δ-intersecting. Then $\chi((\mathcal{H}^-)^{t_1}) > 2$ if and only if $Tr(\mathcal{H}^-) = \mathcal{H}^\Delta|_{t_1}$ where t_1 is the largest member of $\mathbf{F}(\mathcal{H}^-)$.*

Proof. By Theorem 4.2, $\mu \otimes H$ is simply ordered and therefore essentially sequentially simple. Thus the result follows at once from Theorem 4.73. ■

If $\mathcal{H}$ is a fuzzy hypergraph, we define the support of $\mathcal{H}$, $\text{supp}(\mathcal{H})$, to be the set $\text{supp}(\mathcal{H}) = \{\text{supp}(\mu) \mid \mu \in \mathcal{E}\}$.

Theorem 4.75 *Suppose $\mathcal{H}$ is an ordered fuzzy hypergraph. Then the following assertions hold.*

(1) $\mathcal{H}$ is intersecting if and only if $\mathcal{H}^s$ is intersecting.

(2) $\mathcal{H}^-$ is intersecting if and only if $\mathcal{H}^\Delta$ is intersecting.

Proof. It suffices to prove part (1) since $\mathcal{H}^\Delta = (\mathcal{H}^-)^s$ and $\mathcal{H}^-$ is ordered whenever $\mathcal{H}$ is a nontrivial ordered fuzzy hypergraph. Now since $\mathcal{H}$ is ordered, $\text{supp}(\mathcal{H}) = \cup\{\mathbf{E}(H^{r_i}) \mid H^{r_i} \in \mathbf{C}(\mathcal{H})\}$. Thus, $\text{supp}(\mathcal{H}^s) \subseteq \text{supp}(\mathcal{H})$. Furthermore, according to the construction of $\mathcal{H}^s$, every member of the edge set $\mathbf{E}(H^{r_i})$ is either a member or contains a member of $\text{supp}(\mathcal{H}^s)$. Therefore, for any two edges $E, E' \subseteq \text{supp}(\mathcal{H})$ there exist corresponding edges $F, F' \subseteq \text{supp}(\mathcal{H}^s)$ such that $F \subseteq E$ and $F' \subseteq E'$. It is now clear that $\text{supp}(\mathcal{H}^s)$ is intersecting $\Leftrightarrow$ $\text{supp}(\mathcal{H})$ is intersecting. Hence, in view of Theorem 4.53, part (1) is established. ■

Theorem 4.76 *Let $\mathcal{H}$ be a fuzzy hypergraph. Then the following assertions hold.*

(1) If $\mathcal{H}^s$ is intersecting, then $\mathcal{H}$ is strongly intersecting.

(2) If $\mathcal{H}^\Delta$ is intersecting, then $\mathcal{H}^-$ is strongly intersecting.

Proof. We have that every edge E in the core hypergraph $H^{r_i} \in \mathbf{C}(\mathcal{H})$ contains a member of $\text{supp}(\mathcal{H}^s)$ by the construction process in producing $\mathcal{H}^s$ and by the fact that $\mathcal{H}^s$ is elementary. Hence, if $\text{supp}(\mathcal{H}^s)$ is intersecting, then every core hypergraph, H^{r_i}, of $\mathcal{H}$ must be intersecting as well. Thus $\mathcal{H}$ is $\mathcal{L}$-intersecting and so by Theorem 4.59, $\mathcal{H}$ is strongly intersecting. ■

In general, the converse of the implications in Theorem 4.76 do not hold. The following example illustrates this.

Example 4.22 *The incidence matrix for $\mathcal{H}$ is*

$$\begin{array}{c} \\ a \\ b \\ c \\ d \end{array} \begin{array}{c} \begin{array}{ccc} \mu_1 & \mu_2 & \mu_3 \end{array} \\ \begin{pmatrix} 0.9 & 0 & 0 \\ 0.9 & 0.9 & 0 \\ 0 & 0.9 & 0 \\ 0.4 & 0.4 & 0.4 \end{pmatrix} \end{array}.$$

Clearly, $\mathcal{H}$ is strongly intersecting. However, the incidence matrix for $\mathcal{H}^s$ is

$$\begin{array}{c} \\ a \\ b \\ c \\ d \end{array} \begin{array}{c} \begin{array}{ccc} \nu_1 & \nu_2 & \nu_3 \end{array} \\ \begin{pmatrix} 0.9 & 0 & 0 \\ 0.9 & 0.9 & 0 \\ 0 & 0.9 & 0 \\ 0 & 0 & 0.4 \end{pmatrix} \end{array}.$$

Clearly, $\mathcal{H}^s$ is not intersecting.

Theorem 4.77 *Suppose $\mathcal{H}$ is simply ordered and intersecting. Let r_1 be the largest member of $\mathbf{F}(\mathcal{H})$ and let the order of $(H^s)^{r_1} \geq 2$. Then,*

$$\chi(H^{r_1}) > 2 \Leftrightarrow Tr(\mathcal{H}) = \mathcal{H}^s|_{r_1}. \tag{4.58}$$

Proof. We first show that if $\mathcal{H}$ is simply ordered and intersecting and the number of vertices, $n((H^s)^{r_1})$, in the core hypergraph $(H^s)^{r_1} \in \mathbf{C}(\mathcal{H}^s)$ equals or exceeds 2, then $\mathcal{H}$ does not possess a spike edge or an edge with a terminal spike. To prove this assertion notice that whenever $\mathcal{H}$ has either a spike edge or an edge with a terminal spike the process for constructing $\mathcal{H}^s$ produces a spike edge in the edge set, $\mathcal{E}(\mathcal{H}^s)$, of $\mathcal{H}^s$. In addition, whenever $\mathcal{H}$ is simply ordered and intersecting $\mathcal{H}^s$ is intersecting by Theorem 4.75 and $\mathcal{H}^s$ is locally simple, that is, every core hypergraph in $\mathbf{C}(\mathcal{H}^s)$ is simple since $\mathcal{H}$ is sequentially simple. Consequently, every core hypergraph in

$\mathbf{C}(\mathcal{H}^s) = \{(H^s)^{r_j^s} \mid r_m^s < \cdots < r_j^s < \cdots < r_1^s = r_1\}$
is simple and intersecting. Therefore, if $\mathcal{H}^s$ contains a spike, some core hypergraph, $(H^s)^{r_j^s}$, contains a loop and it follows that $n((H^s)^{r_j^s}) = 1$. However this implies $n((H^s)^{r_1^s}) = 1$ which contradicts the hypothesis.

Thus by hypothesis
$\mathcal{H} = \mathcal{H}^-$ and $\mathcal{H}^s = (\mathcal{H}^-)^s = \mathcal{H}^\Delta$.
Therefore, the hypothesis of Theorem 4.77 satisfies the hypothesis of Theorem 4.73, and so statement (4.58) follows from the conclusion of Theorem 4.73. ■

Corollary 4.78 *If $\mathcal{H}$ has the form $\mu \otimes H$ and H is intersecting and has no loops, then (4.58) in Theorem 4.77 holds.*

Proof. We have that $\mathcal{H}$ is simply ordered and intersecting from Theorem 4.2 and from Definition 4.12. Thus $H = H^{r_n}$, where r_n is the smallest member of $\mathbf{F}(\mathcal{H})$). Also notice that $n((H^s)^{r_1}) \geq 2$. For according to the method of constructing $\mathcal{H}^s$, core hypergraph $(H^s)^{r_1}$ of $\mathcal{H}^s$ must contain at least one member of $H^{r_1} \in \mathbf{C}(\mathcal{H})$, where r_1 is the largest member of $\mathbf{F}(\mathcal{H})$. However, as H has no loops, it follows from the structure of $\mu \otimes H$, as described in Definition 4.12, that every crisp edge in core hypergraph H^{r_1} of $\mathbf{C}(\mathcal{H})$ must have cardinality equal or greater than 2. Thus $n((H^s)^{r_1}) \geq 2$. ■

Theorem 4.79 *Suppose $\mathcal{H}$ is a nontrivial fuzzy hypergraph. Let t_1 be the largest member of $\mathbf{F}(\mathcal{H}^-)$ and*

$$Tr(\mathcal{H}^-) = \mathcal{H}^\Delta|_{t_1} \tag{4.59}$$

Then $\mathcal{H}^-$ is strongly intersecting and $\chi((H^-)^{t_1}) > 2$.

Proof. From Theorem 4.25 and Proposition 4.11,

(1) $Tr(\mathcal{H}^s) = Tr(\mathcal{H})$, for every fuzzy hypergraph $\mathcal{H}$, and

(2) the r_1-cut of $Tr(\mathcal{H}) = Tr(H^{r_1})$, where r_1 is the largest member of $\mathbf{F}(\mathcal{H})$.

Hence it follows from statement (4.59) that
$Tr((H^\Delta)^{t_1}) = (H^\Delta)^{t_1}$.

In addition, $(H^\Delta)^{t_1}$ is a (crisp) simple hypergraph of order ≥ 2; this is easily seen from the fact that $\mathcal{H}$ is nontrivial and the construction of $\mathcal{H}^\Delta$ (in particular, see Definitions 4.26, 4.27, 4.52 and 4.55). Thus it follows from Theorem 4.72 that $(H^\Delta)^{t_1}$ is intersecting and $\chi((H^\Delta)^{t_1}) > 2$.

We claim that every core hypergraph $(H^-)^{t_j}$ of $\mathbf{C}(\mathcal{H}^-)$ is a (crisp) intersecting hypergraph.

To prove this claim, consider Definition 4.13. It indicates that each member of the t_j-cut of $Tr(\mathcal{H}^-)$ intersects every member of the edge set $\mathbf{E}((H^-)^{t_j})$. Also, since $\mathcal{H}^\Delta$ is elementary, the t_j-cut of $\mathcal{H}^\Delta|_{t_1}$ is $(H^\Delta)^{t_1}$.

Thus by these two facts and condition (4.59), we have that every member $T \in \left(H^{\Delta}\right)^{t_1}$ intersects each member of $\mathbf{E}((H^{-})^{t_j})$. Therefore, every edge $E \in (H^{-})^{t_j}$ intersects each edge of $T \in \left(H^{\Delta}\right)^{t_1}$. Hence, if two edges E_1 and E_2 of $(H^{-})^{t_j}$ did not intersect, then $\hat{E}_i = E_i \cap \mathbf{V}(\left(H^{\Delta}\right)^{t_1})$ $(i = 1, 2)$ would be non-intersecting transversals of $\left(H^{\Delta}\right)^{t_1}$. This would imply $\chi(\left(H^{\Delta}\right)^{t_1}) = 2$, which contradicts the fact that $\chi(\left(H^{\Delta}\right)^{t_1}) > 2$. Hence, $(H^{-})^{t_j}$ is intersecting and the claim holds. Therefore, by Definition 4.49 and Theorem 4.59, $\mathcal{H}^{-}$ is strongly intersecting. Finally, observe that $\chi((H^{-})^{t_1}) = \chi(\left(H^{\Delta}\right)^{t_1}) > 2$. ■

Corollary 4.80 *Suppose $\mathcal{H}$ is a nontrivial ordered fuzzy hypergraph which satisfies (4.59). Then both $\mathcal{H}^{-}$ and $\mathcal{H}^{\Delta}$ are strongly intersecting.*

Proof. By Theorem 4.79, $\mathcal{H}^{-}$ is intersecting. Thus by Theorem 4.75, $\mathcal{H}^{\Delta}$ is intersecting. Since $\mathcal{H}^{\Delta}$ is ordered, $\mathcal{H}^{\Delta}$ is strongly intersecting. ■

Corollary 4.81 *Suppose $\mathcal{H}$ is a fuzzy hypergraph. Let r_1 be the largest member of $\mathbf{F}(\mathcal{H})$ and*

$$Tr(\mathcal{H}) = \mathcal{H}^s|_{r_1}, \tag{4.60}$$

Then, both $\mathcal{H}$ and $\mathcal{H}^s$ are strongly intersecting.

Proof. Suppose $(H^s)^{r_1}$ has order ≥ 2, where $(H^s)^{r_1} \in \mathbf{C}(\mathcal{H}^s)$. By the first argument presented in the proof of Theorem 4.79, we see that condition (4.60) implies $Tr((H^s)^{r_1}) = (H^s)^{r_1}$. Thus since $(H^s)^{r_1}$ is simple by construction, and since $n((H^s)^{r_1}) \geq 2$ by assumption, it follows from Theorem 4.72 that $\chi((H^s)^{r_1}) > 2$. Accordingly, in conjunction with statement (4.60) or its combination with the fact that $Tr(\mathcal{H}^s) = Tr(\mathcal{H})$, the argument used to prove the claim that every core hypergraph $(H^{-})^{t_j}$ of $\mathbf{C}(\mathcal{H}^{-})$ is a crisp intersecting hypergraph stated in the proof of Theorem 4.79 is sufficient in the present situation to verify that both $\mathcal{H}$ and $\mathcal{H}^s$ are strongly intersecting.

Suppose $(H^s)^{r_1}$ has order 1. Then the only edge in $(H^s)^{r_1}$ is a loop, say, $E = \{x\}$. Therefore $\mathcal{H}^s|_{r_1}$ is a spike σ_x with support $\{x\}$ and height $r_1 = h(\mathcal{H})$. Consequently, in view of statement (4.60), E intersects all edges of each core hypergraph $H^{r_j} \in \mathbf{C}(\mathcal{H})$. Thus, the edge set of each $H^{r_j} \in \mathbf{C}(\mathcal{H})$ is a "star" with common vertex x. By the construction of $\mathcal{H}^s$, it follows that the edge set $\mathcal{E}(\mathcal{H}^s) = \{\sigma_x\}$. Hence, both $\mathcal{H}$ and $\mathcal{H}^s$ are $\mathcal{L}$-intersecting and, therefore, strongly intersecting fuzzy hypergraphs according to Definition 4.49 and Theorem 4.59. ■

Theorem 4.82 *Suppose $\mathcal{H}$ is a nontrivial simply ordered fuzzy hypergraph. Let t_1 be the largest member of $\mathbf{F}(\mathcal{H}^-)$ and $(H^-)^{t_1} \in \mathbf{C}(\mathcal{H}^-)$. Then $Tr(H^-) = H^{\Delta}|_{t_1}$ if and only if $\mathcal{H}$ is essentially intersecting and $\chi((H^-)^{t_1}) > 2$.*

Proof. Suppose that $\mathcal{H}$ is essentially intersecting and $\chi((H^-)^{t_1}) > 2$. By assumption, $\mathcal{H}^-$ is intersecting (see Definition 4.51). Therefore, since $\mathcal{H}$ is ordered, it follows from Theorem 4.75 that $\mathcal{H}^{\Delta}$ is intersecting. Since $\chi((H^-)^{t_1}) > 2$, it now follows from Theorem 4.73 that $Tr(\mathcal{H}^-) = \mathcal{H}^{\Delta}|_{t_1}$.

Conversely, suppose $Tr(\mathcal{H}^-) = \mathcal{H}^{\Delta}|_{t_1}$. Then, by Theorem 4.49, $\mathcal{H}^-$ is intersecting and $\chi((H^-)^{t_1}) > 2$. ■

Theorem 4.83 *Suppose $\mathcal{H}$ is a simply ordered fuzzy hypergraph. Let r_1 be the largest member of $\mathbf{F}(\mathcal{H})$ and $(H^s)^{r_1} \in \mathbf{C}(\mathcal{H}^s)$ and let the order of $(H^s)^{r_1} \geq 2$. Then $Tr(H) = H^s|_{r_1}$ if and only if $\mathcal{H}$ is intersecting and $\chi(H^{r_1}) > 2$.*

Proof. The proof follows from Theorem 4.77 and Corollary 4.81. In particular, Corollary 4.81 together with the property that $Tr(\mathcal{H}) = \mathcal{H}^s|_{r_1}$ imply that $\mathcal{H}$ is intersecting. The hypothesis of Theorem 4.77 now holds and so that $\chi(H^{r_1}) > 2$. Hence the forward implication is established. The converse implication follows at once from Theorem 4.77. ■

$\mathcal{H}$-dominant Transversals

The concept of a *$\mathcal{H}$-dominant fuzzy subset,* introduced in Definition 4.57 below, plays a fundamental role in this section, especially Theorems 4.86 and 4.88 and Corollary 4.87. Theorem 4.88 is used in the proof of Theorem 4.90. This latter theorem provides a characterization of all nontrivial fuzzy hypergraphs, $\mathcal{H}$, for which $\chi(\mathcal{H}) > 2$.

Definition 4.57 *Let $\mathcal{H} = (X, \mathcal{E})$ be a fuzzy hypergraph. A fuzzy subset $\nu \in \mathfrak{F}\wp(X)$ is said to be $\mathcal{H}$-*dominant *if for every $x \in supp(\nu)$, $\nu(x) = \vee\{\mu(x) \mid \mu \in \mathcal{E}\}$. An edge μ of $\mathcal{H}$ is said to be a* dominant edge *of $\mathcal{H}$ if it is $\mathcal{H}$-dominant.*

Note that every member of a fuzzy coloring, Γ, of $\mathcal{H}$ (see Definition 4.43) is a $\mathcal{H}$-dominant fuzzy subset.

Definition 4.58 *Suppose $\mathcal{H} = (X, \mathcal{E})$ is a fuzzy hypergraph and let $\gamma \in \mathfrak{F}\wp(X)$. Then the $\mathcal{H}$-*dominant transform *of γ, denoted by $\gamma^{D(\mathcal{H})}$ (or, simply γ^D), is defined by*

$$\gamma^{D(\mathcal{H})}(x) = \begin{cases} \vee\{\mu(x) \mid \mu \in \mathcal{E}\} & \text{if } x \in supp(\gamma), \\ 0 & \text{otherwise.} \end{cases}$$

*Let $\mathcal{K} = \{\gamma_1, \ldots, \gamma_m\} \subseteq \mathfrak{F}\wp(X)$. Then the $\mathcal{H}$-*dominant transform *of $\mathcal{K}$, denoted by $\mathcal{K}^{D(\mathcal{H})}$ is defined by $\mathcal{K}^{D(\mathcal{H})} = \{\gamma_1^{D(\mathcal{H})}, \ldots, \gamma_m^{D(\mathcal{H})}\}$.*

We some time write $\mathcal{K}^D$ for $\mathcal{K}^{D(\mathcal{H})}$ and γ_i^D for $\gamma_i^{D(\mathcal{H})}, i = 1, 2, \ldots, m$.

Lemma 4.84 *(Berge [1]). Suppose $H = (X, \mathbf{E})$ is a crisp intersecting hypergraph with an edge $E \neq X$, which does not properly contain another edge except possibly for loops. Then, for every $y \in E$, $\{X \setminus E, E \setminus \{y\}, \{y\}\}$ is a 3-coloring of H.* ■

Note that empty colors are acceptable in k-colorings of H. This is the case for color $E \setminus \{y\}$ when E is a loop.

Theorem 4.85 *Let $\mathcal{H} = (X, \mathcal{E})$ be a nontrivial, essentially strongly intersecting fuzzy hypergraph and let $\mathbf{C}(\mathcal{H}^-) = \{(H^-)^{t_j} \mid t_j \in \mathbf{F}(\mathcal{H}^-)\}$. Suppose there is an edge ν of $\mathcal{H}^-$ with the following properties:*

(1) $h(\nu) = h(\mathcal{H}^-)$,

(2) $(\nu^{D(\mathcal{H}^-)})^{t_j} \neq (X^-)^{t_j}$ $(= \mathbf{V}((H^-)^{t_j}))$ for all $t_j \in \mathbf{F}(\mathcal{H}^-)$,

(3) for each $t_j \in \mathbf{F}(\mathcal{H}^-)$ no $\mu^{t_j} \in \mathbf{E}((H^-)^{t_j}) \setminus \nu^{t_j}$ is properly contained in $(\nu^{D(\mathcal{H}^-)})^{t_j}$.

Then, $\chi(\mathcal{H}) \leq 3$.

Proof. Assume without loss of generality, that $\mathbf{V}(\mathcal{H}^-) = X$ and let ν be an edge of $\mathcal{H}^-$ with the above-mentioned properties. Let $y \in \text{supp}(\nu)$ be such that $\nu(y) = h(\mathcal{H}^-)$. Then $P = \{X \setminus \text{supp}(\nu), \text{supp}(\nu) \setminus \{y\}, \{y\}\}$ is a partition of X into three non-empty subsets by property (2) and the fact that ν cannot be a spike since $\nu \in \mathcal{E}(\mathcal{H}^-)$.

We claim that $\Gamma = \{\gamma_1, \gamma_2, \gamma_3\}$, where

$$\gamma_1(x) = \begin{cases} \vee\{\mu(x) \mid \mu \in \mathcal{E}^-\} & \text{if } x \in X \setminus \text{supp}(\nu), \\ 0 & \text{otherwise,} \end{cases}$$

$$\gamma_2(x) = \begin{cases} \vee\{\mu(x) \mid \mu \in \mathcal{E}^-\} & \text{if } x \in \text{supp}(\nu) \setminus \{y\}, \\ 0 & \text{otherwise,} \end{cases}$$

$$\gamma_3(x) = \begin{cases} \vee\{\mu(x) \mid \mu \in \mathcal{E}^-\} & \text{if } x = y, \\ 0 & \text{otherwise,} \end{cases}$$

is a fuzzy coloring of $\mathcal{H}^-$ and $\mathcal{E}^-$ is the fuzzy edge set of $\mathcal{H}^-$.

For each $t_j \in \mathbf{F}(\mathcal{H}^-)$, let

$(H^-)^{t_j} = ((X^-)^{t_j}, (\mathbf{E}^-)^{t_j})$,

where $(\mathbf{E}^-)^{t_j} = \{(\mathbf{E}((H^-)^{t_j}) \setminus \nu^{t_j}) \cup (\nu^{D(\mathcal{H}^-)})^{t_j}\}$. Clearly $(H^-)^{t_j}$ is intersecting and, in view of property (1), $(\nu^{D(\mathcal{H}^-)})^{t_j} \neq \emptyset$ for all $t_j \in \mathbf{F}(\mathcal{H}^-)$. These properties together with properties (2) and (3) together with the choice of y specified earlier imply, according to Lemma 4.84, that the restriction of P, namely Γ^{t_j}, to $(X^-)^{t_j}$ is a coloring of $(H^-)^{t_j}$.

Since $y \notin \nu^{t_j} \subseteq (\nu^{D(\mathcal{H}^-)})^{t_j}$, it is clear that Γ^{t_j} is also a coloring of $(H^-)^{t_j}$; this completes the proof of the claim. Thus, $\chi(\mathcal{H}^-) \leq 3$.

Suppose that $\mathcal{H}^-$ exists. Since $\chi(\mathcal{H}^-) = \chi(\mathcal{H})$, it follows that $\chi(\mathcal{H}) \leq 3$. ■

By Definition 4.58, $T^D(\mathcal{H}) = \{\tau^D \mid \tau \in T(\mathcal{H})\}$ is the set of *$\mathcal{H}$-dominant fuzzy transversals* of $\mathcal{H}$.

Definition 4.59 *A nontrivial fuzzy hypergraph $\mathcal{H}$ is said to be* essentially ordered *if $\mathcal{H}^-$ is ordered.*

Note that if $\mathcal{H}$ is nontrivial and ordered, then $\mathcal{H}$ is essentially ordered; however, even when $\mathcal{H}$ is nontrivial and not ordered, $\mathcal{H}$ may, in some cases, be essentially ordered.

Theorem 4.86 *Let $\mathcal{H} = (X, \mathcal{E})$ be a nontrivial essentially ordered fuzzy hypergraph. Then $\chi(\mathcal{H}) = 2$ if and only if there is a $\mathcal{H}^-$-dominant fuzzy transversal of $\mathcal{H}^-$ which does not contain an edge of $\mathcal{H}^-$.*

Proof. Since $\mathcal{H}$ is nontrivial (see Definition 4.55), $\mathcal{H}^-$ exists.

Suppose that $\chi(\mathcal{H}) = 2$. Since $\chi(\mathcal{H}^-) = \chi(\mathcal{H}) = 2$, there exists a fuzzy bi-coloring $\{\tau, \bar{\tau}\}$ of $\mathcal{H}^-$. Therefore, for every $t_j \in \mathbf{F}(\mathcal{H}^-)$, both τ^{t_j} and $\bar{\tau}^{t_j}$ intersect every edge of $(H^-)^{t_j} \in \mathbf{C}(\mathcal{H}^-)$ since $(H^-)^{t_j}$ contains no loops. Every member of a fuzzy coloring of $\mathcal{H}^-$ is a $\mathcal{H}^-$-dominant fuzzy subset by the note following Definition 4.57. Hence it follows that τ and $\bar{\tau}$ belong to $T^D(\mathcal{H}^-)$ and $\tau \cap \bar{\tau} = \emptyset$. Hence, τ and $\bar{\tau}$ are a disjoint pair of $\mathcal{H}^-$-dominant transversals of $\mathcal{H}^-$, neither of which contain an edge of $\mathcal{H}^-$.

Conversely assume there is a $\mathcal{H}^-$-dominant fuzzy transversal, τ, of $\mathcal{H}^-$ which does not contain an edge of $\mathcal{H}^-$. Let $X^- = \mathbf{V}(\mathcal{H}^-)$, the vertex set of $\mathcal{H}^-$. Then $X^- \setminus \text{supp}(\tau) \neq \emptyset$ for otherwise τ contains all edges of $\mathcal{H}^-$ since τ is $\mathcal{H}^-$-dominant.

Let $\bar{\tau}$ be the $\mathcal{H}^-$-dominant fuzzy subset with $\text{supp}(\bar{\tau}) = X^- \setminus \text{supp}(\tau)$. Since τ is $\mathcal{H}^-$-dominant and contains no edge $\mu \in \mathcal{H}^-$, it follows that $\text{supp}(\mu) \nsubseteq \text{supp}(\tau)$ for every $\mu \in \mathcal{H}^-$. In addition, since τ is a transversal of $\mathcal{H}^-$, $\text{supp}(\mu) \cap \text{supp}(\bar{\tau}) \neq \emptyset$ and $\text{supp}(\mu) \cap \text{supp}(\tau) \neq \emptyset$ for all $\mu \in \mathcal{H}^-$. Hence, τ^{t_m} and $\bar{\tau}^{t_m}$ are transversals of $(H^-)^{t_m} \in \mathbf{C}(\mathcal{H}^-)$, where t_m is the smallest member of $\mathbf{F}(\mathcal{H}^-)$.

Since $\mathcal{H}^-$ is ordered, $\mathbf{E}((H^-)^{t_j}) \subseteq \mathbf{E}((H^-)^{t_m})$ for all $(H^-)^{t_j} \in \mathbf{C}(\mathcal{H}^-)$. Therefore, since $\bar{\tau}^{t_m} \in T((H^-)^{t_m})$ and $\bar{\tau}$ is $\mathcal{H}^-$-dominant, it follows that, for every $t_j \in \mathbf{F}(\mathcal{H}^-)$, $\bar{\tau}^{t_j}$ intersects all edges of $(H^-)^{t_j} \in \mathbf{C}(\mathcal{H}^-)$. Hence $\bar{\tau} \in T^D(\mathcal{H}^-)$.

Consequently, $\{\tau, \bar{\tau}\}$ is a fuzzy bi-coloring of $\mathcal{H}^-$. Thus, $\chi(\mathcal{H}^-) = 2$. Since $\chi(\mathcal{H}) = \chi(\mathcal{H}^-)$, we have that $\chi(\mathcal{H}) = 2$. ■

We state the contrapositive of Theorem 4.86 in the next corollary because of its usefulness.

Corollary 4.87 *Let $\mathcal{H}$ be a nontrivial essentially ordered fuzzy hypergraph. Then $\chi(\mathcal{H}) > 2$ if and only if every $\mathcal{H}^-$-dominant fuzzy transversal of $\mathcal{H}^-$ contains an edge of $\mathcal{H}^-$.*

Theorem 4.88 *Let $\mathcal{H} = (X, \mathcal{E})$ be a nontrivial essentially ordered fuzzy hypergraph. Then $\chi(\mathcal{H}) > 2$ if and only if $T^D(\mathcal{H}^-)$ is intersecting.*

Proof. Suppose $\chi(\mathcal{H}) > 2$. By Corollary 4.87, each member τ' of $T^D(\mathcal{H}^-)$ contains an edge, μ', of $\mathcal{H}^-$. For every member, τ, of $T^D(\mathcal{H}^-)$, $\tau \cap \mu' \neq \emptyset$; therefore, $\tau \cap \tau' \neq \emptyset$ for every pair of members $\{\tau, \tau'\}$ in $T^D(\mathcal{H}^-)$.

Conversely, suppose $\chi(\mathcal{H}) = 2$. Then it follows from Theorem 4.86 that there is a member $\tau \in T^D(\mathcal{H}^-)$ which does not contain an edge of $\mathcal{H}^-$. Therefore, since τ is $\mathcal{H}^-$-dominant, supp(τ) cannot contain the support of any fuzzy edge of $\mathcal{H}^-$. Thus, $X^- \setminus$ supp(τ) intersects the support of every fuzzy edge of $\mathcal{H}^-$, where $X^- = \mathbf{V}(\mathcal{H}^-)$, the vertex set of $\mathcal{H}^-$.

Let $\bar{\tau}$ be the $\mathcal{H}^-$-dominant fuzzy subset with supp$(\bar{\tau}) = X^- \setminus$ supp(τ). Clearly, $\bar{\tau}$ intersects every edge of $\mathcal{H}^-$. Now, by repeating an argument found in the proof of Theorem 4.86, it follows that $\bar{\tau} \in T^D(\mathcal{H}^-)$. Since $\tau \cap \bar{\tau} = \emptyset$, $\{\tau, \bar{\tau}\}$ is a disjoint pair of members in $T^D(\mathcal{H}^-)$. Hence $T^D(\mathcal{H}^-)$ is not intersecting. ■

Recall that $T^D(\mathcal{H}) = \{\tau^D \mid \tau \in T(\mathcal{H})\}$, $Tr^D(\mathcal{H}) = \{\tau^D \mid \tau \in Tr(\mathcal{H})\}$, and supp$(\mathcal{H}) = \{\text{supp}(\mu) \mid \mu \in \mathcal{H}\}$.

Lemma 4.89 *Let $\mathcal{H}$ be a fuzzy hypergraph. Then the following conditions are equivalent.*

(1) $T^D(\mathcal{H})$ is intersecting.

(2) $T(\mathcal{H})$ is intersecting.

(3) $Tr(\mathcal{H})$ is intersecting.

(4) $Tr^D(\mathcal{H})$ is intersecting.

Proof. It has previously been shown that every member of $T(\mathcal{H})$ contains a member of $Tr(\mathcal{H})$; moreover, $Tr(\mathcal{H}) \subseteq T(\mathcal{H})$. Thus, conditions (2) and (3) are equivalent. From Definition 4.48, we see that supp$(T(\mathcal{H}))$ is intersecting if and only if supp$(Tr(\mathcal{H}))$ is intersecting. Moreover, supp$(T^D(\mathcal{H}))$ = supp$(T(\mathcal{H}))$ and supp$(Tr^D(\mathcal{H}))$ = supp$(Tr(\mathcal{H}))$. Thus, supp$(T^D(\mathcal{H}))$ =

supp$(T(\mathcal{H}))$ is intersecting if and only if supp$(Tr(\mathcal{H}))$ = supp$(Tr^D(\mathcal{H}))$ is intersecting. The property

$\mathcal{H}$ is intersecting $\Leftrightarrow$ supp$(\mathcal{H})$ is intersecting

and the preceding statement establish the following chain of equivalences:

$T^D(\mathcal{H})$ is intersecting $\Leftrightarrow$ $T(\mathcal{H})$ is intersecting $\Leftrightarrow$ $Tr(\mathcal{H})$ is intersecting $\Leftrightarrow$ $Tr^D(\mathcal{H})$ is intersecting. ■

For our next result, it is useful to recall the following properties of a nontrivial fuzzy hypergraph $\mathcal{H}$:

(i) $\chi(\mathcal{H}) = \chi(\mathcal{H}^-) = \chi(\mathcal{H}^\Delta)$,

(ii) $Tr(\mathcal{H}^-) = Tr(\mathcal{H}^\Delta)$.

Condition (ii) follows from Theorem 4.25 where it is shown that for every fuzzy hypergraph $\mathcal{H}$, $Tr(\mathcal{H}) = Tr(\mathcal{H}^s)$. Thus, since $\mathcal{H}^\Delta = (\mathcal{H}^-)^s$, (ii) holds. By properties (i) and (ii) stated above together with Theorem 4.88 as applied to $\mathcal{H}^\Delta$, which is ordered, and Lemma 4.89, the following theorem will be established.

Theorem 4.90 *For every nontrivial fuzzy hypergraph $\mathcal{H}$, $\chi(\mathcal{H}) > 2$ if and only if $Tr(\mathcal{H}^-)$ is intersecting.*

Proof. Since $\mathcal{H}$ is nontrivial, $\mathcal{H}^-$ and $\mathcal{H}^\Delta$ exist. Moreover, $\chi(\mathcal{H}) > 2 \Leftrightarrow \chi(\mathcal{H}^\Delta) > 2$

$\Leftrightarrow$ $T^D(\mathcal{H}^\Delta)$ is intersecting (by Theorem 4.88 applied to $\mathcal{H}^\Delta$)

$\Leftrightarrow$ $Tr(\mathcal{H}^\Delta)$ is intersecting (by Lemma 4.89)

$\Leftrightarrow$ $Tr(\mathcal{H}^-)$ is intersecting (by property (ii)).

Note that the application of Theorem 4.88 to $\mathcal{H}^\Delta$ is permissible since $\mathcal{H}^\Delta$ is nontrivial and ordered. Also note that the spike reduced fuzzy hypergraph, $(\mathcal{H}^\Delta)^-$, of $\mathcal{H}^\Delta$ remains $\mathcal{H}^\Delta$. ■

Lemma 4.91 *If H is a crisp hypergraph with no loops (i.e., no singleton edges), then $\chi(H) > 2$ if and only if $Tr(H)$ is intersecting.*

Proof. We have that

$$\chi(H) = 2 \Leftrightarrow \text{some transversal of } H \text{ does not contain an edge of } H. \tag{4.61}$$

In fact, since H has no loops, the two colors of any bi-coloring of H must be a pair of disjoint transversals of H neither of which contains an edge of H. On the other hand, if T is a transversal of H which covers no edge of H, then $T' = V(H) \setminus T$ is also a transversal of H and $\{T, T'\}$ is a bi-coloring of H.

The contrapositive of (4.61) now follows: If H is a crisp hypergraph with no loops then $\chi(H) > 2 \Leftrightarrow$ every transversal of H contains an edge of H. Finally observe that $Tr(H)$ is intersecting if and only if every transversal of H contains an edge of H. To establish this fact, observe that if some transversal, T, of H does not contain an edge of H, then, since H has no loops, $T' = V(H) \setminus T$ is a transversal of H disjoint from T. Now if every transversal, T, contains some edge E of H, then as every transversal, T', of H must intersect E, $T \cap T' \neq \emptyset$ for every pair $\{T, T'\}$, of transversals of H. ■

Note that Lemma 4.91 does not hold if H has a loop $\{x\}$ for then every transversal of H contains x and so $Tr(H)$ is intersecting irrespective of the value of the chromatic number $\chi(H)$.

Theorem 4.92 *For every nontrivial fuzzy hypergraph $\mathcal{H}$, $Tr(\mathcal{H}^-)$ is strongly intersecting if and only if $\chi((H^-)^{t_1}) > 2$, where $(H^-)^{t_1} \in \mathbf{C}(\mathcal{H}^-)$ and t_1 is the largest member of $\mathbf{F}(\mathcal{H}^-)$.*

Proof. Since $\mathcal{H}$ is nontrivial, $\mathcal{H}^-$ exists, and the crisp hypergraph, $(H^-)^{t_1}$, has no loops. Therefore, by Lemma 4.91,

$\chi((H^-)^{t_1}) > 2 \Leftrightarrow Tr((H^-)^{t_1})$ is intersecting.

It has previously been shown that the top cut τ^{t_1} of every minimal transversal $\tau \in Tr(\mathcal{H}^-)$ belongs to $Tr((H^-)^{t_1})$. Therefore, $Tr((H^-)^{t_1})$ is intersecting $\Rightarrow$ $Tr(\mathcal{H}^-)$ is strongly intersecting.

On the other hand, every member of $Tr((H^-)^{t_1})$ is the top cut of some member of $Tr(\mathcal{H}^-)$. Hence, $Tr(\mathcal{H}^-)$ is strongly intersecting $\Rightarrow$ $Tr((H^-)^{t_1})$ is intersecting. ■

4.5 Hebbian Structures

In [14, Ch. 4], Hebb describes collections of neuronal cell-assembles. In [9], Goetschel demonstrates how a collection of such cell-assembles can be constructed as the edges of a fuzzy hypergraph. Hebb explains that cell integration is dependent upon Lorente de Nó's thesis [23] which states that a cell is generally fired non-spontaneously by the simultaneous activity of two or more afferent fibres. Consequently cell assemblies can develop rather specifically through experience with respect to an adaptive underlying neural network.

Cooperation of cells through assemblies is essential within Hebbian analysis. The degree of membership of each cell within an assembly is dependent, in part, upon the strength and complexity of the cell's (pre and post) synaptic connections with other component calls of the assembly. Thus, identifying cell assemblies as fuzzy edges requires that the degree of membership within a given edge be variable.

Consider a cell assembly as a single "conscious content", for example, the recognition of an angle of a triangle (see [14, p. 74, p.84]). The conceptualization of the entire triangle would require the integration of several cell assemblies which together would form the edge set of a fuzzy hypergraph. Hebb identifies the resultant of the integration process as a phase sequence.

An explanation in [9] supports the premise that a phase sequence is a useful schema to support a fuzzy organization which comprises a set of cell assemblies that forms the basis for recognizing the relevant parts of the whole together with a family of fuzzy sequencings of the parts according to emerging or established programs of motor activities.

By combining the activations of $\{a_1, a_2, a_3\}$ into a temporal sequence with repetition Hebb defines a simple execution of a phase sequence which is to be construed as a primary step in the recognition of triangle $\Delta\ A_1$, A_2, A_3. For example, suppose a_1 if activated first (by fixation upon A_1). Central activity in a_1 triggers the arousal of two motor activities of which one becomes liminal causing eye movement. Consequently suppose the eye then fixates upon A_3. Then a_3 becomes aroused through the interfacilitation of sensory activity from A_3 and central activity (induced from a_1) and so on. The phase sequence might then continue on to form a temporal sequence of central activity such as $a_1 - a_3 - a_2 - a_3 - a_2 - a_1 - a_2 -$.

As suggested earlier, the cell assemblies in a phase sequence make up the edge set of a fuzzy hypergraph $\mathcal{H} = (\mathcal{U}, \mathcal{E})$, where $\mathcal{U}$ represents a set of cells and each edge e_j of $\mathcal{E}$, $j = 1, \ldots, n$, corresponds to a given cell assembly in the sequence. Communication between cell assemblies mediated by the motor neurons associated with the assemblies of the phase sequence determine a corresponding "order" fuzzy hypergraph, denoted $O(\mathcal{H})$, on

$$\mathcal{X}_{n \times n} = \{1, \ldots, n\} \times \{1, \ldots, n\}$$

with n edges $O(\mathcal{E}) = \{o_1, \ldots, o_n\}$, where the support of o_j is a subset of

$$\{j\} \times \{1, \ldots, n\}, \quad j = 1, \ldots, n.$$

It is understood that the degree of membership of (j, k) in o_j is determined by the degree to which activation of edge e_j can impose activation upon the edge e_k. It is suggested here that the Hebbian phase sequence PS be interpreted through, or correlated with, the pair of fuzzy hypergraphs:

$$PS = \{\mathcal{H}, O(\mathcal{H})\}.$$

Dialogue between $\mathcal{H}$ and $O(\mathcal{H})$ can follow fuzzy logic rules.

Hebb [14, p.35] likens a phase sequence to a "chain of cortical events with motor links." However, other facilitations between cell assemblies, may eventually develop which may play a significant role in the perception of a gestaltic whole, call it t, which may emerge (sometimes fleetingly) from the processing of a phase sequence. Hebb remarks in [14, pp. 98-99], "When the assembly t [the whole (triangle)] has become organized, psychological evidence indicates that its activity intervenes between the activities of the subordinate assemblies a_1, a_2, and a_3 and does not supersede them. Thus the sequence becomes something like

$a_1 - a_2 - t - a_1 - a_3 - t - a_3 - t - a_2 -$."

Suppose the activity of assembly e_k frequently follows activity of e_j in the execution of phase sequence PS; then, in time, the sequential activity $e_j - e_k$ may produce a fuzzy collection of (axon, soma) pairs, where the axons and somas belong, respectively, to cells in e_j and e_k. More specifically, suppose the phase sequence with cell assemblies $\{e_1, e_2, \ldots, e_n\}$ corresponds to some simple geometric figure. We may suppose the individual strengths of the (axon, soma) connections between e_j and e_k are not especially significant (i.e., weak) when compared to the average strength of the inner connections within either assembly e_j or e_k, this is assuming, of course, that the macular field of vision is probably not significantly enjoined in the development of these interconnections between assemblies; this should be contrasted with the dominant role the macular field of vision must play in stimulating the development of the various assemblies, e_1, e_2 through e_n, of the phase sequence.

The set of (axon, soma) pairs associated with each ordered pair $(j, k) \in I_n \times I_n$, where $j \neq k$ and $I_n = \{1, 2, \ldots, n\}$, generates a detailed fuzzy hypergraph on $\mathcal{U} \times \mathcal{U}$, where $\mathcal{U} = \mathbf{V}(\mathcal{H})$, called the *histologic-order fuzzy hypergraph* of PS. Each edge (j, k) consists of those (axon, soma) pairs generated from the serial activity of the subsequence $e_j - e_k$ in PS; strength of membership is determined by the strength of association between the afferent axon and the postsynaptic soma (on a scale from zero to one). Notationally, if $PS = (\mathcal{H}, O(\mathcal{H}))$, the corresponding histologicorder fuzzy hypergraph is denoted by $\Theta(\mathcal{H})$; thus when considerable detail is required $PS = (\mathcal{H}, O(\mathcal{H}), \Theta(\mathcal{H}))$.

Early learning is a period of time when sensory control within cortical association areas slowly takes the form of structured phase sequences. The formation of early phase sequences provides the foundation for later learning when sensorial control over central associations relaxes in favor of increasing intra-cortical convergences between phase sequences or between groups of phase sequence complexes (related phase sequences organized around a particular concept). Further explanation can be found in [9].

Phase sequence complex organized by a particular precept or concept permits attention to remain for longer periods of time on the concept (since the complex permits longer periods of reverberatory excitation once activated). Thus the environment for insightful associations between concepts represented by complexes is enhanced, especially when the only potential links between concepts would initially be weak and subliminal.

The fuzzy hypergraph representation of PS can be adapted to represent phase sequence complex PSC. Let $PSC = (\mathcal{H}_c, O(\mathcal{H}_c))$ be a fuzzily organized structure of phase sequences:

$\{PS_i = (\mathcal{H}_i, O(\mathcal{H})) \mid i = l, \ldots, m\}$,

where $\mathcal{H}_c$ is a fuzzy hypergraph developed from the collection $\{\mathcal{E}(\mathcal{H}_i) \mid i = 1, \ldots, m\}$ and $O(\mathcal{H}_c)$ is a particularly focused order hypergraph of $\mathcal{H}_c$. The intuitive idea here is to correspond the fuzzy edges of $\mathcal{H}_c$ with the edge sets, $\mathcal{E}(\mathcal{H}_i)$, of the fuzzy hypergraphs $\mathcal{H}_i$, $i = 1, \ldots, m$, and to identify the vertex set of $\mathcal{H}_c$ with the collection of edges $\cup_{i=1}^{m} \mathcal{E}(\mathcal{H}_i)$. In this schema, the vertices play a dual role: on the one hand when it is reasonable to do so they can be treated as "smeared" fuzzy points ν, however, in their formal role as vertices of $\mathcal{H}_c$ they should be treated as crisp points $\{\nu\}$

(i.e., as singleton sets containing ν); thus the vertex set, $\mathbf{V}(\mathcal{H}_c)$, of $\mathcal{H}_c$ is formally defined by

$$\mathbf{V}(\mathcal{H}_c) = \{\{\nu\} \mid \nu \in \mathcal{E}(\mathcal{H}_i),\ i = 1, \ldots, m\}.$$

There is a bijective correspondence between the set of edges, $\mathcal{E}(\mathcal{H}_c)$, of $\mathcal{H}_c$ and the collection of edge sets:

$\{\mathcal{E}(\mathcal{H}_i) \mid \mathcal{H}_i$ is a fuzzy hypergraph representation of the cell assemblies of phase sequence PS_i in the PSC complex $(\mathcal{H}_c, O(\mathcal{H}_c))\}$;

this is true, of course, under the tacit assumption that if $i \neq j$ then $\mathcal{E}(\mathcal{H}_i)$. In addition, there also exists a bijection between $\mathcal{E}(\mathcal{H}_c)$ and the support set $\mathbf{S}(\mathcal{H}_c)$, of $\mathcal{E}(\mathcal{H}_c)$; i.e., the collection of subsets of $\mathbf{V}(\mathcal{H}_c)$ which are the supports of the edges in $\mathcal{E}(\mathcal{H}_c)$. The validity of this latter bijection follows from the above tacit assumption and the property that vertices $\{\nu_1\}$ and $\{\nu_2\}$ are equal if and only if ν_1 and ν_2 represent the same fuzzy subset. Indeed, the bijections are obvious since each member A_i of $\mathbf{S}(\mathcal{H}_c)$ satisfies

$$A_i = \{\{\nu_{i,j}\} \mid \nu_{ij} \in \mathcal{E}(\mathcal{H}_i)\}$$

where, it is assumed that

$$\mathbf{S}(\mathcal{H}_c) = \{A_i \mid i =, \ldots, m\}.$$

Let α_i denote the edge set in $\mathcal{E}(\mathcal{H}_c)$ with support A_i, thus $\text{supp}(\alpha_i) = A_i$ and, of course,

$$\mathcal{E}(\mathcal{H}_c) = \{\alpha_i \mid i = 1, \ldots, m\}.$$

It is reasonable to assume that for each member $\{\nu_{ij}\} \in A_i$, the degree of membership, $\alpha_i(\{\nu_{ij}\})$, in α_i should be derived from a fuzzy logic evaluation which requires information from at least two sources:

(i) A fuzzy measure(ment) of the fuzzy subset ν_{ij}.

(ii) An evaluation of the *degree of cohesiveness* of edge ν_{ij} within $\mathcal{H}_i$, determined by a fuzzy procedure based upon data obtained from the fuzzy order hypergraph $O(\mathcal{H}_i)$ of phase sequence PS_i.

Transversal theory, especially fuzzy minimal transversal theory, is useful in descriptive analysis of the basic phenomena: *short-circuiting of phase sequences (or complexes)*. This is a fundamental phenomena within Hebbian psychological analysis, an analysis based upon physiological properties of neuronal structures. By definition, Hebb states [14, p. 228]:

"The reader will recall that the phase sequence is "recurrent" and "anticipatory", containing cyclical conceptual activities schematized as

$A - B - A - C - B - D - E - F - D - E - G - F - H$, etc,

"Short-circuiting" might cut such a sequence down to

$A - B - D - H$, etc.

That is, (1) on repetition the sequence might touch only the high spots, (2) after some synaptic knobs have deteriorated, however, (or, some cells are in refractory state), D might be no longer able to arouse H directly - only when E, F and G are also aroused."

Hebb's comments on cell assembly, as explained in [9], leads to a fuzzy interpretation of short-circuiting. A cell assembly can be treated like a collection of indivisible modular components, each component acting as a functional unit, an organized collection of cells operating as a transmission unit. One can assume that, once established, an effective transmission unit of simple complexity would tend to be a functional unit in several related cell assemblies as they occur in phase sequences.

This speculation suggests as alternative model for phase sequence $PS = (\mathcal{H}, O(\mathcal{H}))$. We construct a modified pair of fuzzy hypergraphs, denoted $(\mathcal{H}_D, O(\mathcal{H}_D))$. The supports of the functional units derived from the edges $e \in \mathcal{E}(\mathcal{H})$ constitute the vertices of the vertex set $\mathbf{V}(\mathcal{H}_D)$ of $\mathcal{H}_D$. The functional units are recognized as directed linear subchains contained within edges of $\mathcal{H}$; generally, after PS is reasonably well-established, it will be presumed that the members of a chain have equal degrees of membership within any specific containing edge; however, the strength of a chain may vary between any two containing edges of the chain. Hence, the edge set, $\mathcal{E}(\mathcal{H}_D)$, of $\mathcal{H}_D$ represents the edges of $\mathcal{E}(\mathcal{H})$ when viewed as fuzzy assemblies of elementary functional units - as fuzzy subsets of $\mathbf{V}(\mathcal{H}_D)$. Clearly, $O(\mathcal{H}_D) = O(\mathcal{H})$; however, the realization of the histologic-order fuzzy hypergraph, $\Theta(\mathcal{H}_D)$, of $\mathcal{H}_D$ would require some computational effort, perhaps by a fuzzy logic procedure utilizing data from both $\Theta(\mathcal{H})$ and $\mathcal{H}_D$.

More ideas concerning the relationship between Hebbian structures and fuzzy hypergraphs can be found in [9]. Recent material that supports neurophysiological assumptions made by Hebb appears in the interesting work of Lynch, [24], with commentaries by G.M. Shepherd, I.B. Black and H.P. Killackey.

4.6 Additional Applications

In this section, we introduce some concepts such as the strength of an edge, the span of a fuzzy hypergraph and the dual fuzzy hypergraph. The material is from [22]. It is stated in [22] that proposed concepts can be used in system analysis, circuit clustering and pattern recognition, etc. [20 - 22].

In general , a family $\{A_1, A_2, \ldots, A_m\}$ of nonempty subsets of a set X is called a *partition* of X if the following conditions are satisfied.

$\bigcup_{i=1}^{m} A_i = X$,

$A_i \cap A_j = \emptyset, i, j = 1, \ldots, m (i \neq j)$

If the family $\{A_1, A, \ldots, A_m\}$ allows $A_i \cap A_j \neq \emptyset$ for $i \neq j$, it is called a *covering* (or *cover*) of X.

A fuzzy partition of set X is a family $\{\mu_1, \mu_2, \ldots, \mu_m\}$ of nonempty fuzzy subsets of X such that

$$\bigcup_{i=1}^{m} \operatorname{supp}(\mu_i) = X \tag{4.62}$$

$$\sum_{i=1}^{m} \mu_i(x) = 1, \forall x \in X \tag{4.63}$$

We call a family $\{\mu_1, \mu_2, \ldots, \mu_m\}$ a *fuzzy covering* of X if it satisfies only the above condition (4.62), but not (4.63).

Let $H = (V, \mathbf{E})$ be a hypergraph, where $X = \{x_1, \ldots, x_n\}$ and $\mathbf{E} = \{E_1, \ldots, E_m\}$. Two vertices x and y of H are said to be *adjacent* if there exists an edge E_j which contains x and y. Two edges are said to be *adjacent* if their intersection is not empty. The *degree of vertex* x is the number of edges which contain the vertex $(|\{E_j \mid x \in E_j\}|)$. The incidence matrix of the hypergraph H is a matrix $M_H = (a_{ij})_{n \times m}$ with m columns representing the edges and n rows representing the vertices, where the elements a_{ij} are as follows:

$$a_{ij} = \begin{cases} 1 & \text{if } x_i \in E_j \\ 0 & \text{if } x_i \notin E_j \end{cases}$$

For example, consider a hypergraph $H = (V, \mathbf{E})$ such that

$V = \{x_1, x_2, x_3, x_4, x_5\}$

$\mathbf{E} = \{E_1, E_2, E_3\}$

$E_1 = \{x_1, x_2\}$, $E_2 = \{x_2, x_3, x_4\}, E_3 = \{x_4, x_5\}$.

The hypergraph can be described by its incidence matrix as follows:

	E_1	E_2	E_3
x_1	1	0	0
x_2	1	1	0
x_3	0	1	0
x_4	0	1	1
x_5	0	0	1

In general, the edge set is a cover of X. In a hypergraph, if every vertex has degree 1, i.e., $E_i \cap E_j = \phi, i \neq j$, the edge form a partition of X. A hypergraph $H = (X, \mathbf{E})$ can be mapped to a hypergraph $H^* = (X^*, \mathbf{E}^*)$ whose vertices are the points $e_1, e_2, \ldots, e_m$ (corresponding to $E_1, E_2, \ldots, E_m$ respectively), and whose edges are sets $X_1, X_2, \ldots, X_n$ (corresponding to $x_1, x_2, \ldots, x_n$ respectively). Then

$X_j = \{e_i \mid x_j \in E_i, i = 1, \ldots, m\}, j = 1, \ldots, n,$

$X_j \neq \phi,$

$$\bigcup_{j=1}^{n} X_j = X^* = \{e_1, e_2, \ldots, e_m\}$$

The hypergraph H^* is called the *dual hypergraph* of H. The incidence matrix $(a_{ij})_{n \times m}$ of the hypergraph H and that $(b_{ij})_{m \times n}$ of the dual hypergraph H^* are transposed matrices of each other , i.e.,$(a_{ij})^T_{n \times m} = (b_{ij})_{m \times n}$. Thus we have $(H^*)^* = H$. If two vertices x_i and x_j in H are adjacent, the edges X_i and X_j in H^* are adjacent. Similarly, if two edges E_i and E_j in H are adjacent, two vertices e_i and e_j in H^* are adjacent. We can obtain the following dual hypergraph H^* of the hypergraph H given in the above example.

$H^* = (X^*, \mathbf{E}^*)$,

$X^* = \{e_1, e_2, e_3\}$ and $\mathbf{E}^* = \{X_1, X_2, X_3, X_4, X_5\}$, where

$X_1 = \{e_1\}, X_2 = \{e_1, e_2\}, X_3 = \{e_2\}, X_4 = \{e_2, e_3\}, X_5 = \{e_3\}$.

TABLE 4.1 Incidence matrix.

$\mathcal{H}$	μ_1	μ_2	μ_3
x_1	0.8	0	0
x_2	0.5	0.5	0
x_3	0	1	0
x_4	0	0.8	0.8
x_5	0	0	1

The incidence matrix of the hypergraph H^* is as follows:

H^*	X_1	X_2	X_3	X_4	X_5
e_1	1	1	0	0	0
e_2	0	1	1	1	0
e_3	0	0	0	1	1

Given a hypergraph $H = (X, \mathbf{E})$, we associate a graph $G = (X, F)$, called the *corresponding graph* as follows. For any two vertices $x, y \in X$, there is an edge between x and y if and only if there exists at least one $E \in \mathbf{E}$ such that $x, y \in E$. In other words, $F = \{(x, y) \mid \exists E \in \mathbf{E}, x, y \in E\}$. In general, the corresponding graph is a multigraph. The corresponding graph G of the hypergraph H given above is as follows.

$G = (X, F)$

$X = \{x_1, x_2, x_3, x_4, x_5\}$

$F = \{(x_1, x_2), (x_2, x_3), (x_2, x_4), (x_3, x_4), (x_4, x_5)\}$

In the corresponding graph there are three maximal cliques $\{x_1, x_2\}$, $\{x_2, x_3, x_4\}$, and $\{x_4, x_5\}$.

Consider a fuzzy hypergraph $\mathcal{H} = (X, \mathcal{E})$ where $X = \{x_1, x_2, x_3, x_4, x_5\}$ and $\mathcal{E} = \{\mu_1, \mu_2, \mu_3\}$ described by the incidence matrix given in Table 4.1.

In a fuzzy hypergraph, we define the adjacent level between two vertices or between two edges as follows:

adjacent level γ between two vertices x and y $(x \neq y)$ is defined by

$$\gamma(x, y) = \vee\{\mu_j(x) \wedge \mu_j(y) \mid j = 1, \ldots, m\}.$$

adjacent level σ between two edges μ_j and μ_k is defined $\sigma(\mu_j, \mu_k) = \vee\{\mu_j(x) \wedge \mu_k(x) \mid x \in X\}$.

In the above fuzzy hypergraph, the adjacent level $\gamma(x_1, x_2)$ between vertices x_1 and x_2 is 0.5, and the adjacent level $\sigma(\mu_1, \mu_2)$ is 0.5.

Let $t \in [0, 1]$. Define $\mu_{m+1}^t = \{x \in X \mid \mu_j(x) < t, \forall j = 1, \ldots, m\}$. The edge μ_{m+1}^t is added to the set of edges of $\mathcal{H}^t$.

For example, at $t = 0.8$ for the fuzzy hypergraph $\mathcal{H}$ given above, hypergraph $\mathcal{H}^{0.8}$ is given by its incidence matrix M is as follows. In the 0.8-cut hypergraph $\mathcal{H}^{0.8}$, a new edge $\mu_4^{0.8}$ added to contain the element x_2.

$\mathcal{H}^{0.8}$	$\mu_1^{0.8}$	$\mu_2^{0.8}$	$\mu_3^{0.8}$	$\mu_4^{0.8}$
x_1	1	0	0	0
x_2	0	0	0	1
x_3	0	1	0	0
x_4	0	1	1	0
x_5	0	0	1	0

We have seen that the support of a fuzzy edge in a hypergraph corresponds to a clique in the corresponding ordinary graph. Therefore, the edge $\mu_3^{0.8}$ in the hypergraph $\mathcal{H}^{0.8}$ represents the clique $\{x_4, x_5\}$. Again, the edge and the clique correspond to the class in the clustering (partition).

We now consider the strength of an edge.

We see that some edges contains only vertices having high membership degree. For example, in the fuzzy hypergraph $\mathcal{H}$ given above, the edge supp(μ_3) contains the vertices having membership at least 0.8. On the other hand, supp(μ_1) has vertices having membership at least 0.5. We can state that the cohesion in μ_3 is stronger than in μ_1.

Therefore we define the concept of strength of a fuzzy edge. The strength β of a fuzzy edge μ_j is the minimum membership $\mu_j(x)$ of the vertices. That is, $\beta(\mu_j) = \wedge\{\mu_j(x) \mid x \in \text{supp}(\mu_j)\}$. Its interpretation is that the fuzzy edge μ_j groups elements having participation degree at least $\beta(\mu_j)$ in the hypergraph.

For example, in the fuzzy hypergraph $\mathcal{H}$ given above, the strength of each edge is $\beta(\mu_1) = 0.5$, $\beta(\mu_2) = 0.5$ and $\beta(\mu_3) = 0.8$ respectively. In the example, the edge μ_3 is said to be *stronger than* μ_1 and μ_2, since $\beta(\mu_3) > \beta(\mu_1)$ and $\beta(\mu_3) > \beta(\mu_2)$. If we assign $\beta(\mu_j)$ as the membership degree for all x_i such that $\mu_j(x_i) > 0$, we obtain a fuzzy hypergraph which represents the subsets with grouping strength or the *cohesion hypergraph.*

	$\hat{\beta}_1$	$\hat{\beta}_2$	$\hat{\beta}_3$
x_1	0.5	0	0
x_2	0.5	0.5	0
x_3	0	0.5	0
x_4	0	0.5	0.8

We have seen that we can associate an ordinary graph with a hypergraph. A hypergraph can be mapped to a clique(subgraph) in the graph. Similarly, a fuzzy graph can be associated with a fuzzy hypergraph. In this case, a fuzzy edge with its strength β in the fuzzy hypergraph is mapped to clique in the fuzzy graph; all the edges in the clique have the membership degree β. The $\hat{\beta}_i, i = 1, 2, 3$, give the corresponding fuzzy graph to the fuzzy hypergraph H. In the graph, the number attached to the edges represent the membership degree.

We now introduce the concept of a *dual fuzzy hypergraph* for a fuzzy hypergraph. Given a fuzzy hypergraph $\mathcal{H} = (X, \mathcal{E})$, where $X = \{x_1, x_2, \ldots, x_n\}$

TABLE 4.2 Dual fuzzy hypergraph.

$M_{\mathcal{H}^*}$	χ_1	χ_2	χ_3	χ_4	χ_5
e_1	0.8	0.5	0	0	0
e_2	0	0.5	1	0.8	0
e_3	0	0	0	0.8	1

TABLE 4.3 0.8–cut hypergraph.

$M_{(\mathcal{H}^*)^{0.8}}$	$\chi_1^{0.8}$	$\chi_2^{0.8}$	$\chi_3^{0.8}$	$\chi_4^{0.8}$	$\chi_5^{0.8}$
e_1	1	0	0	0	0
e_2	0	0	1	1	0
e_3	0	0	0	1	1

and $\mathcal{E} = \{\mu_1, \mu_2, \ldots, \mu_m\}$, its dual fuzzy hypergraph $\mathcal{H}^* = (X^*; \mathcal{E}^*)$ is defined as follows:

$X^* = \{e_1, e_2, \ldots, e_m\}$: set of vertices corresponding to $\mu_1, \mu_2, \ldots, \mu_m$ respectively.

$\mathcal{E}^* = \{\chi_1, \chi_2, \ldots, \chi_n\}$: set of hyperedges corresponding to $x_1, x_2, \ldots, x_n$ where

$\chi_i(e_j) = \mu_j(x_i), i = 1, 2, \ldots, n; j = 1, 2, \ldots, m.$

For example, consider the fuzzy hypergraph $\mathcal{H}$ of H. Its dual fuzzy hypergraph $\mathcal{H}^*$ is as follows $\mathcal{H}^* = (X^*, \mathcal{E}^*)$, where $X^* = \{e_1, e_2, e_3\}$ and $\mathcal{E}^* = \{\chi_1, \chi_2, \ldots, \chi_5\}$. Its incidence matrix is as given in Table 4.2.

Now, let's cut the fuzzy hypergraph $\mathcal{H}^*$at level 0.8. The 0.8-cut hypergraph $(\mathcal{H}^*)^{0.8}$ is obtained by the incidence matrix $M_{(\mathcal{H}^*)^{0.8}}$ given in Table 4.3.

From the hypergraphs $\mathcal{H}^{0.8}$ and $(\mathcal{H}^*)^{0.8}$, we can see that the dual hypergraph of $\mathcal{H}^{0.8}$ is not equal to $(\mathcal{H}^*)^{0.8}$. That is, the commutativity property is not satisfied between the t-cut operation and the dual operation. However, if the following conditions are satisfied, then the commutativity property is satisfied.

For $\mathcal{H} = (X, \mathcal{E})$,

$\forall x_i \in X, \exists\ \mu_k$ such that $\mu_k(i) \geq t$,

$\forall \mu_i \in \mathcal{E}, \exists x_k$ such that $\mu_i(k) \geq t$.

That is, when f_t is a function which cuts a fuzzy hypergraph at level t and g is a function generating the dual hypergraph of a hypergraph, the composition of the two functions f_t and g is commutative under the above conditions as follows:

$f_t : \mathcal{H} \to \mathcal{H}^t$

$g : \mathcal{H} \to \mathcal{H}^*$

$f_t(\mathcal{H}) = \mathcal{H}^t$

$g(\mathcal{H}^t) = (\mathcal{H}^*)^t$

$g\,(\mathcal{H}) = H^*$

$f_t\,(H^*) = (\mathcal{H}^*)^t$

TABLE 4.4 Fuzzy partition matrix.

H	τ	γ
x_1	0.96	0.04
x_2	1	0
x_3	0.61	0.39
x_4	0.05	0.95
x_5	0.03	0.97

TABLE 4.5 Hypergraph $\mathcal{H}^{0.61}$.

$\mathcal{H}^{0.61}$	τ	γ
x_1	1	0
x_2	1	0
x_3	1	0
x_4	0	1
x_5	0	1

$g \circ f_t = f_t \circ g$

A fuzzy partition can be represented by a fuzzy matrix (a_{ij}) where a_{ij} is the membership degree of element x_i in class j. We see that this matrix can be considered as the incidence matrix of a fuzzy hypergraph. Thus we can represent a fuzzy partition by a fuzzy hypergraph $H = (X, \mathcal{E})$ where $X = \{x_1, x_2, \ldots, x_n\}$, $\mathcal{E} = \{\mu_1, \mu_2, \ldots, \mu_m\}$ $(\mu_j \neq \emptyset, j = 1, 2, \ldots, m)$ and $\sum_{j=1}^{m} a_{ij} = 1$, $i = 1, 2, \ldots, n$, for $a_{ij} = \mu_j(x_i)$, $i = 1, 2, \ldots, n; j = 1, 2, \ldots, m$.

Note that the last condition is added to the fuzzy hypergraph for fuzzy partitions. If the last condition is eliminated, the fuzzy hypergraph can represent a fuzzy covering. Naturally we can apply the t-cut to the fuzzy partition.

Let's consider an example of clustering problem given in [3]. The problem a is typical example of a fuzzy partition in visual image processing. There are five objects and they are classified in to two classes: tank and house. To cluster the elements x_1, x_2, x_3, x_4, x_5 into τ (tank) and γ (house), a fuzzy partition matrix is given as the form of an incidence matrix in Table 4.4 of a fuzzy hypergraph as in [3].

We can apply the t-cut to the fuzzy hypergraph and obtain a hypergraph $\mathcal{H}^t$ which is not fuzzy. Let's denote the edge(class) in the t-cut hypergraph $\mathcal{H}^t$ as A_j^t. This hypergraph $\mathcal{H}^t$ represents generally a covering because the condition $\sum_i \mu_i(x) = 1$, $\forall x \in A_j^t$ is not always guaranteed . The hypergraph $\mathcal{H}^{0.61}$ is shown in Table 4.5.

We can obtain the dual hypergraph $(\mathcal{H}^*)^{0.61}$ of $\mathcal{H}^{0.61}$ as follows (Table 4.6).

TABLE 4.6 Dual fuzzy hypergraph.

$(\mathcal{H}^*)^{0.61}$	X_1	X_2	X_3	X_4	X_5
τ	1	1	1	0	0
γ	0	0	0	1	1

Let's consider the strength of edge(class) A_j^t in the t-cut hypergraph $\mathcal{H}^t$. It is necessary to modify the definition of strength to obtain the strength of edge A_j^t in $\mathcal{H}^t$ as follows: $\beta(A_j^t) = \wedge\{a_{ij} \mid i \text{ such that } x_i \in A_j^t\}$.

The possible interpretations of $\beta(A_j^t)$ are:

1. The edge (class) in the hypergraph (partition) $\mathcal{H}^t$ groups elements having at least β membership, the strength (cohesion) of edge (class) A_j^t in $\mathcal{H}^t$ is β. Thus we can use the strength as the measure of the cohesion or the strength of the class in the partition. For example, the strengths of classes $\tau^{0.61}$ and $\gamma^{0.61}$ at $t = 0.61$ are $\beta(\tau^{0.61}) = 0.61$. $\beta(\gamma^{0.61}) = 0.95$. Thus we say that the class $\gamma^{0.61}$ is stronger than $\tau^{0.61}$ because $\beta(\gamma^{0.61}) > \beta(\tau^{0.61})$. From the above discussion about the hypergraphs $\mathcal{H}^{0.61}$ and $(\mathcal{H}^*)^{0.61}$, we can state that:

2. The fuzzy hypergraph can represent the fuzzy partition visually. The t-cut hypergraph also represents the t-cut partition.

3. The dual hypergraph $(\mathcal{H}^*)^{0.61}$ can represent the elements X_i which can be grouped into a class A_j^t. For example, the edges X_1, X_2, X_3 of the dual hypergraph in Fig. 11 represent the elements x_1, x_2, x_3 that can be grouped into t at level 0.61.

4. In the fuzzy partition, we have $\sum_{j=1}^{n} \mu_j(x) = 1, x \in X_j$. If we t-cut at level $t > 0.5$, there is no element which is grouped into two classes simultaneously. That is if $t > 0.5$, every element is combined in only one class in $\mathcal{H}^t$. Therefore, the hypergraph $\mathcal{H}^t$ represents a partition (if $t \leq 0.5$, the hypergraph may represent a covering).

5. At $t = 0.61$ level, the strength of class $\gamma^{0.61}$ is the highest (0.95), so it is the strongest class. Hence this class can be grouped independently from the other parts. Thus we can eliminate the class γ from the others and continue clustering. Therefore the discrimination of strong classes from the others can allow us to decompose a clustering problem into smaller ones. This strategy allows us to work with the reduced data in a clustering problem.

In 1982, Pawlak introduced the concept of rough set, [30]. This concept is fundamental to the examination of granularity in knowledge. It is a concept which has applications in data analysis. The idea is to approximate a subset of a universal set by a lower approximation and an upper approximation in the following manner. A partition of the universe is given.

The lower approximation is the union of those members of the partition contained in the given subset and the upper approximation is the union of those members of the partition which have a nonempty intersection with the given subset. This framework provides a systematic method for the study on intelligent systems characterized by insufficient or incomplete information. It is well known that a partition induces an equivalence relation on a set and vice versa. The properties of rough sets can thus be examined via either partitions or equivalence relations. The members of the partition (or equivalence classes) can be formally described by unary set-theoretic operators, [37], or by successor functions for upper approximation spaces, [18,19]. This axiomatic approach allows not only for a wide range of areas to be used to describe rough sets. Some examples are topology, (fuzzy) abstract algebra, (fuzzy) directed graphs, (fuzzy) finite state machines, modal logic, interval structures, [18,28,31,37,38,39]. The requirement of a partition (or equivalence relation) seems to be a stringent condition that may limit the application domain of the rough set model. To resolve this problem, many proposals have been made, [28,33,34,35,39], one of which is the replacement of a partition of the universe by a cover of the universe. Consequently the properties of hypergraphs and fuzzy hypergraphs seems to provide an untapped resource for rough set theory.

Let $\mathcal{C}$ be a cover of X. Define L, $U : \mathcal{P}(X) \to \mathcal{P}(X)$ as follows: $\forall S \in \mathcal{P}(X)$,

$$L(S) = \bigcup_{\substack{C \in \mathcal{C} \\ C \subseteq S}} C \text{ and } U(S) = \bigcup_{\substack{C \in \mathcal{C} \\ C \cap S \neq \emptyset}} C.$$

Then $L(S)$ is called the *lower approximation* of S and $U(S)$ is called the *upper approximation* of S. When $\mathcal{C}$ is partition of X, then strong properties hold for L and U. This case has been studied extensively. The case where $\mathcal{C}$ is not a partition has been examined less extensively.

In order to see a connection between hypergraphs and rough sets consider the following result: Let $H = (X, \mathbf{E})$ be a hypergraph. Let L and U be defined on $\mathcal{P}(X)$ as above, where $C = \mathbf{E}$. Then H is intersecting implies $\forall E \in \mathbf{E}, U(E) = X$: H is intersecting $\Leftrightarrow \forall E, E' \in \mathbf{E}, E \cap E' \neq \emptyset$ $\Rightarrow \forall E \in \mathbf{E}, U(E) = X$. Let $H = (X,E)$, where $X = \{1,2,3,4\}$ and $\mathbf{E} = \{\{1,2\},\{1,3\},\{2,4\},\{3,4\}\}$. Then $\forall E \in \mathbf{E}, U(E) = X$, but H is not intersecting.

We now consider fuzzy hypergraphs. Let $\mathcal{H} = (X, \mathcal{E})$ be a fuzzy hypergraph and let be a partition of X. Then $\mathcal{P}$ may be considered to be a primitive coloring or an $\mathcal{L}$−coloring of $\mathcal{H}$. Define $\widetilde{L}, \widetilde{U} : \mathcal{FP}(X) \to \mathcal{FP}(X)$ by $\forall \mu \in \mathcal{E}$,

$$\widetilde{L}(\mu)(x) = \wedge\{\mu(z) | z \in [x]\} \text{ and } \widetilde{U}(\mu)(x) = \vee\{\mu(z) | z \in [x]\}$$

for all $x \in X$, where $[x]$ is the equivalence class of x with respect to the equivalence relation induced by $\mathcal{P}$. It can be shown that the definitions of $\widetilde{L}$ and $\widetilde{U}$ contain the crisp case, i.e., are those for $\widetilde{L}$ and $\widetilde{U}$ when the image of μ is $\{0,1\}$, [26]. With the definitions of $\widetilde{L}$ and $\widetilde{U}$, we have a connection between fuzzy hypergraphs and rough sets.

4.7 References

1. Berge, C., *Graphes et Hypergraphes*, Dunod, Paris, 1972.

2. Berge, C., *Hypergraphs,* North Holland, Amsterdam, 1989.

3. Bezdek, J. C., Partition structure: a tutorial, in *Analysis of Fuzzy Information*, vol. III, CRC press: Boca Raton , pp. 81-108, 1987.

4. Bollobas, B., *Graph Theory on Introductory Course*, Springer -Verlag: New York, 1979.

5. Bouchon, B. and Cohen, G., On fuzzy relations and partitions, in *Advances in Fuzzy Sets, Possibility Theory and its Applications*, P. P. Wang (ed.), pp. 97-105, 1983.

6. Constantin-Paton, M. and Prusky, G., The basis of "recognition" at the synaptic level, in: Price, D.L., Thoenen, H. and Aguayo, A.J., Eds., *Neurodegenerative Disorders, Mechanisms and Prospects for Therapy* (Wiley, Chichester, 1991) 127-144.

7. Das, M., *A note on fuzzy hypergraphs and their duals,* unpublished (1987).

8. Dubois, D., and Prade, H., *Fuzzy Sets and Systems: Theory and Applications,* New York: Academic Press 1980.

9. Goetschel, R., Jr., *Introduction of fuzzy hypergraphs and Hebbian structures,* Fuzzy Sets and Systems **76**:113-130, 1995.

10. Goetschel, R., Jr., Craine, W., Voxman, W., *Fuzzy transversals of fuzzy hypergraphs,* Fuzzy Sets and Systems **84**:235-254,1996.

11. R. Goetschel, Jr., *Fuzzy colorings of fuzzy hypergraphs,* Fuzzy Sets and Systems, **94**:185-204,1998.

12. Goetschel,R. N., Jr., and Voxman, W., *Intersecting fuzzy hypergraphs,* Fuzzy Sets and Systems, **99**:81-96,1998.

13. Graver, J. E., and Watkins, M. E., *Combinatorics With Emphasis on Theory of Graphs*, Springer-Verlag, Berlin, 1977.

14. Hebb, D.O., *The Organization of Behavior* (Wiley, New York, 1949).

15. Ismail, M. A., Soft clustering: algorithm and validity of solutions, in *Fuzzy Computing*, Gupta, M. M., and Yamakawa, T.,(ed.), Elservier Science, North Holland, pp. 445-471, 1988.

16. Kaufmann, A., *Introduction à la Thèorie des Sous-Ensemble Flous*, vol. 1, Masson :Paris 1977.

17. Kaufmann, A., *Introduction à la Thèorie des Sous-Ensemble Flous,* vol. 4, Masson :Paris 1977.

18. Kuroki, N. and Mordeson, J. N., Successor and source functions, *J. Fuzzy Math.,* **5**:173–182, 1997.

19. Kuroki N. and Mordeson, J. N., Structure of rough sets and rough groups, *J. Fuzzy Math.,* **5**:183–191, 1997.

20. Lee-Kwang, H., and Favrel, J., Hierarchical reduction and decomposition method of Petri nets, *IEEE Trans.Syst.Man.Cybern.*, vol. SMC-15, pp. 272-280, 1985.

21. Lee-Kwang, H., and Favrel, J. and Oh, G., Hierarchical decomposition of Petri net languages, *IEEE Trans.Syst.Man.Cybern.*, vol. SMC-17, pp. 877-878, 1987.

22. Lee-Kwang, H., and Lee, K.-M., Fuzzy Hypergraph and Fuzzy Partition, *IEEE Trans. Syst., Man,.and Cybern.,* vol. SMC -25, pp. 196–201, 1995.

23. Lorente de Nó, R., *Synaptic stimulation of motoneurons as a local process,* J. Neurophysiol. **1**:195-206,1938.

24. Lynch, G., *Synapses, Circuits, and the Beginnings of Memory*, MIT Press, Cambridge, MA, 1986.

25. Mordeson, J. N., Rough set theory applied to (fuzzy) ideal theory, *Fuzzy Sets and Systems,* to appear.

26. Mordeson, J. N., Algebraic properties of spaces in rough fuzzy set theory, NAFIPS'99, 56-59, 1999.

27. Morsi, N. N. and Yakout, M. N., Axiomatics for fuzzy rough sets, *Fuzzy Sets and Systems,* **100**:327–342,1998.

28. Orlawska, E., Semantic analysis of inductive reasoning, *Theoretical Computer Science,* **43**:81–89, 1986.

29. Papadimitrion, C. H.and Steiglitz, K., *Combinatorial Optimization: Algorithms and Complexity*, Prentice Hall: Englewood Cliffs, 1982.

30. Pawlak, Z., Rough sets, *Int. J. Comp. Sci.,* **11**:312–356, 1982.

31. Pawlak, Z., *Rough Sets,* Theoretical Aspects About Data, Kluwer Acad. Pub., 1991.

32. Pedrycz, W. and Hirota, K., An application of probabilistic sets to fuzzy clustering, in *Analysis to Fuzzy Information,* vol. III, CRC Press: Boca Raton, pp. 170-181, 1987.

33. Pomykala, J. A., Approximation operations in approximate space, *Bulletin of the Polish Academy of Sciences: Mathematics,* **35**:653–662, 1987.

34. Wasilewska, A., Conditional knowledge representation systems - model for an implementation, *Bulletin of the Polish Academy of Sciences: Mathematics,* **37**:63–69, 1987.

35. Wasilewska, A. and Vigeron, L., On generalized rough sets, preprint.

36. Wybraniec - Skardowska, U., On a generalization of approximation space, *Bulletin of the Polish Academy of Sciences: Mathematics,* **37**:51–61, 1989.

37. Yao, Y. Y., Relational interpretations of neighborhood operators and rough set approximation operators, preprint.

38. Yao, Y. Y., Two views of the theory of rough sets in finite universes, *Int. J. Approximate Reasoning,* **15:**291–317, 1996.

39. Yao, Y. Y. and Lin, T. Y., Generalization of rough sets using modal logics, Intelligent Automation and Soft Computing, **2:**103 - 120, 1996.

40. Zakowski, W., Approximations in the space (U, Π), *Demonstratio Mathematica,* **XVI**:761–769, 1983.

41. Zimmermann, H.-J., *Fuzzy Set Theory and its Applications*, Nijhoff Publishing, Dordrecht, 1985.

LIST OF FIGURES

LIST OF TABLES

LIST OF SYMBOLS

A, B, X, and Y are sets

μ, ν, ξ are fuzzy subsets

$A \cup B$	union of A and B, p. 1
$A \cap B$	intersection of A and B, p. 1
$B \setminus A$	relative complement of A in B, p. 1
A^c	the complement of A in its universal set, p. 1
$x \in A$	x is an element of A, p. 1
$x \notin A$	x is not an element of A, p. 1
$A \subseteq B$	A is contained in B, p. 1
$A \supseteq B$	A contains B, p. 1
$A \subset B$	A is strictly contained in B, p. 2
$B \supset A$	B strictly contains A, p. 2
$\|A\|$	the cardinality of A, p. 2
$\text{card}(A)$	the cardinality of A, p. 2
$\wp(A)$	the power set of A, p. 2
$\emptyset$	the empty set, p. 2
$\mathbb{N}$	the positive integers, p. 2
$\mathbb{Z}$	the integers, p. 2
$\mathbb{R}$	the real numbers, p. 2
(x, y)	ordered pair of x and y, p. 2
$X \times Y$	Cartesian product of sets X and Y, p. 2
X^n	set of ordered n-tuples, p. 2
$\text{Dom}(R)$	domain of the relation R, p. 2
$\text{Im}(R)$	image of the relation R, p. 2

$[x]$	equivalence class determined by x, p. 2
$T \circ R$	composition of relations R and T, p. 2
$f : X \to Y$	function of X into Y, p. 2
μ^t	the $t - cut$ or level set of μ, p. 3
$\text{supp}(\mu)$	the support of μ, p. 3
$\mathfrak{F}\wp(S)$	fuzzy power set of a set S, p. 3
$\wedge$	infimum, p. 3
$\vee$	supremum, p. 3
$\mu \subseteq \nu$	μ is contained ν, p. 3
$\mu \subset \nu$	μ is strictly contained in ν, p. 3
$\mu \cup \nu$	μ union ν, p. 3
$\mu \cap \nu$	μ intersect ν, p.3
μ^c	p. 3
$\cap_{\xi \in \mathcal{S}} \xi$	intersection of those ξ in $\mathcal{S}$, p. 3
$\cup_{\xi \in \mathcal{S}} \xi$	union of those ξ in $\mathcal{S}$, p. 3
χ_A	the characteristic function of A, p. 4
$\rho \circ \varpi$	composition of ρ and ϖ, p. 5
ρ^k	composition of ρ with itself k times, p. 5
ρ^∞	p. 5
ρ^0	p. 5
ρ^c	p. 5
ρ^{-1}	p. 5
F^ρ	p. 12
$F_\epsilon{}^\rho$	p. 12
ϕ_ρ	p. 14
$\phi_\rho{}^\epsilon$	p. 15
(V, R) a graph,	p. 19
(V, μ, ρ)	a fuzzy graph, p. 19
(μ, ρ)	a fuzzy graph, p. 19
P	a path, p. 20
$\text{diam}(x, y)$	p. 21
$\mathbb{Z}_2$	p. 31
∂	boundary operator, p. 31
δ	coboundary operator, p. 31
$C(T)$	set of cycles, p. 32
$m(G)$	cycle rank of a graph G, p. 32
ν_f	p. 34
ρ_E	p. 35
$\mathcal{CS}(\mu, \rho)$	cycle set, p. 37
$\mathcal{FCS}(\mu, \rho)$	fuzzy cycle set, p. 37
$\mathcal{C}o\mathcal{S}(\mu, \rho)$	cocycle set, p. 37
$\mathcal{FC}o\mathcal{S}(\mu, \rho)$	fuzzy cocycle set, p. 37
S_ρ	p. 37
$< \mathcal{S} >$	p. 37

$\mathcal{S}^+$	p. 37
$m(\mu, \rho)$	cycle rank of (μ, ρ), p. 38
$fm(\mu, \rho)$	fuzzy cycle rank of (μ, ρ), p. 38
$m_c(\mu, \rho)$	cocycle rank of (μ, ρ), p. 39
$fm_c(\mu, \rho)$	fuzzy cocycle rank of (μ, ρ), p. 39
$L(G)$	line graph of the graph G, p. 40
$\simeq$	isomorphism of graphs, p. 41
$\mathcal{I}(S)$	intersection graph, p. 41
$D = (V, \mu, \delta)$	fuzzy digraph, p. 45
$h(\alpha)$	height of the fuzzy subset α, p. 45
$\mathrm{fs}(G)$	fundamental sequence of a graph G, p. 45
$\mathrm{Int}(\mathcal{F})$	fuzzy intersection graph of a family $\mathcal{F}$ of fuzzy subsets, p. 46
α_x	p. 46
$\mathcal{I}$	fuzzy interval, p. 47
$\mathcal{K}$	fuzzy clique, p. 49
$\mathcal{J}_x$	p. 50
$< x, y >$	p. 51
ρ_A	p. 53
$<_t$	p. 55
$(G, <)$	p. 58
$\mu_1 \times \mu_2$	p. 62
$\rho_1 \rho_2$	p. 62
$G_1[G_2]$	the composition of graph G_1 with graph G_2, p. 64
$\mu_1 \circ \mu_2$	p. 64
$\rho_1 \circ \rho_2$	p. 64
$G_1 \cup G_2$	union of graphs G_1 and G_2, p. 66
$\mu_1 \cup \mu_2$	p. 67
$\rho_1 \cup \rho_2$	p. 67
$G_1 + G_2$	join of graphs G_1 and G_2, p. 67
$\mu_1 + \mu_2$	p. 67
$\rho_1 + \rho_2$	p. 67
$C_m(\mu, \rho)$	complete fuzzy graph, p. 69
$C_{m,n}(\mu, \rho)$	complete fuzzy bigraph, p. 69
$C(G)$	connectedness level of a fuzzy graph G, p. 72
$h(G, \cdot)$	the cyclomatic function of the fuzzy graph G, p. 73
$S(G)$	acyclic level of the fuzzy graph G, p. 74
$dis(x, y)$	p. 83
$l(P)$	$\rho - length$ of P, p. 83
$\delta(x, y)$	$\rho - distance$ between vertices x and y, p.83
$s(P)$	p. 84
$d(x, y)$	p. 86
M_ρ	fuzzy matrix of a fuzzy graph, p. 87

$h(\mathcal{H})$	height of $\mathcal{H}$, p. 136
$\sigma(a,r)$	p. 136
$H^t = (X^t, \mathbf{E}^t)$	p. 136
$\sqsubseteq$	p. 137
$\sqsubset$	p. 137
$\mathbf{F}(\mathcal{H})$	p. 137
$\mathbf{C}(\mathcal{H})$	p. 137
$\mathbf{E}_i$	p. 137
ν_E	p. 139
$\mu \otimes H$	μ tempered fuzzy hypergraph, p. 139
$Tr(H)$	set of minimal transversals of H, p. 141
$Tr(\mathcal{H})$	set of minimal fuzzy transversals of $\mathcal{H}$, p. 141
$Tr^*(\mathcal{H})$	p. 142
$\mu_{(t)}$	p. 145
$\mu^{(t)}$	p.145
$\mathcal{E}_{(t)}$	p. 145
$\mathcal{E}^{(t)}$	p. 145
$\mathcal{H}_{(t)}$	p. 145
$\mathcal{H}^{(t)}$	p. 145
$\mathbf{S}(\mu)$	p. 146
$\mathbf{C}(\mu)$	p. 146
$\mathbf{E}(\mu)$	p. 146
$\mathbf{J}(\mathcal{H})$	the join of the fuzzy hypergraph $\mathcal{H}$, p. 154
$\mathcal{H}^s$	p. 158
$\hat{\mathbf{C}}(\mathcal{H})$	p. 158
$\hat{\mathbf{F}}(\mathcal{H})$	p. 159
$\chi(H)$	p. 173
$\chi_p(\mathcal{H})$	$p-$chromatic number of $\mathcal{H}$, p. 174
μ^-	spike reduction of μ, p. 175
$\mathcal{H}^-$	p. 175
$\mathcal{H}^\Delta$	the skeleton of $\mathcal{H}^-$, p. 175
$H^\beta(x)$	a β_H star of x, p.179
$d_H{}^\beta(x)$	the $\beta_H - star$ degree of x, p. 179
$\Delta^\beta(H)$	the maximum $\beta_H - degree$ of H, p. 179
$\delta^\beta(H)$	the minimum $\beta_H - degree$ of H, p. 179
H/A	the partial hypergraph circumscribed by A, p. 179
$\mathbf{E}(H/A)$	the edge set of H/A, p. 179
$\mathbf{R}_H{}^\beta(X)$	p. 180
$\mathbf{Q}_H{}^\beta(X)$	p. 180
$\mathbf{L}_H{}^\beta(X)$	p. 180
$H(B)$	p. 182
$\Delta^\beta(H\blacksquare B)$	p. 182
$\delta^\beta(H\blacksquare B)$	p. 182

$M_H(X)$	a linear order of X, p. 184
$\overrightarrow{\mathbf{R}}^{\beta}_{H}(X)$	p. 185
$\overrightarrow{\mathbf{Q}}^{\beta}_{H}(X)$	p. 185
$\overrightarrow{\mathbf{L}}^{\beta}_{H}(X)$	p. 185
$\overleftarrow{\mathbf{R}}^{\beta}_{H}(X)$	p. 185
$\overleftarrow{\mathbf{Q}}^{\beta}_{H}(X)$	p. 185
$\overleftarrow{\mathbf{L}}^{\beta}_{H}(X)$	p. 185
$\gamma_{(\sigma)}$	the elementary center, p. 195
$\Gamma_{(\sigma)}$	the elementary center of Γ, p. 195
$\Lambda_f(\Gamma)$	the f – *chromatic* value of Γ, p. 195
$\bar{\Lambda}_f(\Gamma)$	the f – *chromatic* valuation of Γ, p. 196
$\chi_f(\mathcal{H})$	the Λ_f – *chromatic* number of $\mathcal{H}$, p. 197
$\bar{\chi}_f(\mathcal{H})$	the Λ_f – *chromatic* number of $\mathcal{H}$, p. 197
$\gamma^{D(\mathcal{H})}$	the $\mathcal{H}$ – *dominant* transform of γ, p. 212
$\mathcal{K}^{D(\mathcal{H})}$	the $\mathcal{H}$ – *dominant* transform of $\mathcal{K}$, p. 213
$O(\mathcal{H})$	p. 218
$O(\mathcal{E})$	p. 218
PS	p. 218
$\Theta(\mathcal{H})$	p. 219
PSC	p. 220
$\mathcal{H}_D$	p. 221
H^*	dual hypergraph, p. 222

INDEX

Printing: Weihert-Druck GmbH, Darmstadt
Binding: Buchbinderei Schäffer, Grünstadt